Selected Titles in This Series

240 **Angel Carocca, Víctor González-Aguilera, and Rubí E. Rodríguez, Editors,** Complex geometry of groups, 1999

239 **Jean-Pierre Meyer, Jack Morava, and W. Stephen Wilson, Editors,** Homotopy invariant algebraic structures, 1999

238 **Gui-Qiang Chen and Emmanuele DiBenedetto, Editors,** Nonlinear partial differential equations, 1999

237 **Thomas Branson, Editor,** Spectral problems in geometry and arithmetic, 1999

236 **Bruce C. Berndt and Fritz Gesztesy, Editors,** Continued fractions: From analytic number theory to constructive approximation, 1999

235 **Walter A. Carnielli and Itala M. L. D'Ottaviano, Editors,** Advances in contemporary logic and computer science, 1999

234 **Theodore P. Hill and Christian Houdré, Editors,** Advances in stochastic inequalities, 1999

233 **Hanna Nencka, Editor,** Low dimensional topology, 1999

232 **Krzysztof Jarosz, Editor,** Function spaces, 1999

231 **Michael Farber, Wolfgang Lück, and Shmuel Weinberger, Editors,** Tel Aviv topology conference: Rothenberg Festschrift, 1999

230 **Ezra Getzler and Mikhail Kapranov, Editors,** Higher category theory, 1998

229 **Edward L. Green and Birge Huisgen-Zimmermann, Editors,** Trends in the representation theory of finite dimensional algebras, 1998

228 **Liming Ge, Huaxin Lin, Zhong-Jin Ruan, Dianzhou Zhang, and Shuang Zhang, Editors,** Operator algebras and operator theory, 1999

227 **John McCleary, Editor,** Higher homotopy structures in topology and mathematical physics, 1999

226 **Luis A. Caffarelli and Mario Milman, Editors,** Monge Ampère equation: Applications to geometry and optimization, 1999

225 **Ronald C. Mullin and Gary L. Mullen, Editors,** Finite fields: Theory, applications, and algorithms, 1999

224 **Sang Geun Hahn, Hyo Chul Myung, and Efim Zelmanov, Editors,** Recent progress in algebra, 1999

223 **Bernard Chazelle, Jacob E. Goodman, and Richard Pollack, Editors,** Advances in discrete and computational geometry, 1999

222 **Kang-Tae Kim and Steven G. Krantz, Editors,** Complex geometric analysis in Pohang, 1999

221 **J. Robert Dorroh, Giséle Ruiz Goldstein, Jerome A. Goldstein, and Michael Mudi Tom, Editors,** Applied analysis, 1999

220 **Mark Mahowald and Stewart Priddy, Editors,** Homotopy theory via algebraic geometry and group representations, 1998

219 **Marc Henneaux, Joseph Krasil'shchik, and Alexandre Vinogradov, Editors,** Secondary calculus and cohomological physics, 1998

218 **Jan Mandel, Charbel Farhat, and Xiao-Chuan Cai, Editors,** Domain decomposition methods 10, 1998

217 **Eric Carlen, Evans M. Harrell, and Michael Loss, Editors,** Advances in differential equations and mathematical physics, 1998

216 **Akram Aldroubi and EnBing Lin, Editors,** Wavelets, multiwavelets, and their applications, 1998

For a complete list of titles in this series, visit the AMS Bookstore at **www.ams.org/bookstore/**.

Complex Geometry of Groups

I Iberoamerican Congress on Geometry: "Cruz del Sur"

Names of participants on picture by rows from left to right.

ROW 1 (seated): Herbert Lange; Frederick Gardiner; Martin Chuaqui; Rubén Hidalgo Angel Carocca; Rubí E. Rodríguez; Victor González; Irwin Kra; Bernard Maskit; Clifford Earle.

ROW 2: Antonio Costa; Hugo Caerols; Xavier Gómez Mont; Hershel Farkas; Mabel Rodriguez; Alicia Dickenstein; Leticia Brambila-Paz; Ana Porto; Christina Birkenhake Laura Hidalgo; Linda Keen; Bettye Anne Case; Lori Stone; Fausto Toranzos.

ROW 3: Jonathan Brezin; Peter Buser; Gustavo Labbé; Marcia Lazo; William Harvey Esteban Gómez-González; Sevin Recillas; Armando Sánchez; Yair Minsky; Yaroslav Kopylov.

ROW 4: Gerhard Rosenberger; Alexander Mednykh; Patricia Sipe; Alvaro Paz; Albert Castro; Nikola Lakic; William Abikoff; Andreas Bolsch; Robert Silhol; Jack Quine.

NOT PICTURED: Rodrigo Bamón; Jan Kiwi; Christian Pauly; Gonzalo Riera; Anita Rojas.

CONTEMPORARY MATHEMATICS

240

Complex Geometry of Groups

January 5–11, 1998, Olmué, Chile

Angel Carocca
Víctor González-Aguilera
Rubí E. Rodríguez
Editors

American Mathematical Society
Providence, Rhode Island

This volume contains the proceedings of the I Iberoamerican Congress on Geometry, held in Olmué, Chile on January 5–11, 1998.

1991 *Mathematics Subject Classification*. Primary 30Fxx; Secondary 14Hxx, 14Kxx.

Library of Congress Cataloging-in-Publication Data

Iberoamerican Congress on Geometry (1st : 1998 : Olmué, Chile)

Complex geometry of groups : I Iberoamerican Congress on Geometry, January 5–11, 1998, Chile / [edited by] Angel Carocca, Victor González-Aguilera, Rubí E. Rodríguez.

p. cm. — (Contemporary mathematics, ISSN 0271-4132 ; 240)

Includes bibliographical references.

ISBN 0-8218-1381-1 (alk. paper)

1. Rieman surfaces Congresses. 2. Kleinian groups Congresses. I. Carocca, Angel, 1965– . II. González-Aguilera, Victor, 1950– . III. Rodríguez, Rubí E., 1953– . IV. Title. V. Series: Contemporary mathematics (American Mathematical Society) ; v. 240.

QA333.I24 1998

515'.93—dc21

99-30206

CIP

∞ The paper used in this book is acid-free and falls within the guidelines
established to ensure permanence and durability.
Visit the AMS home page at URL: `http://www.ams.org/`

10 9 8 7 6 5 4 3 2 1   04 03 02 01 00 99

Contents

Preface

In the early eighties Víctor González and Rubí Rodríguez started an informal seminar in Santiago on the subjects of algebraic geometry, Kleinian groups and Riemann surface theory.

The seminar and other related activities soon attracted students and colleagues working at various universities throughout the country, many of whom became interested in beginning research work.

As a consequence of this formative work, some of the students obtained a master's degree in Chile working in this area and continued on with their careers by finishing their doctoral degrees abroad. Presently, there are doctoral dissertations being written in Chile on the subject.

This is how the Chilean Group on Complex Geometry was formed. More researchers have joined the group according as they returned to Chile after their graduate and post-doctoral studies with strong backgrounds on Kleinian groups, complex dynamics, geometric function theory, or group theory. Thus they have enriched the level of discourse and provided new insights into the problems we have been working on and interested in.

International contacts were originally supported enthusiastically by C. H. Clemens and I. Kra, and later on expanded to collaborative work and joint research grants with other colleagues from France, Germany, Spain and the U.S.A.

There is also a great deal of fruitful interaction with the Mexican group working in the area. Besides joint works, grants and research visits to both countries, we have exchanged graduate students.

In September 1995, during an Algebraic Geometry Workshop in Morelia, Mexico, some Iberoamerican mathematicians (from Salamanca, Spain, Santiago and Valparaíso, Chile, Morelia, Guanajuato and Mexico City, Mexico), who were working on the same type of problems, joined efforts and decided to meet periodically. As a result, in July 1996, we held the Workshop on Abelian Varieties and Theta Functions in Morelia, where we also began to plan the present Congress.

Another decisive event in bringing this idea to reality was the Stony Brook 1997 Krafest, in celebration of Irwin Kra, in which participants and the "festejado"supported the idea with great enthusiasm.

The I Iberoamerican Congress on Geometry: *Cruz del Sur* was held in Olmué, Chile during January 5-11, 1998. Its main subject was The Geometry of Groups: Curves, Abelian Varieties, Theoretical and Computational Aspects.

Many people collaborated in making the Congress possible. Participants came from many countries and most of them also helped to procure financial resources. The supporting institutions reflect individual as well as collective efforts in this regard; they were as CONICYT-Chile, CONACYT-Mexico, DAAD-Germany, NSF-USA, FONDECYT, Fundación Andes, Universidad Técnica Federico Santa María and Pontificia Universidad Católica de Chile.

This volume gathers the expanded contributions from most of the participants in the Congress. The articles presented here reflect the subject of the Geometry of Groups in its diversity and unity, through original results and surveys on the areas of curves and Riemann surfaces, abelian varieties, and complex dynamics. The points of view are manifold: Kleinian groups, quasiconformal mappings and Teichmüller spaces, function theory, moduli spaces, automorphism groups, algebraic geometry, and others.

In addition to the authors appearing in this volume, many others have collaborated to its publication, which includes referees with their generous time, Ms. Christine Thivierge from the publication staff of AMS with her helpful advice, and the Congress participants with their enthusiasm for the next one.

We thank them all and hope that this volume will be the the continuation and a prelude to future Iberoamerican meetings.

Angel Carocca
Víctor González-Aguilera
Rubí E. Rodríguez

Santiago, Chile
January, 1999

Contemporary Mathematics
Volume **240**, 1999

Complex Geometry in Chile: Panorama and Perspectives

Rubí E. Rodríguez

Abstract

This is a brief version of a survey conference delivered at the I Iberoamerican Congress on Geometry: *Cruz del Sur*. The main objective of the conference was to present the historical background and an overview of a variety of results obtained by Chilean researchers in complex geometry, specifically on the topics covered by the Congress, namely, The Geometry of Groups: Curves, Abelian Varieties, and Theoretical and Computational Aspects.

1. History

The early seventies were interesting years in Chile. A generation of enthusiastic young people went into science, determined to continue graduate studies abroad and afterwards return to the country to promote drastic changes in a broad spectrum of intellectual aspects of the national life. This was particularly true in mathematics where there was no national policy nor tradition in cultivating the discipline. At the time there were perhaps ten people in the country holding doctoral degrees.

Contributing to this motivation, there was the presence of many foreign mathematicians who had become interested in the process and volunteered to come and teach for a certain period of time. Among these, C. H. Clemens was a pioneer who helped many Chileans become mathematicians as well as instrumental in the creation of a Complex

---

1991 *Mathematics Subject Classification.* Primary 30Fxx, 14Fxx; Secondary 14Kxx, 01Axx.

Partially supported by FONDECYT grant 8970007 and the Presidential Science Chair 1997.

Geometry Group in Chile. In particular, he encouraged Gonzalo Riera and myself into undertaking graduate studies at Columbia where we were inspired by Lipman Bers.

There we met other mathematicians who helped us generously in more ways than one at a difficult time. Some of them were present at the Congress, such as Irwin Kra, Linda Keen, Bernie Maskit, Clifford Earle and the rest of the wonderful people who conform the Ahlfors-Bers Mafia.

While we were in New York Víctor González went to work in Paris under the supervision of J. L. Verdier and F. Norguet, establishing other connections and gaining a different yet related perspective in his mathematical work.

At the beginning of the eighties we returned to Chile and started an informal seminar on the themes of algebraic geometry, Kleinian groups and Riemann surface theory. This weekly seminar is still very much alive. In addition, a yearly national meeting is held and some of you have attended as well as participated in these meetings through the years.

As the group grew, student participation increased and collaborative work with people abroad continued as well as amplified a great deal. Some young researchers obtained their doctoral degrees abroad, while others did so in this country. Concerning those who studied abroad, some have returned to Chile and are now part of this group, others have continued their careers in other countries. Presently, joint work and graduate student exchanges are being carried out with our colleagues and friends from Mexico, Spain, Italy, Germany and the United States.

2. The Mathematics

2.1. The individual work. In this paragraph we briefly describe the individual mathematical interests of the various members of the group.

Jan Kiwi works in the area of complex dynamics. His research is focused on iterations of polynomials of degrees greater than two. His main interest has been understanding how polynomials are organized in parameter space according to their dynamical properties. The emphasis has been in developing a combinatorial classification of higher degree polynomials as well as their relation with the topology of parameter space.

Angel Carocca works in finite group theory and his main interest is in permutability and subnormality criteria. He is presently applying his knowledge on groups to study group actions on Riemann surfaces and abelian varieties.

Martin Chuaqui studies univalent functions in the disk and other simply connected domains through a generalized Schwarzian derivative and a very general sufficient condition for univalence.

This new Schwarzian differs from the classical one in that it depends upon the choice of a conformal metric. This allows one to change the geometry of the domain conveniently thereby recovering, from the general criterion, a vast class of known conditions.

Rubén Hidalgo works mainly on Kleinian groups focused on uniformizations of both noded and non-singular Riemann surfaces. In particular, he has contributed to the understanding of the relationship between automorphisms of Riemann surfaces and their uniformizations by Schottky-type groups. Other topics of interest for him are the geometries of three-orbifolds and generalizations to higher dimensions.

He is also becoming interested in the connections with algebraic geometry through the study of the Jacobians of curves with prescribed automorphisms and is passionately looking for a Torelli-type theorem for general Kleinian groups.

Gonzalo Riera is presently working on Riemann surfaces and abelian varieties with automorphisms. He is currently interested in looking at these objects from the real analytic point of view.

Víctor González-Aguilera works mainly with algebraic curves and surfaces as well as principally polarized abelian varieties with automorphisms. In particular, he has contributed to the understanding of the connection of stable curves with automorphisms and families of principally polarized abelian varieties uniformized by modular curves.

2.2. The group work. We could say that most of the work within the group is concentrated on the geometry of group actions.

The objects under study are Riemann surfaces and abelian varieties, in particular the spaces parameterizing them as well as their degenerations.

Many different tools are used to accomplish these goals, such as complex analysis, algebraic and hyperbolic geometry, and groups and their representations, among others.

We let $\mathfrak{M}_g$ denote the Riemann moduli space of (isomorphism classes of) compact Riemann surfaces of genus g, and we will let $\mathcal{A}_g$ denote the moduli space of (isomorphism classes of) principally polarized abelian varieties of dimension g. Let $\Pi : \mathfrak{M}_g \longrightarrow \mathcal{A}_g$ denote the Abel-Jacobi period map. Also $\overline{\mathfrak{M}_g}$ denotes the (Mumford-Deligne) compactification of $\mathfrak{M}_g$ by stable curves. It is known that both $\mathfrak{M}_g$

and $\mathcal{A}_g$ are analytic spaces of dimensions $3g-3$ and $g(g+1)/2$, for g at least two, respectively.

We believe that our main scientific contribution has been to conjugate harmoniously different points of view from different areas of knowledge as well as propose and attack problems on the subject.

The **main problems** that have been undertaken by the group are the following.

A. **The understanding of the singular loci of $\mathfrak{M}_g$ and its degeneration towards the boundary.**

This singular locus corresponds to the Riemann surfaces with automorphisms. We have developed a theory that allows us to describe their uniformizations in various ways: for instance, by means of Fuchsian groups or Schottky groups and, unlike the classical case, we obtain uniformizations which reflect the action of the automorphism group. Here the main ingredients involved are hyperbolic geometry, used to prove the existence of suitable fundamental polygons, and group theory, from which we use of adequate presentations and Cayley diagrams. For example, we have applied it to give complete geometric-analytic descriptions of the sub-varieties of the singular locus of $\mathfrak{M}_g$ that correspond to the different groups acting for low values of g.

This theory has also proved to be useful in studying the deformations of the non-singular curves with automorphisms as they approach the corresponding stable curves on $\overline{\mathfrak{M}_g} - \mathfrak{M}_g$. In this setting the general intricate problem of giving explicit variations towards the limiting values of the parameters may be solved.

B. **The study of principally polarized abelian varieties with automorphisms.**

We have developed a real theory of principally polarized abelian varieties (p.p.a.v.'s) which generalizes ideas of H. Weyl and A. Albert. This real theory is more suitable than the classical one for the study of p.p.a.v.'s with automorphisms and it includes the development of real theta functions which are invariant under automorphisms. This has been applied to, for instance, the explicit determination of all p.p.a.v.'s admitting a cyclic group of prime order as automorphisms and we have also characterized those corresponding to Jacobians.

Again, the theory mentioned in item A of above provides a concrete way of finding symplectic representations for the automorphism groups of curves, allowing the computation of the fixed Riemann matrices that give rise to the corresponding principally polarized abelian varieties. For instance, we have used these methods we have found, for each g larger than one, a family of irreducible p.p.a.v.'s of dimension g isomorphic to a product of elliptic curves. For the case $g = 4$, one of the elements of the family is the Jacobian of Bring's curve, which gives a positive answer to a decomposability question posed by Ekedahl and Serre.

We have also studied families of p.p.a.v.'s generated by geometric objects with symmetries, such as prisms and pyramids, which give general insight into the role played by special properties of the groups involved. For instance, this study allowed us to prove that when the automorphism group is complete its normalizer inside the symplectic group is a finite extension of the group and the map from Siegel space to $\mathcal{A}_g$ is well-behaved.

C. **The study of the decomposition of Jacobians with group actions.**

Here the objective is to find isogenous equivariant decompositions of Jacobians into factors of geometric significance by taking advantage of the automorphism group involved. In strong collaboration with the Mexican group and using elements from algebraic geometry and Riemann surface theory, we have obtained partial results: a generalization of Recillas' trigonal construction to the case of curves admitting an S_3-action, where the decomposition is given by Prym varieties and Jacobians of appropriate intermediate quotients, and a similar decomposition for tetragonal curves. We have thus answered a question proposed by J. L. Verdier.

A long-standing problem in the area was *Schottky's problem* of characterizing the image of $\mathfrak{M}_g$ inside $\mathcal{A}_g$. It was solved by Arbarello and De Concini in 1984, and also by Shiota around the same time. Essentially, they proved that a principally polarized abelian variety A is the Jacobian of a Riemann surface if and only if the theta function of A (a certain function associated naturally to any principally polarized abelian variety) is a solution to a specific partial differential equation of type KP.

A more geometric description has been widely sought, but not found, and conjectures abound. For example, the trisecant conjecture which relates Jacobians to trisecants of the associated Kummer variety.

In this connection, and together with the Mexican and Spanish groups in the area, we have proposed the following problem.

PROBLEM (The Schottky problem with automorphisms). *Characterize geometrically the image of the singular locus of* $\mathfrak{M}_g$ *in* $\mathcal{A}_g$ *and describe the locus in* $\mathcal{A}_g$ *which consists of principally polarized abelian varieties with nontrivial automorphisms.*

FACULTAD DE MATEMÁTICAS, PONTIFICIA UNIVERSIDAD CATÓLICA DE CHILE, CASILLA 306, CORREO 22, SANTIAGO, CHILE
E-mail address: `rubi@riemann.mat.puc.cl`

Contemporary Mathematics
Volume **240**, 1999

Quasiconformal Conjugation of Möbius transformations

William Abikoff

ABSTRACT. In this note we examine the change in translation length of a Möbius transformation when it is conjugated into another Möbius transformation by a quasiconformal mapping. A sharp bound is given in terms of the dilatation of the mapping. Also we consider the conjugation of a Möbius transformation γ by powers of a quasiconformal mapping, all of which conjugate γ into transformations γ_n lying in the Möbius group. In particular, we are interested in bounding the growth in the relative spacings between successive pairs of axes of the γ_n.

The purpose of this note is to examine two questions. The first is how does the complex translation length of a Möbius transformation change under conjugation by a K-quasiconformal mapping. We thereby generalize a result of Teichmüller, later rediscovered by Wolpert, from the hyperbolic plane to hyperbolic 3-space. In the second case, we consider conjugations, by powers of a single K-quasiconformal mapping, of a fixed Möbius transformation into the Möbius group, The question then is to find some estimate of the spacing between the axes of the conjugates. We show that the spacings grow at most exponentially with K. The estimate is local in the sense that, if two successive axes are closer than M, then the next spacing differs from KM by a constant not depending on the specific iterate, but only on K. We shall refer to that result as the *spreading out of axes* of the γ_n.

We can compare the case of conjugation by a Möbius transformation; there the spacings remain constant. Perhaps the most important consequence of the growth estimate lies in the realm of discrete groups. Our theorem implies that the closed geodesics, covered by these axes, in the quotient orbifold, are regularly distributed. The proof given here is based on an expression for the complex distance using the cross ratio; the form of the expression is standard in two dimensions but not commonly used in dimension 3.

The spreading out of axes is key to a (Riemann) moduli space, rather than Teichmüller space, approach to Thurston's fibration theorem. The latter gives the existence of hyperbolic structures on the mapping tori of irreducible diffeomorphisms of Riemann surfaces of finite area. That proof will be given elsewhere.

1991 *Mathematics Subject Classification.* 30F40,30C62,30F60.

1. Some preliminaries

We model hyperbolic 3-space, $\mathcal{H}^3$, on the upper half-space $\mathbb{R}^3_+ := \{w = (z, y) : z \in \mathbb{C}, y \in \mathbb{R}_+\}$ with the metric $ds^2 := y^{-2}|dw|^2$. Here $|dw|^2$ is the square of the Euclidean norm. The orientation-preserving isometries are then the complex Möbius transformations, $\mathcal{M}\text{öb}_{\mathbb{C}}$, extended to $\mathcal{H}^3$ using the standard method which goes back to Poincaré.

Supppose L_1 and L_2 are hyperbolic lines in $\mathcal{H}^3$ with distinct endpoints. Then, using a Möbius transformation we may move the lines so that L_1 has endpoints ± 1 and L_2 has endpoints at e^β and $-e^\beta$. (It is an elementary exercise that any ordered set of four distinct points in $\widehat{\mathbb{C}} := \mathbb{C} \cup \{\infty\}$ can be so moved by a Möbius transformation.) The distance between L_1 and L_2 is achieved along the vertical ray from 0. It is obtained by integrating y^{-1} from $y = 1$ to $y = |e^\beta|$, hence the distance is $|\log|e^\beta|| = |\Re\beta|$. The angle may be defined invariantly as follows: if the distance between the lines is positive and Π is the unique plane containing L_1 and the common perpendicular to the two lines, the angle is the angle between L_2 and Π. Otherwise measure the angle in the plane containing both lines. The angle between the two lines is then $\pm \arg e^\beta$ taken $\mod \pi$. (The ambiguity in sign may be removed by considering oriented angles.) Thus, up to orientation of lines and angles, the angle between L_1 and L_2 is $\Im\beta$ which we may take to be between $\pm\pi/2$. Then we define the *complex distance* between L_1 and L_2 to be β. Note that the angle between noncoplanar L_1 and L_2 is also the angle between the vertical plane through L_1 and L_2. If they are coplanar measure the angle in that plane.

Given four distinct points z_i for $i = 1, \dots, 4$ in $\widehat{\mathbb{C}} := \mathbb{C} \cup \{\infty\}$, we denote the cross-ratio of the z_i by

$$[z_1, z_2, z_3, z_4] = \left(\frac{z_1 - z_3}{z_1 - z_4}\right)\left(\frac{z_2 - z_4}{z_2 - z_3}\right). \tag{1.1.1}$$

The cross-ratio is invariant under Möbius transformations in the sense that

$$[\gamma(z_1), \gamma(z_2), \gamma(z_3), \gamma(z_4)] = [z_1, z_2, z_3, z_4] \tag{1.1.2}$$

when $\gamma \in \mathcal{M}\text{öb}_{\mathbb{C}}$.

It is immediate to show

LEMMA 1.1. *Normalize L_1 to have endpoints ± 1 and L_2 to have endpoints $\pm e^\beta$. Then the cross-ratio of the endpoints of the directed lines is*

$$[e^\beta, -e^\beta, 1, -1] = \tanh^2(\beta/2).$$

Further, if the endpoints of the L_i are translated by any Möbius transformation to points (given in order) ζ_i for $i = 1, \dots, 4$ then, if ζ_3, ζ_4 and β are known, ζ_1 and ζ_2 satisfy a quadratic equation.

We have noted above that β is not uniquely defined, however $\tanh^2(\beta/2)$ is well-defined.

We will denote by K-QC the K-quasiconformal mappings of the extended complex plane.

2. Quasiconformal conjugation of Möbius transformations

Our first theorem provides an estimate for the change in the translation length of a Möbius transformation induced by conjugation by a quasiconformal mapping f. If γ is conjugate in the Möbius group to $w \mapsto \lambda w$ then we say that γ has *complex translation length* $\ell_{\mathbb{C}}(\gamma) := \log \lambda$. Parabolic transformations have complex translation length zero.

THEOREM 2.1. *Suppose* γ *is a Möbius transformation and* $f \in$ K-QC. *Further assume that* $\eta := f \circ \gamma \circ f^{-1}$ *is also a Möbius transformation. Then*

(i) *if* γ *is parabolic, so is* η,
(ii) *if* γ *is loxodromic with complex translation distance* $\ell_{\mathbb{C}}(\gamma)$, *then* η *is also loxodromic and its complex translation distance* $\ell_{\mathbb{C}}(\eta)$ *is bounded by a constant depending only on* K *and* $\ell_{\mathbb{C}}(\gamma)$; *as* K *approaches 1,* $\ell_{\mathbb{C}}(\eta) \to \ell_{\mathbb{C}}(\gamma)$,
(iii) *if* γ *is elliptic of finite order, then* η *is elliptic of the same order.*
(iv) *if* γ *elliptic then* η *is elliptic.*

PROOF. Since f is a homeomorphism of $\widehat{\mathbb{C}}$, η has the same number of fixed points as γ; further, if γ is periodic then η is periodic of the same order. This proves both (i) and (iii). (iv) has been proved if γ is of finite order. If it is elliptic of infinite order, then it has two fixed points and other orbits are dense in circles. Orbits under loxodromic transformations only cluster at the fixed points and (iv) is proved.

To prove (ii), we may assume that both γ and η have their fixed points at 0 and ∞. We may also assume that f fixes $0, 1$ and ∞. Then f maps $\gamma(1)$ to $\eta(1)$. Among all quasiconformal mappings of $\widehat{\mathbb{C}}$ mapping the quadruple $(0, 1, \infty, \gamma(1))$ to $(0, 1, \infty, \eta(1))$, there is one of minimal dilatation because of the weak compactness of the normalized K-QC maps. Therefore a K-quasiconformal mapping cannot map $\gamma(1)$ too close to 0 or ∞ hence it cannot distort complex translation length in an unbounded manner. As K nears one, f must become close to the identity because of the normalization. (ii) is proved. □

The proof of part (ii), while quite efficient, does not give specific bounds. We recall the corresponding result for quasiconformal mappings of hyperbolic 2-space which conjugate a purely hyperbolic group into another purely hyperbolic group. It is a result of Teichmüller, rediscovered later by Wolpert and frequently attributed to the latter alone (cf. Abikoff [**1**] and others), that a K-QC map cannot increase the multiplier of a hyperbolic transformation by more than the factor K. The sharp form of (ii) is given by an appeal to the deformation theory of tori in the following fashion. Suppose $\gamma : z \mapsto \lambda z$ with $|\lambda| < 1$. Then γ identifies the unit circle with the circle of radius $|\lambda|$ usually non-radially. The identification forms a torus and the cover group for that torus may be taken as having generators $2\pi i$ and $\log \lambda$. We need only define the branch of the logarithm. Set $\tau = \log \lambda / 2\pi i$. Then our choice of $|\lambda|$ forces $\Im \tau > 0$ and we may choose $-1 < \Re \tau \leq 1$, hence force a choice of the branch of $\log \lambda$. We base our choice of the branch of $\log \mu$ on the choice of the branch of $\log \lambda$ as below.

THEOREM 2.2. *Suppose* $\gamma \in \mathcal{M}\text{öb}_{\mathbb{C}}$ *is loxodromic with multiplier* λ *and* $f \in$ K-QC *conjugates* γ *into the Möbius transformation* η *with multiplier* μ. *Then*

$$\frac{|\log \lambda \bar{\mu}| + |\log \frac{\lambda}{\mu}|}{|\log \lambda \bar{\mu}| - |\log \frac{\lambda}{\mu}|} \leq K. \tag{2.2.1}$$

with $\arg \mu$ *chosen so that* $|\Im \mu - \Im \lambda| \leq \pi$.

PROOF. By conjugating γ and η by Möbius transformations, we may assume that both fix 0 and ∞, with the latter acting as repelling fixed point. Let G

(respectively, H) be the group generated by γ (respectively η) and let $\dot{\mathbb{C}} := \mathbb{C}\setminus\{0\}$. Then $\lambda = \gamma(1)$ and $\mu = \eta(1)$ and both have modulus less than one. As above, $E_1 := \dot{\mathbb{C}}/G$ (respectively, $E_2 := \dot{\mathbb{C}}/H$) is a torus represented by the point τ_1 (respectively, τ_2) in the upper half-plane U. Since the Teichmüller distance d in U between the marked surfaces represented by the τ_i is identical to the hyperbolic distance and since f induces a K-quasiconformal mapping between these surfaces, we have, following Beardon [**2**],

$$\log\frac{|\tau_1-\bar{\tau}_2|+|\tau_1-\tau_2|}{|\tau_1-\bar{\tau}_2|-|\tau_1-\tau_2|}\le \log K \tag{2.2.2}$$

or

$$\frac{|\tau_1-\bar{\tau}_2|+|\tau_1-\tau_2|}{|\tau_1-\bar{\tau}_2|-|\tau_1-\tau_2|}\le K. \tag{2.2.3}$$

τ_1 is determined as in the remarks preceeding the theorem as is $\Im\tau_2$. $\Re\tau_2$ is determined by the hypothesis of the theorem. It is convenient that $\overline{\log z} = \log\bar{z}$. Substituting into Equation (2.2.3), we obtain

$$\frac{|\log\lambda+\log\bar{\mu}|+|\log\lambda-\log\mu|}{|\log\lambda+\log\bar{\mu}|-|\log\lambda-\log\mu|}\le K$$

and the theorem follows. □

As a Corollary we get a simpler estimate suggested by Fred Gardiner.

COROLLARY 2.3. *Under the hypotheses of the Theorem,*

$$\frac{1}{K}\le\frac{\log|\lambda|}{\log|\mu|}\le K.$$

PROOF. It follows from Equation (2.2.2), that

$$\frac{1}{K}\le\frac{|\tau_1-\bar{\tau}_2|+|\tau_1-\tau_2|}{|\tau_1-\bar{\tau}_2|-|\tau_1-\tau_2|}\le K.$$

We may apply the triangle inequality separately to the numerator and denominator of the middle term to obtain

$$\frac{2|\Im\tau_1|}{2|\Im\tau_2|}\le K.$$

The left-hand side of the desired inequality follows from the definitions of the τ_i as does the right-hand side. □

Our next result estimates the distribution of axes of loxodromic transformations obtained by conjugating a fixed transformation by successive powers of a quasiconformal mapping.

THEOREM 2.4. *Let γ be a loxodromic Möbius transformation and $\phi\in$ K-QC. If $\gamma_n=\phi^n\circ\gamma\circ\phi^{-n}$ is Möbius, then there is a constant $C>0$ so that:*

$$|K\,|\Re\beta_{n+1}|-|\Re\beta_n||\le C$$

where β_n is the complex distance between the axes L_n of γ_n and L_{n+1} of γ_{n+1} in hyperbolic space.

PROOF. Let r_n (respectively, a_n) be the repelling (respectively, attracting) fixed point of γ_n. $|\Re\beta_n|$ is the hyperbolic distance between L_n and L_{n+1}. Since we will only use β_n in the squares of hyperbolic functions, i.e. as arguments of even functions, we may assume that $\Re\beta_n$ is non-negative. We may further assume that $\Im\beta_n$ has absolute value bounded by $\pi/2$. Lemma 1.1 tells us that, when the fixed points are ordered properly,

$$\tanh^2(\beta_n/2) = [a_{n+1}, r_{n+1}, a_n, r_n] \tag{2.2.4}$$

where the righthand side denotes the cross-ratio. For convenience, we may set $w_n = \tanh^2(\beta_n/2)$. Choose $\eta_n \in \mathcal{M}\text{öb}_{\mathbb{C}}$, so that η_n maps the ordered triple r_{n+1}, a_n, r_n to $0, 1, \infty$. Further choose Möbius transformations θ_n so that $\psi_n := \theta_n \circ \phi \circ \theta_n^{-1}$ fixes 0,1 and ∞. Since $\phi \in$ K-QC, so is ψ_n. Observe that $\phi(a_n) = a_{n+1}$ and $\phi(r_n) = r_{n+1}$. Because cross-ratios are invariant under Möbius transformations, we have

$$w_n = [w_n, 0, 1, \infty] = [\eta_n(a_{n+1}), \eta_n(r_{n+1}), \eta_n(a_n), \eta_n(r_n)].$$

It follows that $w_n = \eta_n(a_{n+1})$ and $\phi \circ \eta_n^{-1}$ maps the ordered 4-tuple $w_n, 0, 1, \infty$ to the ordered 4-tuple $a_{n+2}, r_{n+2}, a_{n+1}, r_{n+1}$. Consequently ψ_n fixes $0, 1, \infty$ and maps w_n to w_{n+1}. Since $\psi_n \in$ K-QC, it is classical (cf. [**3**]) that

$$|w_{n+1} - 1| \le C|w_n - 1|^{1/K} \tag{2.2.5}$$

for some constant C and the bound is valid on compact subsets of $\mathbb{C}$.

We note that, for any $z \in \mathbb{C}$,

$$|\tanh^2(z) - 1| = 4e^{-2\Re z}|1 + e^{-2z}|^{-2}.$$

Therefore, in terms of β_n, Equation (2.2.5) may be rewritten as

$$\exp[\Re(\beta_n/K - \beta_{n+1})] \le C\frac{|1 + \exp(-\beta_{n+1})|^2}{|1 + \exp(-\beta_n)|^{2/K}} \tag{2.2.6}$$

where $C = 4^{(K+1)/K}$. Since we assume that $\Re\beta_n \ge 0$ and $|\Im\beta_n| \le \pi/2$ for all n, we may take logarithms in the previous Equation to obtain

$$\Re\left(\frac{\beta_n}{K} - \beta_{n+1}\right) \le C' + 2\log|1 + \exp(-\beta_{n+1}| - \frac{1}{K}\log|1 + \exp(-\beta_n)|$$

where $C' = \log C$ and each of the logarithms is bounded above by 2 and below by zero. This follows from our assumptions on the real and imaginary parts of β_n. The desired result obtains. □

References

[1] W[illiam] Abikoff, *The real analytic theory of Teichmüller space*, Lecture Notes in Mathematics, vol. 820, Springer-Verlag, 1980.

[2] A[lan] F. Beardon, *The geometry of discrete groups*, Springer-Verlag, 1983.

[3] O. Lehto and K. I. Virtanen, *Quasiconformal mappings in the plane*, Springer-Verlag, 1973.

UNIVERSITY OF CONNECTICUT

E-mail address: abikoff@math.uconn.edu

Contemporary Mathematics
Volume **240**, 1999

A family of n-dimensional differential equations with Lorenz-like attractors.

Rodrigo Bamón*

Abstract

We construct a family of degree-3 polynomial vector fields in $\boldsymbol{R}^n$ which yield an ample supply of attractors when parameters are varied. Most of these look like the Lorenz attractor.

1 Introduction.

The flow of the Lorenz equation

$$\begin{cases} \dot{x} &= -10x + 10y \\ \dot{y} &= 28x - y - xz \\ \dot{z} &= -8/3z + xy \end{cases}$$

and the geometric Lorenz attractor of Guckenheimer and Williams share an important dynamical property. Namely, that the whole unstable manifold of a saddle point passes near its stable manifold. Of course, this does not say very much, but it is a fundamental ingredient to show the existence of attracting basins.

With this in mind, we first find a degree-3 polynomial family of vector fields in $\boldsymbol{R}^n$ so that each of these has a saddle point at the origin with an $n-2$ dimensional unstable manifold which ends in a saddle-node point. Further, this saddle-node point lies in the closure of the stable manifold of the origin and it attracts every point of $\boldsymbol{R}^n$ except the stable manifolds of a finite number of saddle points.

Afterwords, we introduce some small perturbations which are well-suited for producing a good number of attractors and bifurcations.

In this note we restrict ourselves to presenting and explaining the construction of this family of vector fields. We produce figures showing some of the attractors and we also illustrate some of the bifurcations which occur.

It seems to us that this family merits a detailed study. At this moment we are studying some of the bifurcations suggested by the figures, as well as analyzing some of the attractors.

In the last section of this paper we outline a similar construction for diffeomorphisms of the plane.

*Partially supported by Fondecyt, Proyecto 1960848

2 A family of differential equations.

The vector fields in the 2-parameter family

$$(1) \qquad \begin{cases} \dot{x} &= ax - \lambda xy \\ \dot{y} &= x^2 - \mu y(y-1)^2 \end{cases}$$

with $0 < \mu < a < \lambda < 3a$ have one of the two following phase portraits:

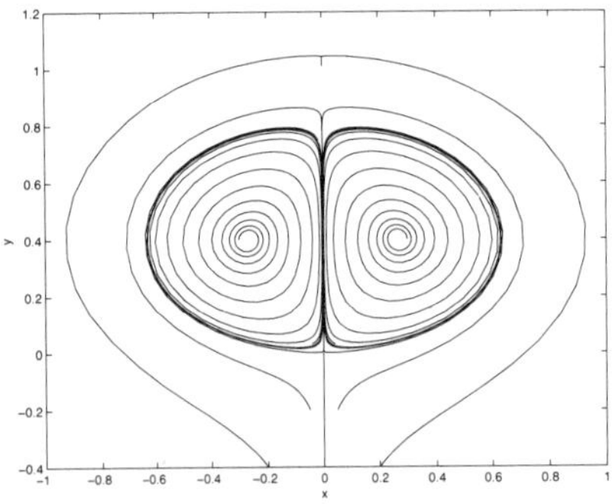

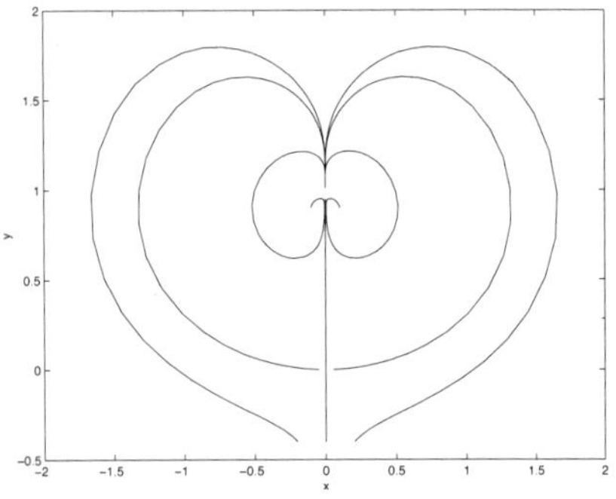

In one of these cases the saddle-node point at $(0, 1)$ attracts an open dense set of $\boldsymbol{R}^2$. For these vector fields the complete unstable manifold of the saddle point at the origin ends in the saddle-node point. In the other case there are two big periodic orbits that attract the unstable manifold of the origin.

One can easily construct similar families in higher dimensions. For example, we can simply copy (1) at each coordinate plane containing the axis with the saddle-node point. In dimension 3 we can also consider:

$$\begin{cases} \dot{x} &= a_1 x - \lambda_1 xz \\ \dot{y} &= a_2 y - \lambda_2 yz \\ \dot{z} &= x^2 + y^2 - \mu z(z-1)^2 \end{cases}$$

with similar relations among the parameters.

If $a = a_1 = a_2$ and $\lambda = \lambda_1 = \lambda_2$, these vector fields copies (1) in each plane containing the $z-$axis.

We can extend this procedure to higher dimensions and obtain explicit vector fields with a saddle point at the origin whose unstable manifold is of dimension $n-1$ and ends in a saddle-node point. However, we do not do this here. Instead, we proceed with our construction for dimensions 3 and 4.

To any of the above vector fields we can add an attractive direction, say:

$$\dot{w} = -bw, \qquad b > 0$$

We obtain vector fields with a hyperbolic saddle point at the origin with an $(n-2)$-dimensional unstable manifold which ends in a saddle-node point. Moreover, this saddle-node point attracts an open dense set of R^n. This saddle-node point also lies in the closure of the stable manifold of the saddle point at the origin.

These vector fields involve only two degeneracies, namely, the saddle-node point and the non-transversal intersection between the stable manifold of the saddle point and the unstable manifold of the saddle-node point. All these vector fields have an axial symmetry with respect to the axis that contains the saddle-node point. In the space of vector fields having this symmetry, the given ones have only one degeneracy, namely the saddle-node point.

When the saddle-node point splits into two singularities one obtains a structurally stable vector field (in compact regions) with trivial nonwandering sets. The interesting dynamics appears when the saddle-node point desappears.

Let us now proceed to introduce perturbations. Consider the 3-dimensional case:

$$\begin{cases} \dot{x} &= ax - \lambda xy \\ \dot{y} &= x^2 - \mu y(y-1)^2 \\ \dot{w} &= -bw \end{cases}$$

with $0 < \mu < a < \lambda < 3a$ and $b > 0$.

The first parameter we consider is the one that unfolds the saddle-node point. Thus we add a component $\dot{y} = -\epsilon\, y$ with ϵ small and positive. If we only do this, then there will appear two attracting periodic orbits in the xy-plane. The stable manifold of the origin coincides with the plane $x = 0$. Each branch of $W^u(O)$ remains in one of the half-spaces $x < 0$ or $x > 0$.

In order to make the branches of $W^u(O)$ cross to a different side of $W^s(O)$ we can add a rotation along the y-axis, say: $\dot{x} = -\alpha\, w$; $\dot{w} = \alpha\, x$ with α small (as a first step). We can also add a different rotation at each plane $y =$ constant, say: $\dot{x} = -\alpha y w$; $\dot{w} = \alpha y x$. We then obtain the following family of vector fields:

$$\text{(2)} \qquad \begin{cases} \dot{x} &= ax - \lambda xy - \alpha yw \\ \dot{y} &= x^2 - \mu y(y-1)^2 - \epsilon y \\ \dot{w} &= -bw + \alpha yx \end{cases}$$

where all parameters involved are positive, a, b, λ and μ satisfy (1), and α and ϵ are small.

Since before introducing these perturbations the saddle-node point is a global attractor, almost every perturbation will produce some kind of attractor set. The perturbations that we have chosen seem to us to be the most natural ones, but probably these are not versal perurbations, even among those systems that verify the indicated symmetry. As we will see in our final remark below, α is not really a perturbation.

Figures 1 and 2 below show several attractors which were obtained from this family. Information about these figures is given in the last section.

With the same principles as before we obtain the following family of vector fields in $\boldsymbol{R}^4$:

$$\text{(3)} \qquad \begin{cases} \dot{x} &= a_1 x - \lambda_1 xz - \alpha_1 zy - \alpha_2 zw \\ \dot{y} &= a_2 y - \lambda_2 yz + \alpha_1 zx \\ \dot{z} &= x^2 + y^2 - \mu z(z-1)^2 - \epsilon z \\ \dot{w} &= -bw + \alpha_2 zx \end{cases}$$

where all parameters are positive, $a_1, a_2, b, \lambda_1, \lambda_2$ and μ satisfy (1) and ϵ, α_1 and α_2 are small.

To plot the phase portrait of these vector fields we have to carry out two projections. This makes it hard to "see" the attractors. In each column of Figure 3 below we fix all parameters and we show different projections of a same attractor.

Final Remark: If in equation (2) we take $\epsilon = 0$ and we consider a general rotation depending on each y-plane we obtain the following family of vector fields:

$$\begin{cases} \dot{x} &= ax - \lambda xy - g(y)w \\ \dot{y} &= x^2 - \mu y(y-1)^2 \\ \dot{w} &= -bw + g(y)x \end{cases}$$

If the parameters verifies the relations in (1) and if $g(0) < \sqrt{ab}$ the origin is a saddle point with a one dimensional unstable manifold which ends at the saddle-node point $(0,1,0)$. This saddle-node point lies in the boundary of the stable manifold of the saddle point at the origin and attracts and open dense set of $\boldsymbol{R}^3$. Every vector field in this family has the symmetry $\varphi_* X = X$, where $\varphi(x,y,w) = (-x, y, -w)$. Among the systems with this symmetry the ones in the above family have one degeneracy, namely the saddle-node point. A natural perturbation is then $\dot{y} = -\epsilon\, y$.

3 Towards a discrete 2-dimensional Lorenz-like attractor.

In this brief section, we speculate on how to apply the previous ideas to diffeomorphisms of the plane.

The diffeomorphism

$$F(x,y) = ((D(y-y_0)^2 + c)x, g(y) + bx^2),$$

where

$D = (1-c)/(\alpha - y_0)^2$ and $g(y) = y + ay(y+y_m)(y-1)^2(y-y_M)$,

has the phase portrait of Figures 4a) below when the parameters are positive and:

$c < 1;\ 1/3 < \alpha < 1 < y_M < 2y_0 - \alpha;\ \ y_M - y_m < 2\alpha;\ a < 1/(y_m y_M)$.

In order to have local expansiveness near the saddle point at the origin, we can also consider parameters for which $\det(DF_{(0,0)}) > 1$.

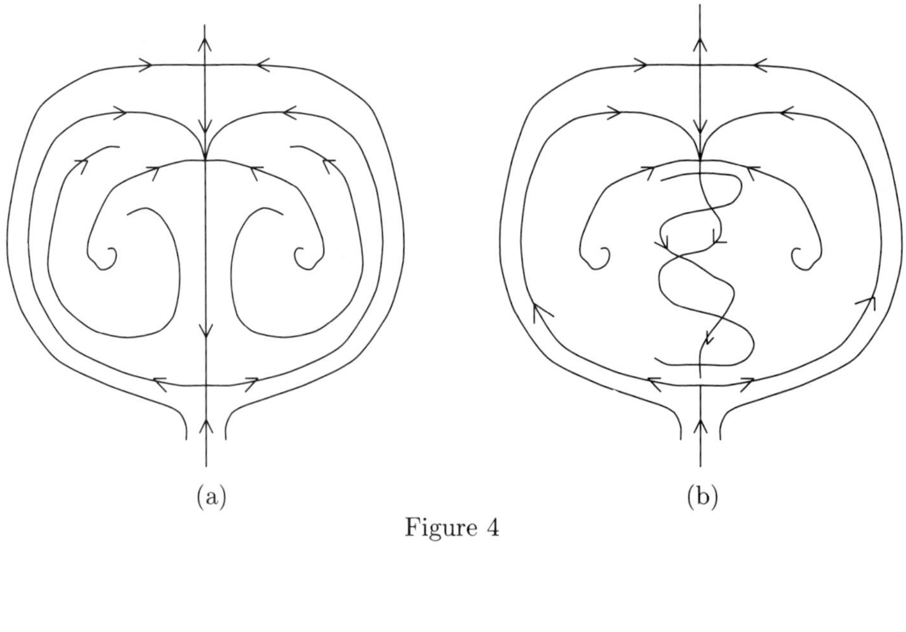

(a) (b)

Figure 4

The saddle-node point at $(0,1)$ attracts an open set of R^2 which contains the unstable manifold of the saddle point at the origin and it is in the closure of the stable manifold of the same saddle point. If we consider a situation as in Figure 4b), where the unstable manifold of the saddle-node point is transversal to the stable manifold of the saddle point, we obtain a diffeomorphism having a saddle-node point as a unique degeneracy.

Considering peturbations, we naturally add a term of the form $(0, -\epsilon\ y)$ with ϵ small. With an extra parameter we would like to produce transversal homoclinic orbits with both branches of the saddle point at the origin. Unfortunately, up to now we haven't been able to find such a good perturbation. Neither have we been able to obtain a perturbation with a phase portrait such as that of Figure 4b). With respect to this last configuration, it seems interesting to study the dynamics when the saddle-node point disappears.

4 About the Figures

Figure 1 spreads out on three pages. It shows some attractors of equation (2). The initial point of integration is close to the unstable manifold of the saddle point at the origin. For the parameters values $\alpha = .18$, .52, .58, .60, 1.0, 1.2 and 1.3 each branch of this unstable manifold converges to a different attractor. In this case two initial points of integration are chosen, one for each branch. In the figure both attractors overlap. A large transient has been omitted. The following parameters have been fixed:

$a = 1.8$; $\lambda = 2$; $\mu = .5$; $b = .6$; and $\epsilon = .09$.

The parameter α varies and takes the values indicated on each small subfigure.

Figure 2 spreads out on two pages. It shows some other attractors of equation (2) corresponding to the following fixed parameters:

$a = .8$; $\lambda = 2$; $\mu = .5$; $b = .4$; and $\epsilon = .1$.

The parameter α varies and takes the values indicated on each small subfigure. The remarks that we made for figure 1 apply in this case as well.

Each column of Figure 3 shows an attractor of equation (3) from different angles. All parameters have been fixed as follows:

$a_1 = 1.9$; $a_2 = 1.1$; $\lambda_1 = \lambda_2 = 2$; $b = 2$; $\mu = .3$; $\alpha_1 = 1.5$; $\alpha = .6$.

For the left column $\epsilon = .05$; for the right column $\epsilon = .09$.

The attractors showed in the figures coexists with other attractors. The showed ones are generated by a numerical integration starting from a point which is close to the unstable manifold of the saddle point at the origin.

For the numerical integration we have used Matlab 5.1

Rodrigo Bamón.
Departamento de Matemática.
Facultad de Ciencias.
Universidad de Chile.
Casilla 653, Santiago.
rbamon@abello.dic.uchile.cl

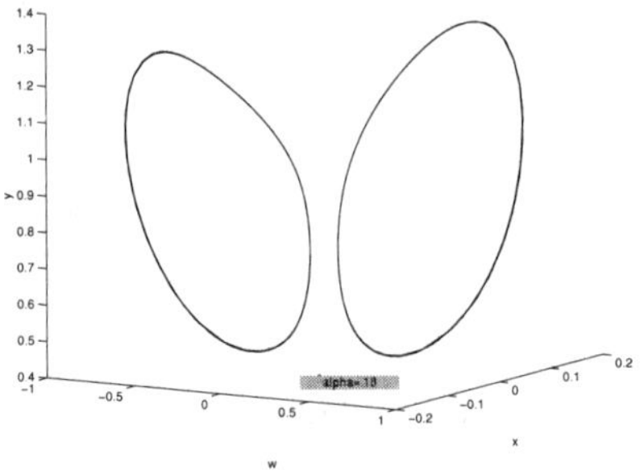

$\alpha = .18$

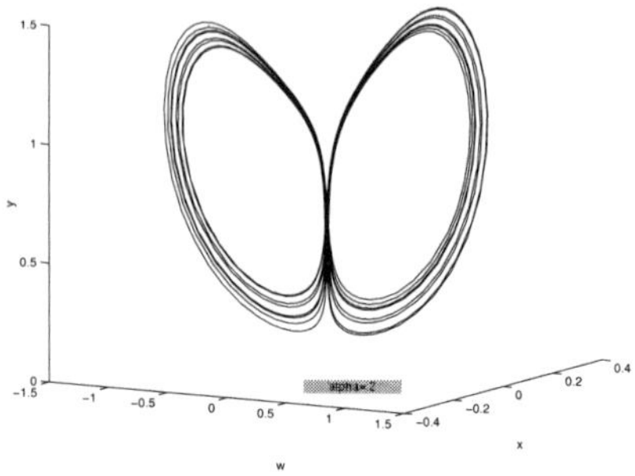

$\alpha = .20$

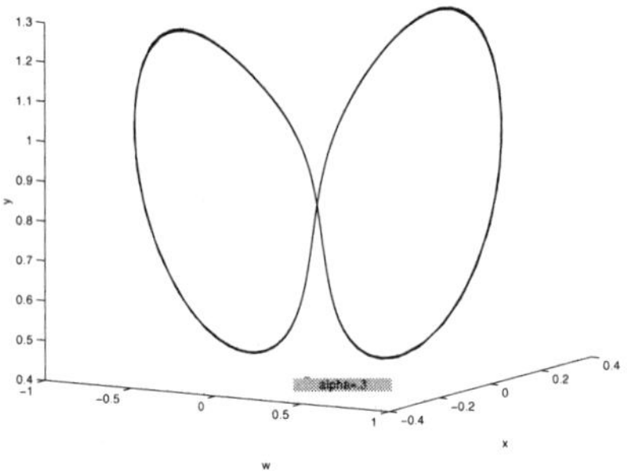

$\alpha = .3$

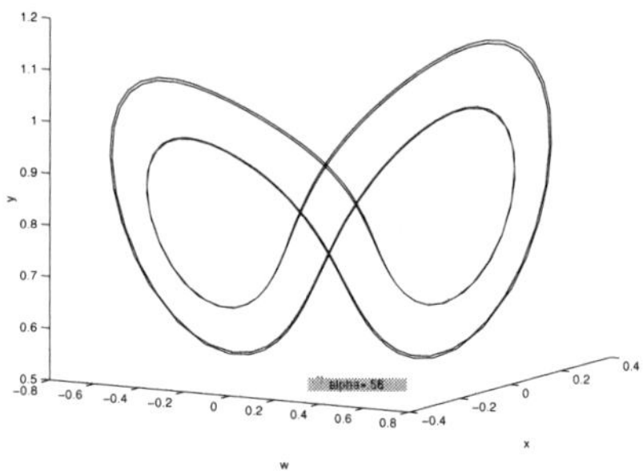

$\alpha = .56$

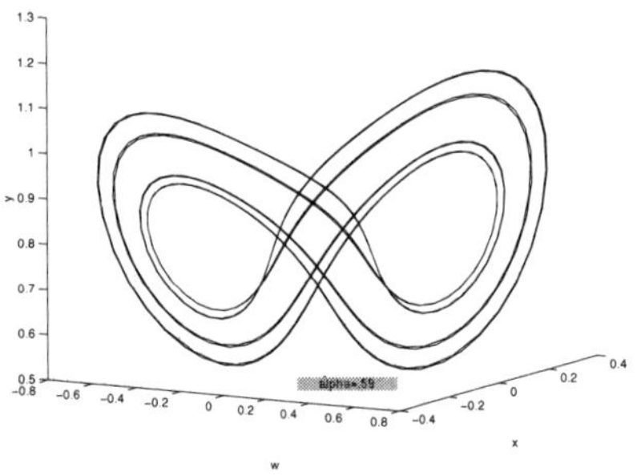

$\alpha = .59$

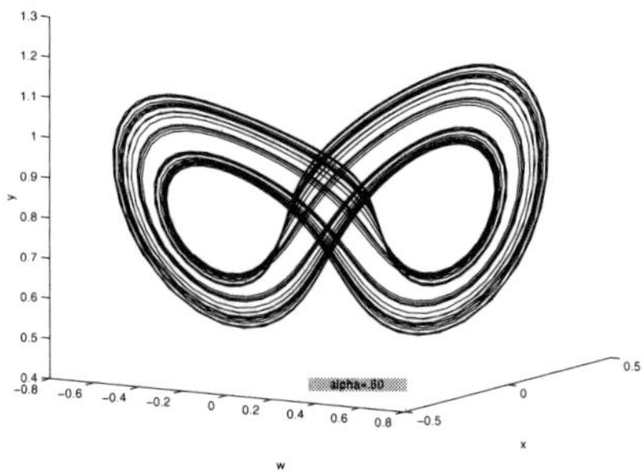

$\alpha = .6$

Figure 1

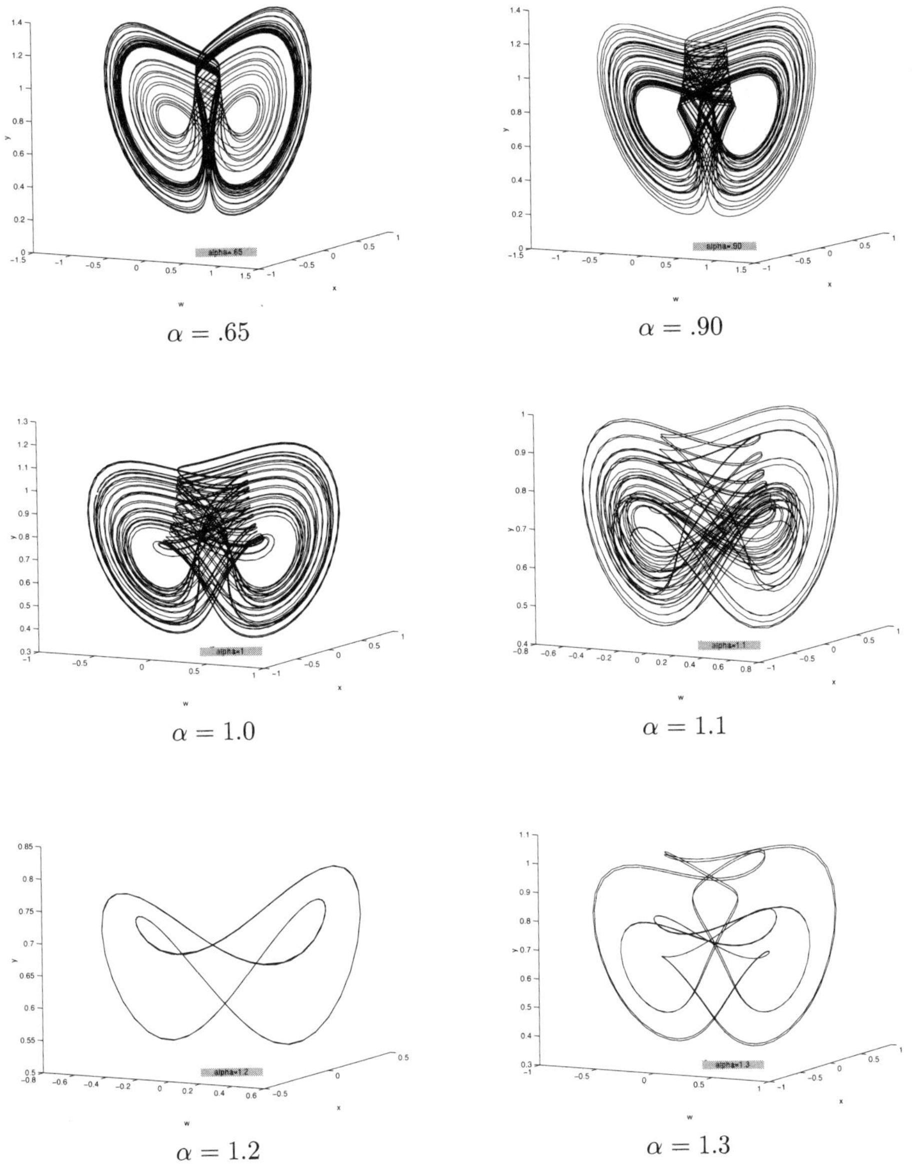

Figure 1 (continuation)

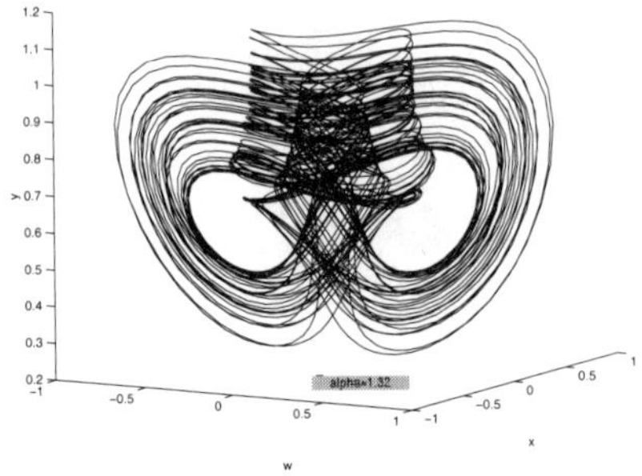

$\alpha = 1.32$

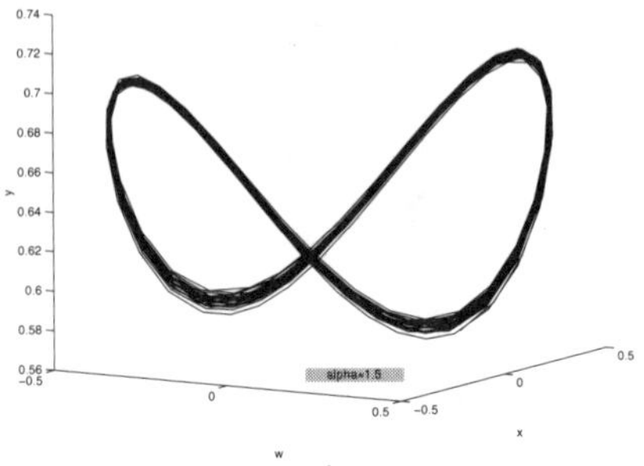

$\alpha = 1.5$

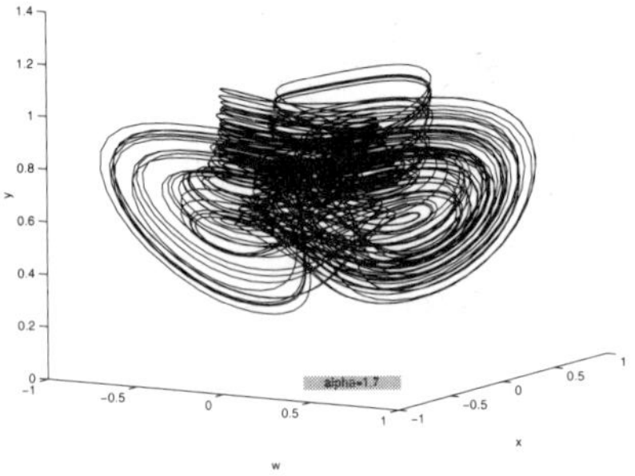

$\alpha = 1.7$

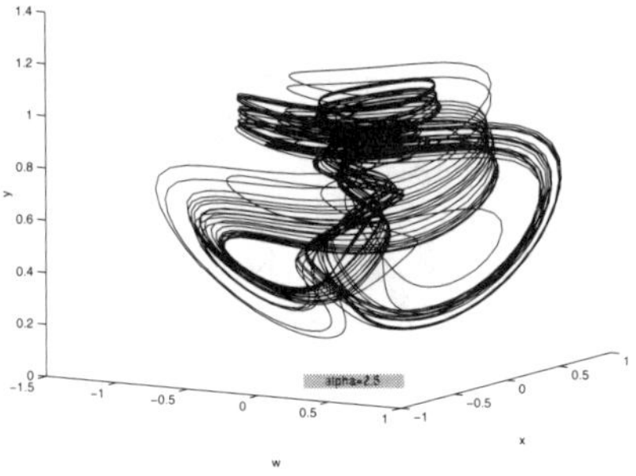

$\alpha = 2.5$

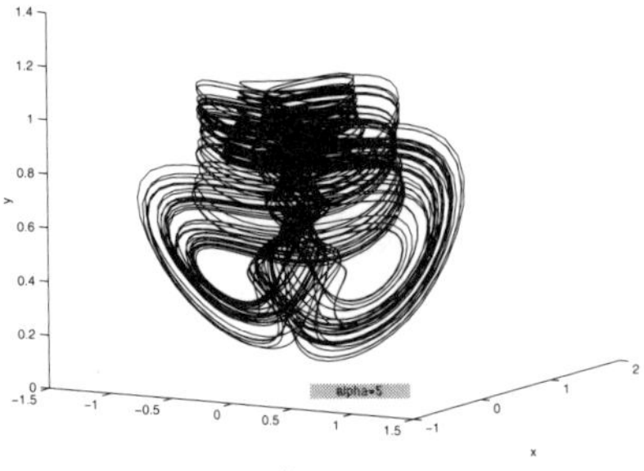

$\alpha = 5.0$

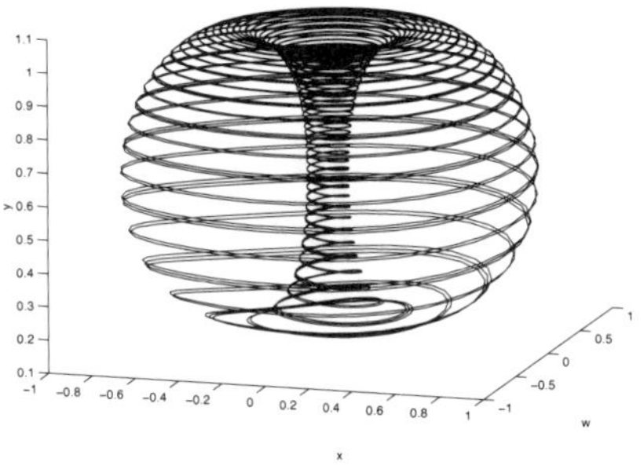

$\alpha = 20.0$

Figure 1 (continuation)

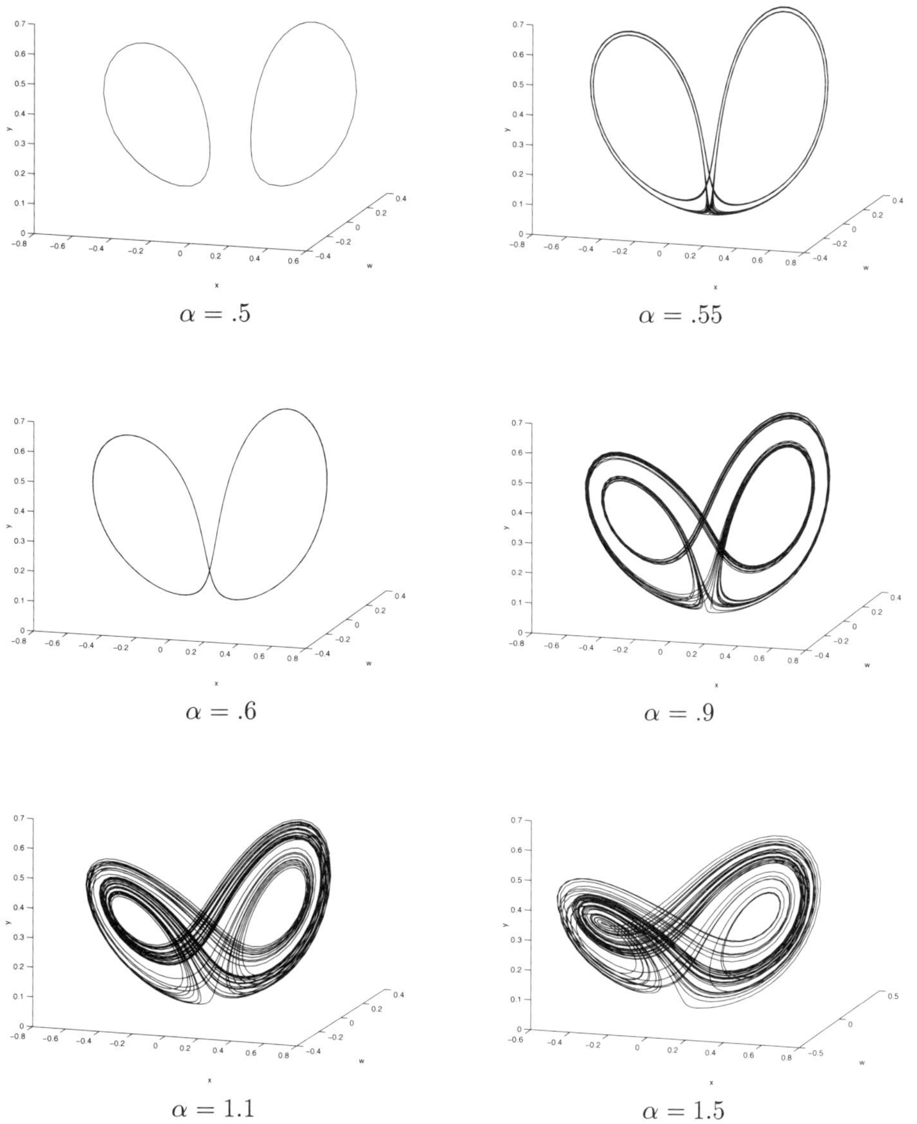

$\alpha = .5$ $\alpha = .55$

$\alpha = .6$ $\alpha = .9$

$\alpha = 1.1$ $\alpha = 1.5$

Figure 2

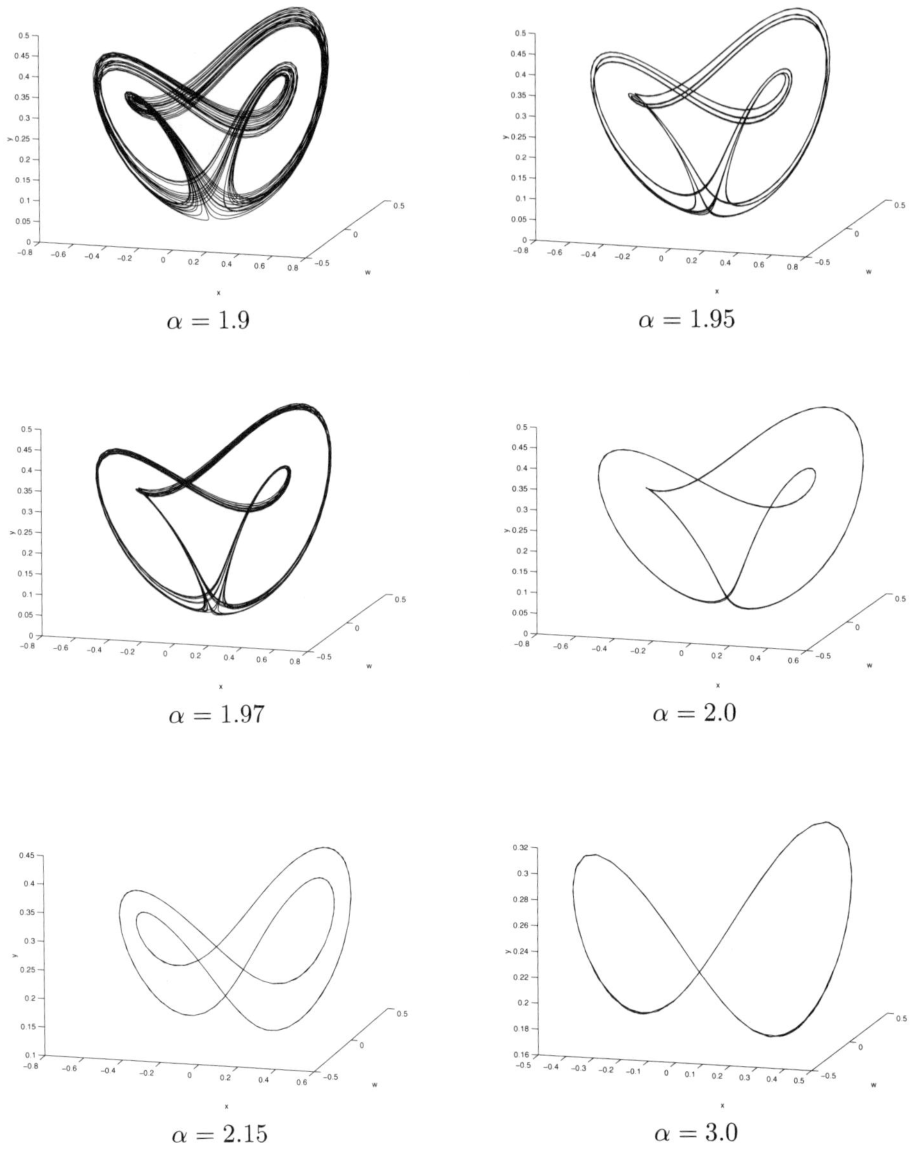

Figure 2 (continuation)

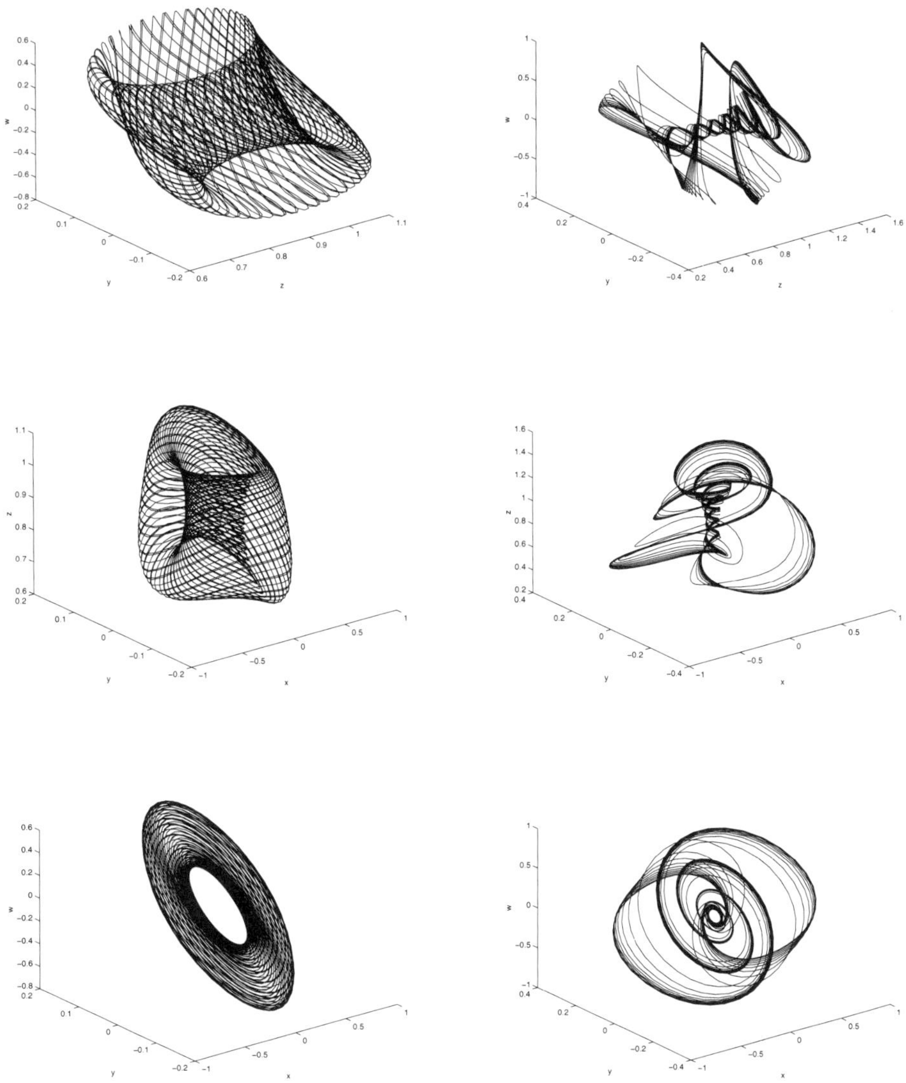
w
0.6 0.4 0.2 0 −0.2 −0.4 −0.6 −0.8
y
0.2 0.1 0 −0.1 −0.2
z
0.6 0.7 0.8 0.9 1 1.1
w
1 0.5 0 −0.5 −1
y
0.4 0.2 0 −0.2 −0.4
z
0.2 0.4 0.6 0.8 1 1.2 1.4 1.6
z
1.1 1 0.9 0.8 0.7 0.6
y
0.2 0.1 0 −0.1 −0.2
x
−1 −0.5 0 0.5 1
z
1.6 1.4 1.2 1 0.8 0.6 0.4 0.2
y
0.4 0.2 0 −0.2 −0.4
x
−1 −0.5 0 0.5 1
w
0.6 0.4 0.2 0 −0.2 −0.4 −0.6 −0.8
y
0.2 0.1 0 −0.1 −0.2
x
−1 −0.5 0 0.5 1
w
1 0.5 0 −0.5 −1
y
0.4 0.2 0 −0.2 −0.4
x
−1 −0.5 0 0.5 1

Figure 3

Contemporary Mathematics
Volume **240**, 1999

Automorphisms of 3-dimensional Abelian Varieties

Ch. Birkenhake, V. González-Aguilera, H. Lange

Abstract. We give a classification of automorphisms of 3-dimensional abelian varieties, completing the results of Roan (see [**RO**]).

Introduction

A complex torus of dimension g is by definition a quotient $X = \mathbb{C}^g/\Lambda$ of the vector space $\mathbb{C}^g$ modulo a lattice Λ of maximal rank. X is called an *abelian variety* if it admits a *polarization*, that is a positive definite hermitian from H on $\mathbb{C}^g$ whose imaginary part is integer valued on Λ. According to the elementary divisor theorem Λ admits a basis such that the matrix of $\operatorname{Im} H$ is of the form $\left(\begin{smallmatrix} 0 & D \\ -D & 0 \end{smallmatrix}\right)$ with a diagonal matrix $D = \operatorname{diag}(d_1, \cdots, d_g)$ and positive integers d_i with $d_i | d_{i+1}$ for $i = 1, \cdots, g$. The g-tupel $(d_1, \cdots, d_g)$ or the matrix D is called the *type* of the polarization H. A polarization of type $(1, \cdots, 1)$ is called a *principal polarization.* The pair (X, H) is called a *polarized abelian variety.* An *automorphism* of X is by definition a biholomorphic map $\alpha : X \to X$ which respects the group law of X. The automorphisms of X form a group. This group is denoted by $\operatorname{Aut}(X)$. In general it is an infinite group. For example if $X = E \times \cdots \times E$, the n-fold product of a general elliptic curve E with itself, then $\operatorname{Aut}(X) = GL_n(\mathbb{Z})$.

An automorphism α of $X = \mathbb{C}^g/\Lambda$ induces an automorphism of the vector space $\mathbb{C}^g$, the *analytic representation* $\rho_a(\alpha)$ of α. We say that α *respects the polarization* H of X if $\rho_a(\alpha)^* H = H$. In that case α is called an *automorphism of the polarized abelian variety* (X, H). The automorphisms of (X, H) form a group denoted by $\operatorname{Aut}_H(X)$, which is always a finite group [**CAV**]. It is well known that for an elliptic curve E and any polarizarion H on E we have $\operatorname{Aut}(E) = \operatorname{Aut}_H(E) = \mathbb{Z}_d$, the cyclic group of order d with $d = 2, 4$ or 6.

Much is known about automorphisms as well as the groups $\operatorname{Aut}(S)$ and $\operatorname{Aut}_H(S)$ in the case of an abelian surface S (see [**FU**]). In [**BGL**] we study the group $\operatorname{Aut}(X)$ of an *abelian threefold* X, that is an abelian variety X of dimension 3. To be more precise we determine the maximal finite subgroups of $\operatorname{Aut}(X)$ in this case or, slightly more generally, for any complex torus of dimension 3. It is the aim of this note to present some results on automorphisms as well as on the group $\operatorname{Aut}_H(X)$ in the case of polarized abelian threefolds (X, H). In order to be as complete as possible we also include some apparently unpublished results of Roan [**RO**], of the

---

1991 *Mathematics Subject Classification.* Primary: 14K10; Secondary: 14K99.

Supported by Conicyt, DGIP U.Santa María and DAAD.

thesis of D. Schmidt [**SCH**] (written under the supervision of the first and third author), and other authors, sometimes without giving explicit attributions.

There are several reasons why one is interested in automorphisms of polarized abelian threefolds:

(1) As for every geometric structure the group of automorphisms is an important invariant. An ultimate goal would be to give a stratification of the moduli space $\mathcal{A}_D$ of polarized abelian varieties of type D.
(2) For any pair (X, H) as above there is an ample line bundle L with first Chern class H. Any two such line bundles differ by a translation of X. Let $\varphi_L : X \to \mathbb{P}_N = P(H^0(L)^*)$ denote the rational map defined by the complete linear system $|L|$. It is often hard to study geometric properties of the map φ_L, in particular in the threefold case. Sometimes one can use automorphisms for these questions. See for example [**BL**].
(3) If C is a smooth projective curve of genus 3 over the field of complex numbers, its Jacobian $(J(C), \Theta)$ is a principally polarized abelian threefold. Torelli's theorem implies that there is a close relationship between the groups $\mathrm{Aut}\,(C)$ and $\mathrm{Aut}_{\Theta}(J(C))$.
    (i) In some cases one can use the automorphisms to determine an explicit period matrix for $J(C)$ (see e.g. [**RAU**]).
    (ii) One can use the automorphism groups to determine a splitting of $J(C)$ (see e.g. [**RIE**]).
(4) The desingularization of the quotient of an abelian threefold modulo a finite group of automorphisms may lead to interesting threefolds. In some cases these varieties are known to be Calabi-Yau threefolds. One should determine all groups of automorphisms which lead to Calabi-Yau threefolds in this way.

This paper is organized as follows: In Section 1 we recall some versions of the fixed point formula and prove some preliminary results. Section 2 contains the decomposition theorem. In Section 3 we study abelian threefolds of CM-type. Section 4 contains the classification of polarized abelian threefolds with an automorphism α of order d such that $X_{<\alpha>} := \cup_{k=1}^{d-1} \mathrm{Fix}\,(\alpha^k)$ is finite. Finally in Section 5 we study abelian threefolds X with an automorphism α such that $\dim X_{<\alpha>} \geq 1$.

1. The Fixed Point Formula

Let $X = \mathbb{C}^g/\Lambda$ be an abelian variety of dimension g. Denote by $\mathrm{End}\,(X)$ the ring of endomorphisms $f : X \to X$. In these terms the group $\mathrm{Aut}\,(X)$ is just the groups of units in $\mathrm{End}\,(X)$. For any integer n different from zero we denote by $n_X : X \to X$ the endomorphism defined by $n_X(x) = nx$. Its kernel X_n is called the group of n-division points.

Let $f \in \mathrm{End}\,(X)$ be any endomorphism. Denote by $\mathrm{Fix}\, f$ the subset of X of fixed points of f:

$$\mathrm{Fix}\, f := \ker\,(\mathrm{id}_X - f)$$

Clearly $\mathrm{Fix}\, f$ is a subgroup either of positive dimension or finite. Denote $\#\mathrm{Fix}\, f := \deg(\mathrm{id}_X - f)$, i.e.

$$\#\mathrm{Fix}\, f = \begin{cases} \text{cardinality of } \ker(\mathrm{id}_X - f) & \text{if } \dim \mathrm{Fix}\, f = 0 \\ 0 & \text{if } \dim \mathrm{Fix}\, f > 0 \end{cases}.$$

In order to state the Fixed Point Formula we first need to introduce the analytic and rational representation of End (X): Any endomorphism $f \in \mathrm{End}\,(X)$ lifts to a $\mathbb{C}$-linear map $\rho_a(f) : \mathbb{C}^g \to \mathbb{C}^g$ in a unique way. Choose an isomorphism $\varphi : \mathbb{Z}^{2g} \to \Lambda$. Then the restriction of $\rho_a(f)$ to Λ defines a $\mathbb{Z}$-linear map

$$\rho_r(f) := \varphi^{-1} \cdot (\rho_a(f)|\Lambda) \cdot \varphi : \mathbb{Z}^{2g} \to \mathbb{Z}^{2g}.$$

The assignments $f \mapsto \rho_a(f)$ and $f \mapsto \rho_r(f)$ define representations

$$\rho_a : \mathrm{End}\,(X) \to GL_g(\mathbb{C})$$

and

$$\rho_r : \mathrm{End}\,(X) \to GL_{2g}(\mathbb{Z})$$

called the *analytic* and the *rational representation* of End (X). By construction $\rho_a(f)$ and $\rho_r(f)$ fit into the following diagram of short exact sequences.

$$\begin{array}{ccccccccc} 0 & \longrightarrow & \mathbb{Z}^{2g} & \xrightarrow{\varphi} & \mathbb{C}^g & \longrightarrow & X & \longrightarrow & 0 \\ & & \downarrow{\scriptstyle \rho_r(f)} & & \downarrow{\scriptstyle \rho_a(f)} & & \downarrow{\scriptstyle f} & & \\ 0 & \longrightarrow & \mathbb{Z}^{2g} & \xrightarrow{\varphi} & \mathbb{C}^g & \longrightarrow & X & \longrightarrow & 0 \end{array}$$

Using this notation we have

LEMMA 1.1 (Fixed Point Formula, 1. Version). a) Fix f *is a closed subgroup of* X *of dimension equal to the multiplicty of* 1 *as an eigenvalue of* $\rho_a(f)$.
b) $\#\mathrm{Fix}\, f = \det(\mathbf{1}_{2g} - \rho_r(f)) = |\det(\mathbf{1}_g - \rho_a(f))|^2$

PROOF. The first statement is obvious. For b) we recall that the rational representation $\rho_r(g)$ of an endomorphism g of X is equivalent (over $\mathbb{C}$) to the direct sum of representations $\rho_a(g) \oplus \overline{\rho_a(g)}$ (see [**CAV**], Section 1.2). Moreover the above diagram shows that

$$\ker g = \ker(\rho_a(g))/\varphi(\ker(\rho_r(g))) = \ker(\rho_a(g))/\Lambda \cap \ker(\rho_a(g)).$$

Hence

$$\deg g = \det \rho_r(g) = |\det(\rho_a(g))|^2.$$

Now apply this equation to $g = \mathrm{id}_X - f$. □

There is a more subtle version of for Fixed Point Formula which involves the special structure of the endomorphism algebra of X: According to Poincare's Reducibility Theorem there is an isogeny

$$X \to X_1^{n_1} \times X_2^{n_2} \times \cdots \times X_r^{n_r}$$

with simple and pairwise non-isogenous abelian varieties X_i of dimension g_i. Denote by

$$\mathrm{End}_{\mathbb{Q}}(X) := \mathrm{End}\,(X) \otimes_{\mathbb{Z}} \mathbb{Q}$$

the endomorphism algebra of $\mathrm{End}_{\mathbb{Q}}(X)$. Then the above isogeny induces a $\mathbb{Q}$-algebra isomorphism

$$(*) \qquad \mathrm{End}_{\mathbb{Q}}(X) \approx \oplus_{i=1}^r M_{n_i}(\mathrm{End}_{\mathbb{Q}}(X_i)).$$

Recall that the endomorphism algebra of a simple abelian variety is a skew-field of finite dimension over $\mathbb{Q}$ (see [**CAV**]). Denote $D_i = \mathrm{End}_{\mathbb{Q}}(X_i)$, $K_i := \mathrm{center}(D_i)$, $e_i := [K_i : \mathbb{Q}]$, $d_i^2 := [D_i : K_i]$ and $N_i : D_i \to \mathbb{Q}$ the reduced norm map. Note that

N_i is well defined even in the non-commutative case. Moreover for $f \in \operatorname{End}_{\mathbb{Q}}(X)$ let

$$f = f_1 + \cdots + f_r$$

be the decomposition (*), i.e., $f_i \in M_{n_i}(\operatorname{End}_{\mathbb{Q}}(X_i))$. With this notation we have

LEMMA 1.2 (Fixed Point Formula). *The number of fixed points of* $f = \sum_{i=1}^{r} f_i \in \operatorname{End}_{\mathbb{Q}}(X) = \oplus_{i=1}^{r} M_{n_i}(\operatorname{End}_{\mathbb{Q}}(X_i)))$ *is*

$$\#\operatorname{Fix} f = \prod_{i=1}^{r} N_i(\det(1 - f_i))^{\frac{2g_i}{d_i e_i}}.$$

PROOF. In view of the first version of the Fixed Point Formula and the decomposition (*) it suffices to show that the restriction of the map $\det \circ \rho_r : \operatorname{End}_{\mathbb{Q}}(X) \to \mathbb{Q}$ to $\operatorname{End}_{\mathbb{Q}}(X_i^{n_i}) = M_{n_i}(D_i)$ is $N_i(\det(\cdot))^{\frac{2g_i}{d_i e_i}}$. To see this note that $\det \circ \rho_r$ is of degree $2g_i n_i$ on $\operatorname{End}_{\mathbb{Q}}(X_i^{n_i})$, X_i being of dimension g_i. On the other hand $N_i \circ \det : M_{n_i}(D_i) \to \mathbb{Q}$ is of degree $n_i d_i e_i$. But any two norm maps of the same degree coincide (see [**MUM**] Lemma p. 179), we get $(\det \circ \rho_r)^{d_i e_i} = (N_i \cdot \det)^{2g_i}$. This implies the assertion. □

REMARK 1.3. Although the components f_i of f depend on the choice of the isogeny $X \to \prod_{i=1}^{n} X_i^{n_i}$, the norm $N_i(\det(1 - f_i))$ does not.

Next we apply the Fixed Point Formula to automorphisms and groups of automorphisms of the abelian variety X.

First let us fix some notation: Suppose $G \subset \operatorname{Aut}(X)$ is a finite group of automorphisms. Denote

$$X_G := \bigcup_{\mathrm{id}_X \neq \alpha \in G} \operatorname{Fix} \alpha.$$

According to Lemma 1 X_G is a closed algebraic subset of X. It was introduced by Roan in [**RO**]. Given an automorphism α denote by d_α its order. An immediate consequence of the Fixed Point Formula is:

PROPOSITION 1.4. *X_G is finite if and only if for all* $\mathrm{id}_X \neq \alpha \in G$ *all eigenvalues of $\rho_a(\alpha)$ (or equivalently those of $\rho_r(\alpha)$) are primitive d_α-th roots of unity.*

REMARK 1.5. In the case of an abelian group it suffices to check the if condition of the Proposition for any set of generators of G. In particular, given a cyclic group $G = \langle \alpha \rangle$ of order d the set X_G is finite if and only if all eigenvalues of $\rho_a(\alpha)$ are primitive d-th roots of unity.

Suppose now that α is an automorphism of X of order d and let $G = \langle \alpha \rangle$ be the associated cyclic group. Assume X_G is finite. Then the fixed points of α are division points. To be more precise we have

PROPOSITION 1.6. *Suppose $X_{<\alpha>}$ is finite. Then*

$$\operatorname{Fix} \alpha \subset \bigcap_{n | d_n \geq 2} X_n.$$

PROOF. All eigenvalues of $\rho_a(\alpha)$ are primitive d-th roots of unity, X_G being finite. Let $n \geq 2$ be a divisor of d. If $\beta = \alpha^{\frac{d}{n}}$, all eigenvalues of $\rho_a(\beta)$ are primitive n-th roots of unity. Hence $\sum_{i=0}^{n-1} \beta^i = 0$ and for any $x \in \operatorname{Fix}\beta$ we have

$$nx = nx - \left(\sum_{i=0}^{n-1} \beta^i\right)(x) = nx - \sum_{i=0}^{n-1} x = 0.$$

Now the assertion follows from the fact that $\operatorname{Fix}\alpha \subseteq \operatorname{Fix}\beta$. □

COROLLARY 1.7. *$\#\operatorname{Fix}\alpha \geq 2$ implies that $d = p^k$ for some prime number p. Moreover $\operatorname{Fix}\alpha \subset X_p$ in this case.*

Denote by φ the Euler function counting the number of primitive roots of unity of a given order d. Then we have

PROPOSITION 1.8. *Suppose α is an automorphism of order d of the g-dimensional abelian variety X such that $X_{\langle\alpha\rangle}$ is finite, then*

a) *$\varphi(d)$ divides $2g$.*
b) *The set $\Phi_\alpha := \{$eigenvalues of $\rho_a(\alpha)\}$ contains $\frac{\varphi(d)}{2}$ different pairwise non complex conjugate primitive d-th roots of unity.*
c) *If $d = p^k$ with p prime, $\#\operatorname{Fix}\alpha = p^{\frac{2g}{\varphi(d)}}$.*

PROOF. According to Proposition 1.4 all eigenvalues of $\rho_r(\alpha)$ are primitive d-th roots of unity. Thus the characteristic polynomial of $\rho_r(\alpha)$,

$$P_\alpha^r(\lambda) := \det(\lambda \mathbf{1}_{2g} - \rho_r(\alpha)),$$

is a multiple of the d-th cyclotomic polynomial μ_d. Hence $\varphi(d) = \deg \mu_d$ must divide $\deg P_\alpha^r = 2g$. This proves a).
The set $\Phi_\alpha \cup \overline{\Phi_\alpha}$ is the set of all eigenvalues of the rational representation $\rho_r(\alpha)$, since ρ_r and $\rho_a \oplus \overline{\rho}_a$ are equivalent representations. But all $\varphi(d)$ d-th roots of unity occur as an eigenvalues of $\rho_r(\alpha)$, since $\mu_d | P_\alpha^r$. This implies b). Comparing degrees we have $P_\alpha^r = \mu_d^{\frac{2g}{\varphi(d)}}$. Hence by the Fixed Point Formula

$$\#\operatorname{Fix}\alpha = \det(1_{2g} - \rho_r(\alpha)) = P_\alpha^r(1) = \mu_\alpha(1)^{\frac{2g}{\varphi(d)}}.$$

Now c) follows from the well known fact that for $d = p^k$:

$$\mu_\alpha(1) = N_{\mathbb{Q}(\xi_d)/\mathbb{Q}}(1 - \xi_d) = p$$

where ξ_d is a primitive d-th root of unity and $N_{\mathbb{Q}(\xi_d)/\mathbb{Q}}$ is the norm map. □

EXAMPLE 1.9. Let α be an automorphism of an elliptic curve of order $d > 1$. Then we have the following possibilities

d	2	3	4	6
$\#\operatorname{Fix}\alpha$	4	3	2	1

PROOF. According to Proposition 1.8 a) the Euler number of d satisfies $\varphi(d)|2$, hence $d \in \{2, 3, 4, 6\}$ and Proposition 1.8 c) and Corollary 1.7 give the assertion. □

It is a well-known fact that these automorphisms can be realized by the elliptic curves with j-invariant 0 and 1728 .

EXAMPLE 1.10. Let α be an automorphism of an abelian surface with $X_{\langle\alpha\rangle}$ finite. Then

d	2	3	4	5	6	8	10	12
$\#\mathrm{Fix}\,\alpha$	16	9	4	5	1	2	1	1

The proof is again an easy computation using Propositions 1.8 and 1.4. □

2. The Decomposition Theorem

Suppose X is an abelian variety of dimension g and α is an automorphism of X of order d. Denote by $G = \langle\alpha\rangle$ the cyclic group defined by α. In this section we study the case that the set X_G is of positive dimension. We will see that G defines a decomposition of X up to isogeny into subtori of X invariant with respect to the action of G. To be more precise we show the following theorem due to Roan.

THEOREM 2.1 (Decomposition Theorem). *Let $G = \langle\alpha\rangle$ be a cyclic group of automorphisms of X. Suppose $1 \leq d_1 < d_2 < \ldots < d_r$ are the orders of the eigenvalues of $\rho_a(\alpha)$. Then there are G-stable abelian subvarieties $X_1, \cdots, X_r$ of X such that*

a) *$\alpha_i = \alpha|X_i$ is of order d_i*

b) $(X_1)_{\langle\alpha_1\rangle} = \begin{cases} X_1 & \text{if } d_1 = 1 \\ \text{finite} & \text{if } d_1 > 1 \end{cases}$ *and $(X_i)_{\langle\alpha_i\rangle}$ is finite for $i > 1$.*

c) *The addition map*

$$\mu : X_1 \times X_2 \times \cdots \times X_r \to X$$

is an isogeny.

REMARK 2.2. One can describe the subtori X_i in the Decomposition Theorem explicitly: Let the notation be as in the Theorem. Define a filtration of X into G-stable abelian subvarieties

$$0 = Y_r \subset Y_{r-1} \subset \cdots \subset Y_1 \subset Y_0 = X$$

by

$$Y_i := \mathrm{im}\left(\mathrm{id} - \alpha^{d_i}\right)\big| Y_{i-1}.$$

Then

$$X_i := \ker\left((\mathrm{id} - \alpha^{d_i})\big| Y_{i-1}\right)_0 = \mathrm{Fix}\left(\alpha^{d_i}\big| Y_{i-1}\right)_0$$

where the index 0 stands for the component of the kernel containing 0.

PROOF OF THE DECOMPOSITION THEOREM. Let the notation be as in the Remark. Obviously the abelian subvarieties Y_i as well as X_i are G-stable. By construction all the eigenvalues of $\rho_a(\alpha|X_i)$ are primitive of order d_i, this implies a) and b).

As for c): Denote for abbreviation $f_i := (\mathrm{id} - \alpha^{d_i})|Y_{i-1}$ for $i = 1, \cdots, r$. So f_i is an endomorphism of Y_{i-1} and $Y_i = \mathrm{im} f_i$ and $X_i = (\ker f_i)_0$ are abelian subvarieties of Y_{i-1}. In particular this shows that the addition map $X_i \times Y_i \to Y_{i-1}$ is an isogeny, for $i = 1, \cdots, r$. Putting all these addition maps together we get a sequences of isogenies

$$Y_r \times X_r \times X_{r-1} \times \cdots \times X_1 \to \cdots \to Y_2 \times X_2 \times X_1 \to Y_1 \times X_1 \to Y_0 = X.$$

This completes the proof of c), since $Y_r = 0$. □

PROPOSITION 2.3. *Suppose $G \subset \operatorname{Aut} X$ is a finite abelian group of automorphisms of an abelian variety X such that X_G is finite. Then G is a cyclic group.*

PROOF. The analytic representation induces a linear action of G on the $\mathbb{Q}$-vector space $H_1(X, \mathbb{Q})$. Let

$$H_1(X, \mathbb{Q}) = V_1 \oplus \cdots \oplus V_s$$

be its decomposition into irreducible G-modules and denote by G_i the image of G in $GL(V_i)$. Since G and thus also G_i is abelian, the group G_i is even contained in the center of $GL_{G_i}(V_i)$. But $GL_{G_i} V_i$ is the group of units of $\operatorname{End}_{G_i} V_i$, which is a skew field by Schur's Lemma. So G_i is cyclic, being a finite subgroup of the group of units of a skew field.

We have to show that G is cyclic too. To see this consider the restriction map: $\operatorname{res}_i : G \to G_i$. By construction res_i is onto. Suppose $\alpha \in \ker \operatorname{res}_i \subset G$, i. e., $\alpha | V_i$ is the identity. By Proposition 1.4 $\alpha = \operatorname{id}_X$, so res_i is an isomorphism. This completes the proof. □

3. Abelian Threefolds of CM-type

In this section we study some abelian threefolds of CM-type with automorphism α of order 7 and 9. We give the period matrices, the rational and the analytic representations of α in some normalized form. Moreover we prove a theorem due to Roan, which is essential for the classification of abelian threefolds with automorphism of the next section.

A *CM-field* K is by definition a totally imaginary quadratic extension of a totally real number field, say of degree g over $\mathbb{Q}$. A CM-type of K is a set $\Phi = \{\sigma_1, \cdots, \sigma_g\}$ of embeddings $K \mapsto \mathbb{C}$, pairwise not complex conjugate. An abelian variety $X = \mathbb{C}^g/\Lambda$ is called to be of *CM-type* (K, Φ), if there is an embedding $\rho : K \mapsto \operatorname{End}_{\mathbb{Q}}(X)$, such that:

$$\rho_a \circ \rho \sim_{\mathbb{C}} \operatorname{diag}(\sigma_1, \cdots, \sigma_g) : K \mapsto M_g(\mathbb{C}).$$

Here "$\sim_{\mathbb{C}}$" means equivalence of representations over $\mathbb{C}$.

To every CM-type Φ of K one can associate an abelian variety $X(K, \Phi)$ in a canonical way: The tensor product $K \otimes_{\mathbb{Q}} \mathbb{R}$ is an $\mathbb{R}$-vector space of dimension $2g$. The CM-type $\Phi = (\sigma_1, \cdots, \sigma_g)$ induces a complex structure on $K \otimes_{\mathbb{Q}} \mathbb{R}$ via the $\mathbb{R}$-linear isomorphism

$$(\sigma_1, \cdots, \sigma_g) : K \otimes_{\mathbb{Q}} \mathbb{R} \xrightarrow{\sim} \mathbb{C}^g.$$

The ring of integers $\mathcal{O}$ of K is a lattice of rank $2g$ in $K \otimes_{\mathbb{Q}} \mathbb{R}$. Hence the quotient

$$X(K, \Phi) := K \otimes_{\mathbb{Q}} \mathbb{R}/\mathcal{O}$$

is a complex torus of dimension g. According to [**MUM**] p. 213 $X(K, \Phi)$ is an abelian variety. By construction we have

$$\mathcal{O} \subset \operatorname{End}(X(K, \Phi))$$

implying

$$\mathcal{O}^* \subset \operatorname{Aut}(X(K, \Phi)).$$

There are exactly two cyclotomic fields leading to abelian threefolds in this way, namely $\mathbb{Q}(\xi_7)$ and $\mathbb{Q}(\xi_9)$. The cyclotomic field $\mathbb{Q}(\xi_7)$ admits 2 different CM-types, namely $\Phi_1^7 = \{\xi_7, \xi_7^2, \xi_7^3\}$ and $\Phi_2^7 = \{\xi_7, \xi_7^2, \xi_7^4\}$. Similarly $\mathbb{Q}(\xi_9)$ admits two different CM-types $\Phi_1^9 = \{\xi_9, \xi_9^2, \xi_9^4\}$ and $\Phi_2^9 = \{\xi_9, \xi_9^4, \xi_9^7\}$.

PROPOSITION 3.1. *There exist 4 abelian threefolds $X(\mathbb{Q}(\xi_d), \Phi_i^d)$, for $d = 7$ and $d = 9$ and $i = 1, 2$, admiting automorphisms ξ_7 and $-\xi_7$ (respectively ξ_9 and $-\xi_9$) of order 7 and 14 (respectively of order 9 and 18) on $X(\mathbb{Q}(\xi_7), \Phi_i^7)$, (respectively $X(\mathbb{Q}(\xi_9), \Phi_i^9)$). The matrices Π_i^d below are period matrices for $X(\mathbb{Q}(\xi_d), \Phi_i^d)$:*

$$\Pi_1^7 = \begin{pmatrix} -1+2\xi_7+\xi_7^3+\xi_7^6 & 1+\xi_7^3+\xi_7^5 & 1+\xi_7^5-\xi_7^6 & 1 & 0 & 0 \\ 1+\xi_7^3+\xi_7^5 & \xi_7^2 & -\xi_7-\xi_7^3-\xi_7^5 & 0 & 1 & 0 \\ 1+\xi_7^5-\xi_7^6 & -\xi_7-\xi_7^3-\xi_7^5 & -\xi_7^5 & 0 & 0 & 1 \end{pmatrix}$$

$$\Pi_2^7 = \begin{pmatrix} 3\frac{\xi_7+\xi_7^2+\xi_7^4}{4} & \frac{-2+\xi_7+\xi_7^2+\xi_7^4}{4} & \frac{-4-\xi_7-\xi_7^2-\xi_7^4}{4} & 1 & 0 & 0 \\ \frac{-2+\xi_7+\xi_7^2+\xi_7^4}{4} & \frac{-2+\xi_7+\xi_7^2+\xi_7^4}{4} & \frac{-10+\xi_7+\xi_7^2+\xi_7^4}{4} & 0 & 1 & 0 \\ \frac{-4-\xi_7-\xi_7^2-\xi_7^4}{4} & \frac{-10+\xi_7+\xi_7^2+\xi_7^4}{4} & 7\frac{-4+\xi_7+\xi_7^2+\xi_7^4}{4} & 0 & 0 & 7 \end{pmatrix}$$

$$\Pi_1^9 = \begin{pmatrix} \begin{smallmatrix} \frac{1}{397}(174+139\xi_9 \\ +54\xi_9^2+135\xi_9^3 \\ +101\xi_9^4-95\xi_9^5) \end{smallmatrix} & \begin{smallmatrix} \frac{1}{397}(-515-3\xi_9 \\ -64\xi_9^2-160\xi_9^3 \\ -105\xi_9^4+142\xi_9^5) \end{smallmatrix} & \begin{smallmatrix} \frac{1}{397}(-549-199\xi_9 \\ -143\xi_9^2-159\xi_9^3 \\ +181\xi_9^4-241\xi_9^5) \end{smallmatrix} & 1 & 0 & 0 \\ \begin{smallmatrix} \frac{1}{397}(-515-3\xi_9 \\ -64\xi_9^2-160\xi_9^3 \\ -105\xi_9^4+142\xi_9^5) \end{smallmatrix} & \begin{smallmatrix} \frac{1}{397}(1125+180\xi_9 \\ +267\xi_9^2+72\xi_9^3 \\ +345\xi_9^4-183\xi_9^5) \end{smallmatrix} & \begin{smallmatrix} \frac{1}{397}(783+30\xi_9 \\ +243\xi_9^2+12\xi_9^3 \\ -141\xi_9^4+168\xi_9^5) \end{smallmatrix} & 0 & 3 & 0 \\ \begin{smallmatrix} \frac{1}{397}(-549-199\xi_9 \\ -143\xi_9^2-159\xi_9^3 \\ +181\xi_9^4-241\xi_9^5) \end{smallmatrix} & \begin{smallmatrix} \frac{1}{397}(783+30\xi_9 \\ +243\xi_9^2+12\xi_9^3 \\ -141\xi_9^4+168\xi_9^5) \end{smallmatrix} & \begin{smallmatrix} \frac{1}{397}(1917+402\xi_9 \\ +636\xi_9^2+399\xi_9^3 \\ -222\xi_9^4+822\xi_9^5) \end{smallmatrix} & 0 & 0 & 3 \end{pmatrix}$$

$$\Pi_2^9 = \begin{pmatrix} 1+\xi_9^3 & 0 & 0 & 1 & 0 & 0 \\ 0 & 1+\xi_9^3 & 0 & 0 & 1 & 0 \\ 0 & 0 & 1+\xi_9^3 & 0 & 0 & 1 \end{pmatrix}$$

Π_1^7 and Π_2^9 define principal polarizations, Π_2^7 defines a polarization of type $(1, 1, 7)$ and Π_1^9 defines a polarization of type $(1, 3, 3)$.

PROOF. Theorem 4.2 below implies that $X(\mathbb{Q}(\xi_d), \Phi_i^d)$ are the only abelian threefolds on which ξ_d induces an automorphism of order d for $d = 7$ and $d = 9$. Hence it suffices to show that there are matrices $\rho_a^i(\xi_d) \in M_3(\mathbb{C})$ and $\rho_r^i(\xi_d) \in M_6(\mathbb{Z})$ of order d such that

$$(*) \qquad \rho_a^i(\xi_d)\Pi_i^d = \Pi_i^d \rho_r^i(\xi_d)$$

These matrices are then the analytic and rational representation of the automorphism ξ_d. It is an immediate computation to check equation (*) with the following pairs of matrices:

$$\rho_a^1(\xi_7) = \begin{pmatrix} -1+2\xi_7+\xi_7^3+\xi_7^6 & -1+\xi_7-\xi_7^2+\xi_7^3-\xi_7^4 & 1+2\xi_7+2\xi_7^3+2\xi_7^5 \\ 1+\xi_7^3+\xi_7^5 & 1+\xi_7^2+\xi_7^3+\xi_7^5 & 1-\xi_7+\xi_7^2 \\ 1+\xi_7^5-\xi_7^6 & 2+\xi_7^2+\xi_7^4+\xi_7^5 & 1+\xi_7^2+\xi_7^4 \end{pmatrix}$$

and

$$\rho_r^1(\xi_7) = \begin{pmatrix} 0 & 0 & 1 & 1 & 1 & 1 \\ 0 & 0 & 1 & 0 & 1 & 1 \\ 0 & 1 & 0 & 0 & 0 & 1 \\ -1 & 0 & 0 & 0 & 0 & 0 \\ 1 & -1 & 0 & 0 & 0 & 0 \\ 0 & 0 & -1 & 0 & 0 & -1 \end{pmatrix},$$

$$\rho_a^2(\xi_7) = \begin{pmatrix} 1+3\frac{\xi_7+\xi_7^2+\xi_7^4}{4} & \frac{-2-\xi_7-\xi_7^2-\xi_7^4}{4} & \frac{5+3(\xi_7+\xi_7^2+\xi_7^4)}{4} \\ \frac{2+\xi_7+\xi_7^2+\xi_7^4}{4} & \frac{-2-\xi_7-\xi_7^2-\xi_7^4}{4} & \frac{1+2(\xi_7+\xi_7^2+\xi_7^4)}{7} \\ -1-\frac{\xi_7+\xi_7^2+\xi_7^4}{4} & \frac{10-\xi_7-\xi_7^2-\xi_7^4}{4} & \frac{-1+\xi_7+\xi_7^2+\xi_7^4}{2} \end{pmatrix}$$

and

$$\rho_r^2(\xi_7) = \begin{pmatrix} 0 & 0 & 0 & 1 & 0 & 0 \\ 0 & -3 & -7 & 0 & -1 & 7 \\ 0 & 0 & -1 & 0 & 0 & 1 \\ -1 & -3 & -7 & 1 & -1 & 7 \\ 0 & -2 & -7 & 1 & -1 & 7 \\ 0 & -1 & -4 & 0 & 0 & 3 \end{pmatrix},$$

$$\rho_a^1(\xi_9) = \begin{pmatrix} \begin{matrix}\frac{1}{397}(78+76\xi_9 \\ -99\xi_9^2-49\xi_9^3 \\ +278e_7^4-289e_7^5)\end{matrix} & \begin{matrix}\frac{1}{1191}(-381+209\xi_9 \\ -173\xi_9^2-234\xi_9^3 \\ +169\xi_9^4-100\xi_9^5)\end{matrix} & \begin{matrix}\frac{1}{1191}(642-199\xi_9 \\ -143\xi_9^2-159\xi_9^3 \\ +181\xi_9^4-241\xi_9^5)\end{matrix} \\ \begin{matrix}\frac{1}{397}(-313+204\xi_9 \\ +382\xi_9^2-236\xi_9^3 \\ -6\xi_9^4+269\xi_9^5)\end{matrix} & \begin{matrix}\frac{1}{397}(77+187\xi_9 \\ +284\xi_9^2-84\xi_9^3 \\ +193\xi_9^4+15\xi_9^5)\end{matrix} & \begin{matrix}\frac{1}{397}(-136+10\xi_9 \\ +81\xi_9^2+4\xi_9^3 \\ -47\xi_9^4+56\xi_9^5)\end{matrix} \\ \begin{matrix}\frac{1}{397}(411+34\xi_9 \\ +593\xi_9^2+93\xi_9^3 \\ -\xi_9^4+508\xi_9^5)\end{matrix} & \begin{matrix}\frac{1}{397}(79-35\xi_9 \\ +312\xi_9^2-14\xi_9^3 \\ -34\xi_9^4+201\xi_9^5)\end{matrix} & \begin{matrix}\frac{1}{397}(-155+134\xi_9 \\ +212\xi_9^2+133\xi_9^3 \\ -74\xi_9^4+274\xi_9^5)\end{matrix} \end{pmatrix}$$

and

$$\rho_r^1(\xi_9) = \begin{pmatrix} 0 & 0 & 0 & 2 & 3 & 0 \\ -1 & 2 & 1 & 1 & 3 & 0 \\ 0 & 0 & 0 & 1 & 1 & 1 \\ -2 & 3 & 3 & 2 & 3 & 3 \\ 1 & -2 & -1 & -1 & -2 & -1 \\ 1 & -2 & -2 & -1 & -2 & -2 \end{pmatrix},$$

$$\rho_a^2(\xi_9) = \begin{pmatrix} 0 & 0 & \xi_9^3 \\ 1 & 0 & 0 \\ 0 & 1 & 0 \end{pmatrix} \quad \text{and} \quad \rho_r^2(\xi_9) = \begin{pmatrix} 0 & 0 & 0 & 0 & 0 & 1 \\ 1 & 0 & 0 & 0 & 0 & 0 \\ 0 & 1 & 0 & 0 & 0 & 0 \\ 0 & 0 & -1 & 0 & 0 & -1 \\ 0 & 0 & 0 & 1 & 0 & 0 \\ 0 & 0 & 0 & 0 & 1 & 0 \end{pmatrix}$$

□

Let X be an abelian variety of dimension g and α an automorphism of order d of X with $X_{\langle\alpha\rangle}$ finite. According to Proposition 1.8 $\varphi(d)$ divides $2g$. In [**BGL**], Proposition 3.3 we showed that if $\varphi(d) = 2g$, then $(\mathbb{Q}(\xi_d), \Phi_\alpha)$ is a CM-field and X is an abelian variety of CM-type $(\mathbb{Q}(\xi_d), \Phi_\alpha)$. The following theorem due to Roan (see [**RO**]) is a generalization of this.

THEOREM 3.2 (Roan). *Let X be an abelian variety of dimension g and α an automorphism of X of order $d \geq 3$ with $X_{\langle\alpha\rangle}$ finite and $\#\Phi_\alpha = \frac{\varphi(d)}{2}$. Then $\varphi(d)|2g$ and $(\mathbb{Q}(\xi), \Phi_\alpha)$ is a CM-field, there are $k = \frac{2g}{\varphi(d)}$ abelian varieties $X_1, \cdots, X_k$ of CM-type $(\mathbb{Q}(\xi_d), \Phi_\alpha)$ such that*

$$X \cong X_1 \times \cdots \times X_k,$$

and α decomposes into a product of automorphisms of the X_ν induced by primitive d-th roots of unities.

PROOF. According to the proof of Proposition 3.1 the automorphism α induces a $\mathbb{Q}(\xi_d)$-vector space structure on $H_1(X, \mathbb{Q})$. Moreover $H_1(X, \mathbb{Z})$ is a torsion free $\mathbb{Z}[\xi_d]$-module of rank k, since all eigenvalues of $\rho_r(\alpha)$ are primitive, $\text{rk}_\mathbb{Z}\, \mathbb{Z}[\xi_d] = \varphi(d)$, and $\text{rk}_\mathbb{Z} H_1(X, \mathbb{Z}) = 2g$. Now $\mathbb{Z}[\xi]$ is a Dedekind ring. Hence there are ideals $I_1, \cdots, I_k \subseteq \mathbb{Z}[\xi_d]$ such that

$$H_1(X, \mathbb{Z}) \cong I_1 \oplus \cdots \oplus I_k.$$

Denote $V_\nu := I_\nu \otimes_\mathbb{Z} \mathbb{Q} = \mathbb{Q}(\xi_d)$. Then

$$H_1(X, \mathbb{Q}) = V_1 \oplus \cdots \oplus V_k$$

is a decomposition of $H_1(X, \mathbb{C})$ into one-dimensional $\mathbb{Q}(\xi_d)$-vector spaces, inducing an isomorphism of real tori:

$$X \cong X_1 \times \cdots \times X_k$$

with $X_\nu = V_\nu \otimes_{\mathbb{Q}} \mathbb{R}/I_\nu$. The analytic representation induces on the lattice the rational representation, implying $\rho_a(\alpha)I_\nu = \rho_r(\alpha)I_\nu \subseteq I_\nu$ and hence $\rho_a(\alpha)V_\nu \subseteq V_\nu$ for $\nu = 1, \cdots, k$. Hence by construction $\rho_a(\alpha)$ stabilizes the real tori X_ν.

According to Proposition 1.8 b) $(\mathbb{Q}(\xi_d), \Phi_\alpha)$ is a CM-field, since $\#\Phi_\alpha = \frac{\varphi(d)}{2} =: s$. The set Φ_α defines a complex structure J_ν on $V_\nu \otimes_{\mathbb{Q}} \mathbb{R} = \mathbb{Q}(\xi_d) \otimes_{\mathbb{Q}} \mathbb{R}$. It remains to show that the complex structure on $V_\nu \otimes_{\mathbb{Q}} \mathbb{R}$ induced by the complex structure of $H_1(X, \mathbb{R})$ coincides with J_ν.

Denoting $\Phi_\alpha = \{\lambda_1, \cdots, \lambda_s\}$, the complex structure J_ν on $V_\nu \otimes_{\mathbb{Q}} \mathbb{R} = \mathbb{Q}(\xi_\alpha) \otimes_{\mathbb{Q}} \mathbb{R}$ is defined by the following commutative diagram

$$\begin{array}{ccc} V_\nu \otimes_{\mathbb{Q}} \mathbb{R} = \mathbb{Q}(\xi_d) \otimes_{\mathbb{Q}} \mathbb{R} & \xrightarrow{\phi_\nu} & \mathbb{C}^s \\ {\scriptstyle J_\nu}\downarrow & & \downarrow{\scriptstyle i\mathbf{1}_s} \\ V_\nu \otimes_{\mathbb{Q}} \mathbb{R} = \mathbb{Q}(\xi_d) \otimes_{\mathbb{Q}} \mathbb{R} & \xrightarrow{\phi_\nu} & \mathbb{C}^s \end{array}$$

where ϕ_ν is defined by $(\lambda_1, \cdots, \lambda_s)$. Since $1, \xi_d, \cdots, \xi_d^{d-1}$ generate the $\mathbb{R}$-vector space $\mathbb{Q}(\xi) \otimes_{\mathbb{Q}} \mathbb{R}$, there are real numbers $r_0, \cdots, r_{d-1}$ such that the image of $1 \otimes 1 \in \mathbb{Q}(\xi_d) \otimes_{\mathbb{Q}} \mathbb{R} = V_\nu \otimes_{\mathbb{Q}} \mathbb{R}$ under J_ν is of the form

$$J_\nu(1 \otimes 1) = \sum_{j=0}^{d-1} \xi_d^j \otimes r_j$$

This implies

$$\begin{pmatrix} i \\ \vdots \\ i \end{pmatrix} = i\,\phi_\nu(1 \otimes 1) = \phi_\nu J_\nu(1 \otimes 1)$$

$$= \begin{pmatrix} \lambda_1 J_\nu(1\otimes 1) \\ \vdots \\ \lambda_s J_\nu(1\otimes 1) \end{pmatrix} = \begin{pmatrix} \lambda_1\left(\sum_{j=1}^{d-1} \xi_d^j \otimes r_j\right) \\ \vdots \\ \lambda_s\left(\sum_{j=1}^{d-1} \xi_d^j \otimes r_j\right) \end{pmatrix} = \sum_{j=1}^{d-1} r_j \begin{pmatrix} \lambda_1^j \\ \vdots \\ \lambda_s^j \end{pmatrix}$$

On the other hand $\lambda_1, \cdots, \lambda_s$ are exactly the eigenvalues of $\rho_a(\alpha)$. Hence this equation implies

$$i \cdot \mathrm{id}_{H_1(X,\mathbb{R})} = \sum_{j=0}^{d-1} r_j \rho_a(\alpha)^j$$

where $H_1(X, \mathbb{R})$ is considered as a complex vector space. This gives the assertion. □

As an immediate consequence we obtain

COROLLARY 3.3. *Suppose α is an automorphism of an abelian variety of order $d \geq 3$ with $\rho_a(\alpha) = \xi_d \cdot id$. Then $d = 3, 4$ or 6 and*

$$X = E \times \cdots \times E$$

where E denotes the elliptic curve admitting an automorphism of order d.

4. Abelian Threefolds with $\dim X_{\langle\alpha\rangle} = 0$

In this section we determine all pairs (X,α) where X is an abelian threefold and α an automorphism of X of order $d \geq 3$ such that the set $X_{\langle\alpha\rangle}$ is finite.

Suppose (X,α) is such a pair. According to Proposition 1.8 the number $\varphi(d)$ divides 6, hence

$$d = \operatorname{ord}\alpha \in \{3,4,6,7,9,14,18\}.$$

Moreover by loc. cit. the set of eigenvalues Φ_α of α consists of at least $\frac{\varphi(d)}{2}$ primitive non-complex conjugate d-th roots of unity. Replacing eventually α by a suitable power this gives:

Lemma 4.1. *The pair (d,Φ_α) is contained in the following list:*

d	Φ_α
3	$\{\rho\}$
	$\{\rho,\rho^2\}$
4	$\{i\}$
	$\{i,-i\}$
6	$\{-\rho\}$
	$\{-\rho,-\rho^2\}$

7	$\{\xi_7,\xi_7^2,\xi_7^3\} = \Phi_1^7$
	$\{\xi_7,\xi_7^2,\xi_7^4\} = \Phi_2^7$
14	$\{-\xi_7,-\xi_7^2,-\xi_7^3\}$
	$\{-\xi_7,-\xi_7^2,-\xi_7^4\}$
9	$\{\xi_9,\xi_9^2,\xi_9^4\} = \Phi_1^9$
	$\{\xi_9,\xi_9^4,\xi_9^7\} = \Phi_2^9$
18	$\{-\xi_9,-\xi_9^2,-\xi_9^4\}$
	$\{-\xi_9,-\xi_9^4,-\xi_9^7\}$

Here $\rho := \xi_3$, a primitive third root of unity.

If $\#\Phi_\alpha = \frac{\varphi(d)}{2}$ (that is $d \in \{3,4,6\}$ and $\#\Phi_\alpha = 1$ or $d \in \{7,9,14,18\}$), there is a uniquely determined abelian threefold X admitting an automorphism α with eigenvalues Φ_α as in the lemma. To be more precise, according to the results of section 3 we have

Theorem 4.2. *Let (X,α) be an abelian threefold with an automorphism α of order d such that $X_{\langle\alpha\rangle}$ is finite and $\#\Phi_\alpha = \frac{\varphi(d)}{2}$. Then X admits a polarization H invariant with respect to α and replacing eventually α by a suitable power the data $(X,\Phi_\alpha,$ type of $H)$ are contained in the following list*

d =*ord* α	X	Φ_α	*type of* H
3	$E_\rho \times E_\rho \times E_\rho$	$\{\rho\}$	$(3,3,3)$
4	$E_i \times E_i \times E_i$	$\{i\}$	$(4,4,4)$
6	$E_\rho \times E_\rho \times E_\rho$	$\{-\rho\}$	$(3,3,3)$
7	$X(\mathbb{Q}(\xi_7), \Phi_1^7)$	$\Phi_1^7 = \{\xi_7, \xi_7^2, \xi_7^3\}$	$(1,1,1)$
	$X(\mathbb{Q}(\xi_7), \Phi_2^7)$	$\Phi_2^7 = \{\xi_7, \xi_7^2, \xi_7^4\}$	$(1,1,7)$
14	$X(\mathbb{Q}(\xi_7), \Phi_1^7)$	$-\Phi_1^7$	$(1,1,1)$
	$X(\mathbb{Q}(\xi_7), \Phi_2^7)$	$-\Phi_2^7$	$(1,1,7)$
9	$X(\mathbb{Q}(\xi_9), \Phi_1^9)$	$\Phi_1^9 = \{\xi_9, \xi_9^2, \xi_9^4\}$	$(1,3,3)$
	$X(\mathbb{Q}(\xi_9), \Phi_2^9)$	$\Phi_2^9 = \{\xi_9, \xi_9^4, \xi_9^7\}$	$(1,1,1)$
18	$X(\mathbb{Q}(\xi_9), \Phi_1^9)$	$-\Phi_1^9$	$(1,3,3)$
	$X(Q(\xi_9), \Phi_9^2)$	$-\Phi_2^9$	$(1,1,1)$

Here E_τ *denotes the elliptic curve* $\mathbb{C}/\tau\mathbb{Z} + \mathbb{Z}$.

PROOF. This an immediate consequence of Theorem 3.2. For an explicit description of the of the abelian threefolds of CM-type (cases $d = 7, 14, 9$ or 18) in terms of period matrices see Section 3. □

We now turn to the remaining case, $\#\Phi_\alpha > \frac{\varphi(d)}{2}$: Suppose (X, H, α) is a polarized abelian threefold with an automorphism α of order $d \in \{3,4,6\}$, a finite set $X_{\langle\alpha\rangle}$ and $\#\Phi_\alpha = 2$. Then the automorphism α induces an embedding $\mathbb{Q}(\xi_d) \mapsto \operatorname{End}_{\mathbb{Q}}(X)$, i.e. X admits complex multiplication by $\mathbb{Q}(\xi_d)$. Such abelian threefolds were constructed by Shimura (see [**CAV**] Section 9.6). For the sake of completeness we sketch this construction:

Suppose $d = 3, 4$ or 6. Let $\mathcal{M} \subset \mathbb{Q}(\xi_d)^{\oplus 3} \subset \mathbb{C}^3$ be a free $\mathbb{Z}$-submodule of rank 6 and $T \in GL_3(\mathbb{Q}(\xi_d))$ a nondegenerate matrix satisfying

$$tr_{\mathbb{Q}(\xi_d)/\mathbb{Q}}\left({}^t\lambda T \mu\right) \in \mathbb{Z}$$

for all $\lambda, \mu \in \mathcal{M}$. Suppose moreover that T is skew hermitian of signature $(1,2)$, i.e. ${}^t\overline{T} = -T$ and

$$T = {}^tW \begin{pmatrix} i & 0 & 0 \\ 0 & -i & 0 \\ 0 & 0 & -i \end{pmatrix} \overline{W}$$

for some $W \in GL_3(\mathbb{C})$. (Below we give examples of such pairs $(\mathcal{M}, T)$).

Consider the complex manifold

$$\mathcal{H} := \left\{ \begin{pmatrix} z_1 \\ z_2 \end{pmatrix} \in \mathbb{C}^2 \mid |z_1|^2 + |z_2|^2 < 1 \right\}$$

A point $z = \begin{pmatrix} z_1 \\ z_2 \end{pmatrix} \in \mathcal{H}$ defines an isomorphism of $\mathbb{R}$-vector spaces

$$J_z : \mathbb{C}^3 \to \mathbb{C}^3,\ v \mapsto \begin{pmatrix} (z, \mathbf{1}_2) & 0 \\ 0 & (1, {}^tz) \end{pmatrix} \begin{pmatrix} Wv \\ \overline{Wv} \end{pmatrix} = \begin{pmatrix} (z, \mathbf{1}_2)Wv \\ (1, {}^tz)\overline{Wv} \end{pmatrix}$$

as well as an hermitian form $H_z : \mathbb{C}^3 \times \mathbb{C}^3 \to \mathbb{C}$ by the matrix

$$2\begin{pmatrix} (\mathbf{1} - \overline{z}^t z)^{-1} & 0 \\ 0 & (1 - {}^t\overline{z}z)^{-1} \end{pmatrix} = 2\begin{pmatrix} 1-|z_1|^2 & \overline{z}_1 z_2 & 0 \\ z_1\overline{z}_2 & 1-|z_2|^2 & 0 \\ 0 & 0 & 1-|z_1|^2-|z_2|^2 \end{pmatrix}^{-1}$$

PROPOSITION 4.3. *$(X_z = \mathbb{C}^3/J_z(\mathcal{M}), H_z)$ is a polarized abelian threefold admitting an embedding $\iota : \mathbb{Q}(\xi_d) \mapsto \operatorname{End}_{\mathbb{Q}}(X_z)$ such that*

i) *$\rho_a\iota(a)$ is equivalent to $\left(\begin{smallmatrix} a\mathbf{1}_2 & 0 \\ 0 & \overline{a} \end{smallmatrix}\right)$ for all $a \in \mathbb{Q}(\xi_d)$;*

ii) *the Rosati involution with respect to H_z restricts to complex conjugation on $\mathbb{Q}(\xi_d)$.*

In particular X_z admits an automorphism α of order $d \in \{3,4,6\}$ such that $(X_z)_{<\alpha>}$ is finite, $\#\Phi_\alpha = 2$ and H_z is invariant with respect to α.

In more sophisticated terms this means that the polarized abelian variety (X_z, H_z) admits multiplication by $\mathbb{Q}(\xi_d)$ of signature $(1,2)$.

PROOF. Obviously X_z is a complex torus, $\mathcal{M}$ being a lattice in $\mathbb{C}^3$ and J_z being an isomorphism of $\mathbb{R}$-vector spaces. Moreover by definition of $\mathcal{H}$, H_z is a positive definite hermitian form. In order to show that H_z defines a polarization on X_z it remains to show that its imaginary part Im H_z takes only integer values on the lattice $J_z(\mathcal{M})$. To see this we need the following identity:

$$\begin{pmatrix} {}^tz \\ \mathbf{1}_2 \end{pmatrix} (\mathbf{1}_2 - \overline{z}\,{}^tz)^{-1}(\overline{z}, \mathbf{1}_2) - \begin{pmatrix} 1 \\ \overline{z} \end{pmatrix} \left(1 - {}^tz\overline{z}\right)^{-1} (1, {}^tz) = \begin{pmatrix} -1 & 0 \\ 0 & \mathbf{1}_2 \end{pmatrix}. \qquad (*)$$

For the proof of (*) we first observe that $\overline{z}(1 - {}^tz\overline{z})^{-1} = (\mathbf{1}_2 - \overline{z}{}^tz)^{-1}\overline{z}$ and equivalently $(1 - {}^tz\,\overline{z})^{-1}\,{}^tz = {}^tz\,(\mathbf{1}_2 - \overline{z}{}^tz)^{-1}$. Using this we have:

$$\begin{aligned}
\text{left hand side of } (*) &= \begin{pmatrix} {}^tz(\mathbf{1}_2-\overline{z}{}^tz)^{-1}\overline{z} & {}^tz(\mathbf{1}_2-\overline{z}{}^tz)^{-1} \\ (\mathbf{1}_2-\overline{z}{}^tz)^{-1}\overline{z} & (\mathbf{1}_2-\overline{z}{}^tz)^{-1} \end{pmatrix} - \begin{pmatrix} (1-{}^tz\,\overline{z})^{-1} & (1-{}^tz\,\overline{z})^{-1}\,{}^tz \\ \overline{z}(1-{}^tz\overline{z})^{-1} & \overline{z}(1-{}^tz\overline{z})^{-1}\,{}^tz \end{pmatrix} \\
&= \begin{pmatrix} (1-{}^tz\overline{z})^{-1} & 0 \\ 0 & (\mathbf{1}_2-\overline{z}{}^tz)^{-1} \end{pmatrix} \left[\begin{pmatrix} {}^tz\overline{z} & {}^tz \\ \overline{z} & \mathbf{1}_2 \end{pmatrix} - \begin{pmatrix} 1 & {}^tz \\ \overline{z} & \overline{z}{}^tz \end{pmatrix}\right] \\
&= \begin{pmatrix} -1 & 0 \\ 0 & \mathbf{1}_2 \end{pmatrix} = \text{right hand side of } (*)\,.
\end{aligned}$$

Now we have for all $\lambda, \mu \in \mathcal{M}$:

$$\begin{aligned}
\text{Im}\quad & H_z(J_z(\lambda), J_z(\mu)) = \\
&= 2\ \text{Im}\ {}^t\begin{pmatrix} W\lambda \\ \overline{W\lambda} \end{pmatrix} \begin{pmatrix} \binom{{}^tz}{\mathbf{1}_2}(\mathbf{1}_2-\overline{z}{}^tz)^{-1}(\overline{z},\mathbf{1}_2) & 0 \\ 0 & \binom{1}{z}(1-{}^t\overline{z}z)(1,{}^t\overline{z}) \end{pmatrix} \begin{pmatrix} \overline{W\mu} \\ W\mu \end{pmatrix} \\
&= 2\ \text{Im}\ \left[{}^t\lambda{}^tW\left[\binom{{}^tz}{\mathbf{1}_2}(\mathbf{1}_2 - \overline{z}{}^tz)^{-1}(\overline{z}, \mathbf{1}_2) - \binom{1}{\overline{z}}(1 - {}^tz\overline{z})(1, {}^tz)\right]\overline{W}\overline{\mu}\right] \\
&\overset{(*)}{=} 2\ \text{Im}\ \left[{}^t\lambda{}^tW\begin{pmatrix} -1 & 0 \\ 0 & \mathbf{1}_2 \end{pmatrix}\overline{W}\overline{\mu}\right] \\
&= 2\ \text{Im}\ [i{}^t\lambda T\overline{\mu}] \\
&= 2\text{Re}\ [{}^t\lambda T\overline{\mu}] \\
&= tr_{\mathbb{Q}(\xi_d)/\mathbb{Q}}({}^t\lambda T\overline{\mu}) \in \mathbb{Z}
\end{aligned}$$

This proves the first statement of the Proposition. As for the second statement consider the following commutative diagram

$$\begin{array}{ccc}
\mathbb{Q}(\xi_d)^{\oplus 3} & \xrightarrow{J_z} & \mathbb{C}^3 \\
{\scriptstyle a\mathbf{1}_3}\downarrow & & \downarrow{\scriptstyle \left(\begin{smallmatrix} a\mathbf{1}_2 & 0 \\ 0 & \overline{a} \end{smallmatrix}\right)} \\
\mathbb{Q}(\xi_d)^{\oplus 3} & \xrightarrow{J_z} & \mathbb{C}^3
\end{array}$$

where $a \in \mathbb{Q}(\xi_d)$. This shows that scalar multiplication by $\mathbb{Q}(\xi_d)$ induces an embedding $\iota : \mathbb{Q}(\xi_d) \mapsto \text{End}_{\mathbb{Q}}(X_z)$ satisfying property i).

Finially using the block matrix structure of H_z and the above diagram one immediately sees that for all $a \in \mathbb{Q}(\xi_d)$ and $v, w \in \mathbb{C}^3$:

$$H_z(\iota(a)v, w) = H_z(v, \iota(\overline{a})w).$$

Hence the Rosati-involution, which is the adjoint operator on $\operatorname{End}_{\mathbb{Q}}(X)$ with respect to H_z, restricts to complex conjugation on $\mathbb{Q}(\xi_d)$. This completes the proof. □

Note that the choice of the pair $(\mathcal{M}, T)$ determines the type of the polarization H_y. It is easy to see that every type (d_1, d_2, d_3) occurs at least once.

According to [**CAV**], Proposition 9.6.5 any polarized abelian threefold with multiplication by $\mathbb{Q}(\xi_d)$ of signature $(1, 2)$ is isomorphic to (X_z, H_z) for some pair $(\mathcal{M}, T)$ as above and some $z \in \mathcal{H}$. Hence we conclude

THEOREM 4.4. *Let (X, H, α) be a polarized abelian threefold with an automorphism α of order $d = 3, 4$ or 6 such that $X_{\langle\alpha\rangle}$ is finite, $\#\Phi_\alpha = 2$ and H is invariant with respect to α. Then there is a $\mathbb{Z}[\xi_d]$-module $\mathcal{M} \subset \mathbb{Q}(\xi_d)^{\oplus 3}$ of rank 6 and a nondegenerate skew-hermitian matrix $T \in GL_3(\mathbb{Q}(\xi_d))$ of signature $(1, 2)$ satisfying $tr_{\mathbb{Q}(\xi_d)/\mathbb{Q}}({}^t a T \bar{b}) \in \mathbb{Z}$ for all $a, b \in \mathcal{M}$, such that (with the notation of Proposition 4.3)*

$$(X, H, \alpha) \sim (X_z = \mathbb{C}^3/J_z(\mathcal{M}),\ H_z,\ \iota(\xi_d)).$$

PROOF. By what we have said above (X, H) is isomorphic to (X_z, H_z) for some pair $(\mathcal{M}, T)$ as above and some $z \in \mathcal{H}$. According to Proposition 4.3 there is an embedding $\iota : \mathbb{Q}(\xi_d) \mapsto \operatorname{End}_{\mathbb{Q}}(X_z) = \operatorname{End}_{\mathbb{Q}}(X)$ such that $\iota(\xi_d)$ admits exactly two complex conjugate eigenvalues. We have to show that $\alpha = \iota(\xi_d) \in \operatorname{End}(X)$. But thus is equivalent to $\mathcal{M}$ being a $\mathbb{Z}[\xi_d]$-module. Finally using again the block-matrix structure of H_z it is easy to see that $\alpha^* H_z = H_z$. □

EXAMPLES 4.5. a) *Case $d = 3$ or 6:*
Choose $\mathcal{M} = \mathbb{Z}[\xi_d]^{\oplus 3} = \mathbb{Z}[\rho]^{\oplus 3} \subset \mathbb{Q}(\rho)$, $T = \begin{pmatrix} (\rho-\rho^2) & 0 \\ 0 & (\rho^2-\rho)\mathbf{1}_2 \end{pmatrix}$, and $W = \sqrt[4]{3}\,\mathbf{1}_3$. Then

$${}^tW \begin{pmatrix} i & 0 \\ 0 & -i\mathbf{1}_2 \end{pmatrix} \overline{W} = \begin{pmatrix} \sqrt{3}i & 0 \\ 0 & -\sqrt{3}i\mathbf{1}_2 \end{pmatrix} = \begin{pmatrix} \rho-\rho^2 & 0 \\ 0 & (\rho^2-\rho)\mathbf{1}_2 \end{pmatrix} = T.$$

Suppose $z = \binom{z_1}{z_2} \in \mathcal{H}$. An immediate computation shows that $J_z(\mathcal{M}) = \pi_3\mathbb{Z}^6$ with

$$\pi_3 = \sqrt[4]{3}\begin{pmatrix} z_1 & 1 & 0 & \rho z_1 & \rho & 0 \\ z_2 & 0 & 1 & \rho z_2 & 0 & \rho \\ 1 & z_1 & z_2 & \rho^2 & \rho^2 z_1 & \rho^2 z_2 \end{pmatrix}$$

Hence $(X = \mathbb{C}^3/\pi_3\mathbb{Z}^6, H_z)$ is a polarized abelian threefold with automorphism $\alpha = \begin{pmatrix} \rho\mathbf{1}_2 & 0 \\ 0 & \rho^2 \end{pmatrix}$ of order 3 and an automorphism $-\alpha$ of order 6.

b) *Case $d = 4$:* Choose $\mathcal{M} = \mathbb{Z}[i]^{\oplus 3} \in \mathbb{Q}(i)$, and $T = \begin{pmatrix} i & 0 \\ 0 & -i\mathbf{1}_2 \end{pmatrix}$. For $z = \binom{z_1}{z_2} \in \mathcal{H}$ we have $J_z(\mathcal{M}) = \pi_4\mathbb{Z}^6$ with

$$\pi_4 = \begin{pmatrix} z_1 & 1 & 0 & iz_1 & i & 0 \\ z_2 & 0 & 1 & iz_2 & 0 & i \\ 1 & z_1 & z_2 & -i & -iz_1 & -iz_2 \end{pmatrix}.$$

Hence $(X = \mathbb{C}^3/\pi_4\mathbb{Z}^6, H_z)$ is a polarized abelian threefold with automorphism $\alpha = \begin{pmatrix} i\mathbf{1}_2 & 0 \\ 0 & -i \end{pmatrix}$ of order 4.

5. Abelian Threefolds X with $\dim X_{\langle\alpha\rangle} > 0$

In this section we study abelian threefolds X admitting an automorphism α of order $d \geq 3$ with $\dim X_G \geq 1$, where G denotes the cyclic group $\langle\alpha\rangle$ generated α. According to the Decomposition Theorem we have to distinguish 2 cases:

Case I: Exactly 2 of the 3 eigenvalues of $\rho_a(\alpha)$ have the same order. Let the orders of the eigenvalues be $d_1 \neq d_2$, where d_2 occurs twice. In this case there are G-stable abelian subvarieties E (of dimension 1) and A (of dimension 2) of X such that the restriction $\alpha_E := \alpha|E$ is of order d_1 and $\alpha_A := \alpha|A$ is of order d_2 and such that the addition map $\mu : E \times A \to X$ is an isogeny.

Case II: The orders d_1, d_2 and d_3 of the 3 eigenvalues of $\rho_a(\alpha)$ are pairwise different. In this case there are G-stable abelian subvarieties E_1, E_2, and E_3 (of dimension 1) in X such that $\alpha_{E_i} := \alpha|E_i$ is of order d_i for $i = 1, 2, 3$ and such that the addition map $\mu : E_1 \times E_2 \times E_3 \to X$ is an isogeny.

Case I

Let the notation be as in Case I above. Moreover let e denote the greatest common divisor of d_1 and d_2. The kernel of the isogeny

$$\mu : E \times A \to X$$

has the following properties:

PROPOSITION 5.1. *(i)* $\mathrm{Ker}(\mu) \subseteq \mathrm{Fix}\,\alpha_E^e \times \mathrm{Fix}\,\alpha_A^e$
(ii) The projections $p : \mathrm{Ker}(\mu) \to E$ *and* $q : \mathrm{Ker}(\mu) \to A$ *are injective. Moreover* $\mathrm{Ker}(\mu)$ *is* $(\alpha_E \times \alpha_A)$*-stable.*

PROOF. For $(x, y) \in \mathrm{Ker}(\mu)$ we have $y = -x$ (note that we consider E and A as subvarieties of X). Hence the projections p and q are injective. Moreover $0 = \mu((\alpha_E \times \alpha_A)^{d_1}(x, -x)) = x - \alpha^{d_1}(x)$ and $0 = \mu(\alpha_E \times \alpha_A)^{d_2}(x, -x)) = \alpha^{d_2}(x) - x$ imply

$$x \in \mathrm{Fix}\,(\alpha^{d_1}) \cap \mathrm{Fix}\,(\alpha^{d_2}) \cap E = \mathrm{Fix}\,(\alpha^e) \cap E = \mathrm{Fix}\,\alpha_E^e.$$

Similarly $y \in \mathrm{Fix}\,\alpha_A^e$. The last assertion is trivial. □

PROPOSITION 5.2. *Let the notation be as above. Suppose*

(i) $\frac{d_1}{e}$ *is not a prime power or*
(ii) $A_{\langle\alpha_A\rangle}$ *is finite and* $\frac{d_2}{e}$ *is not a prime power or*
(iii) $E_{\langle\alpha_E^e\rangle}$ *and* $A_{\langle\alpha_A^e\rangle}$ *are finite and* $d_1 \neq e \neq d_2$,

then $\mu : E \times A \to X$ *is an isomorphism.*

PROOF. Suppose $\frac{d_1}{e}$ is not a prime power. Then α_E^e is of order $\frac{d_1}{e} \neq 1$ and hence $E_{\langle\alpha_E^e\rangle}$ is finite. Thus Corollary 1.7 implies $\#\mathrm{Fix}\,\alpha_E^e = 1$. Now Proposition 5.1 implies assertion (i). The same proof works for assertion (ii). Finally in case (iii)

$$\mathrm{Ker}(\mu) \simeq p(\mathrm{Ker}(\mu)) \subseteq \mathrm{Fix}\,(\alpha_E^e) \subseteq E_{\frac{d_1}{e}} \simeq \left(\mathbb{Z}/\tfrac{d_1}{e}\mathbb{Z}\right)^2$$

and

$$\mathrm{Ker}(\mu) \simeq q(\mathrm{Ker}(\mu)) \subseteq \mathrm{Fix}\,(\alpha_A^e) \subseteq A_{\frac{d_2}{e}} \simeq \left(\mathbb{Z}/\tfrac{d_2}{e}\mathbb{Z}\right)^4$$

Since $\gcd\left(\frac{d_1}{e}, \frac{d_2}{e}\right) = 1$, this implies $\#\mathrm{Ker}(\mu) = 1$. Hence μ is an isomorphism. □

In order to construct abelian threefolds X with an automorphism α with $\dim X_{\langle\alpha\rangle} \geq 1$ we use the following proposition which may be considered as a sort of converse statement to the Decomposition Theorem in this case.

PROPOSITION 5.3. *Let E be an elliptic curve with an automorphism α_E of order d_1 and A an abelian surface with an automorphism α_A of order d_2 and assume $d_1 \neq d_2$. Denote $e = \gcd(d_1, d_2)$ and let K be a finite subgroup of $E \times A$ with*

$$K \subseteq \mathrm{Fix}\,(\alpha_E^e) \times \mathrm{Fix}\,(\alpha_A^e) \ \ \textit{and}\ \ (\alpha_E \times \alpha_A)K \subseteq K,$$

such that the projections $p : K \to E$ and $q : K \to A$ are injective. Then the abelian variety $X = E \times A/K$ admits a unique automorphism α such that E and A, considered as abelian subvarieties of X, are α-stable with $\alpha|E = \alpha_E$ and $\alpha|A = \alpha_A$. The isogeny $E \times A \to X = E \times A/K$ is given by the addition map. Moreover $\mathrm{ord}(\alpha) = \mathrm{lcm}(d_1, d_2)$ *and* $\dim X_{\langle\alpha\rangle} \geq 1$.

The proof is immediate, so we omit it. □

According to Example 1.9 the degrees of the nontrivial automorphisms of an elliptic curve are 2, 3, 4 and 6. According to Example 1.10 the degrees of the the nontrivial automorphisms α of an abelian surface X with $X_{\langle\alpha\rangle}$ finite are $2, 3, 4, 5, 6, 8, 10$ and 12. Hence the possible degrees d_1 and d_2 are $d_1 \in \{1, 2, 3, 4, 6\}$ and $d_2 \in \{1, 2, 3, 4, 5, 6, 8, 10, 12\}$. Moreover we may assume $d_E \neq d_A$ and $\mathrm{lcm}(d_1, d_2) \geq 3$, since we are interested in automorphisms of degree $d \geq 3$. Hence if $E, \alpha_{E_1}, A, \alpha_A$ and K are given and X and α are constructed as in Proposition 5.3, we have for the degree d of α

$$d \in \{3, 4, 5, 6, 8, 10, 12, 15, 20, 24, 30\}.$$

We may apply Proposition 5.2 to deduce that in some cases the subgroup K in Proposition 5.3 is necessarily trivial, implying $X = E \times A$ and $\alpha = \alpha_E \times \alpha_A$.

This is the case for:

$$\begin{aligned}(d_1, d_2) \ \in \ \{&(1,6),(1,10),(1,12),(2,3),(2,5),(2,12),(3,2),(3,4),\\ &(3,5),(3,8),(3,10),(4,3),(4,5),(4,6),(4,10),(6,1),\\ &(6,4),(6,5),(6,8),(6,10),(6,12)\}\end{aligned}$$

It remains to consider the cases of the following table. We will see below that apart from the trivial case $K = \{(0,0)\}$ there are also nontrivial subgroups $K \subset E \times A$ such that $\alpha_E \times \alpha_A$ induces an automorphism α on $X = E \times A/K$.

As for the notation in Table 5.4 below: E denotes an arbitrary elliptic curve (so $E = \mathbb{C}/\tau\mathbb{Z} + \mathbb{Z}$ for some $\tau \in \mathfrak{H}_1$), A an arbitrary abelian surface (so $A = \mathbb{C}^2/(Z, \delta)\mathbb{Z}^4$ with $Z = \left(\begin{smallmatrix} z_1 & z_2 \\ z_2 & z_3 \end{smallmatrix}\right) \in \mathfrak{H}_2$ and some type of polarization $\delta = \left(\begin{smallmatrix} \delta_1 & 0 \\ 0 & \delta_2 \end{smallmatrix}\right)$), E_j denotes the elliptic curve with j-invariant j, and $X(\mathbb{Q}(\xi_d))$ one of the abelian surfaces $X(\mathbb{Q}(\xi_d), (\xi_d^{a_1}, \xi_d^{a_2})) = \mathbb{Q}(\xi_d) \otimes_{\mathbb{Q}} \mathbb{R}/\mathcal{O}$ of CM-type $(\xi_d^{a_1}, \xi_d^{a_2})$ for $d = 5, 8, 12$. It is well known (see e.g. [**FU**]) that the elliptic curves and abelian surfaces in the 3rd and 4th column of Table 5.4 below are the only abelian varieties X of dimension 1 and 2 admitting an automorphism α_E and α_A of the indicated order with $E_{\langle\alpha_E\rangle}$ and $A_{\langle\alpha_A\rangle}$ finite. By a slight abuse of notation we denote by ξ_d the automorphism on E and A induced by multiplication by a dth primitive root of unity. Finally recall that $e = \gcd(d_1, d_2)$ and $d = \mathrm{ord}(\alpha) = \mathrm{lcm}(d_1, d_2)$ is the order of the automorphism α on $X = E \times A/K$.

TABLE 5.4.

No	(d_1, d_2)	E	A	α_E	α_A	e	d
1	$(1,3)$	E	$E_0 \times E_0$	id	ρ	1	3
2	$(1,4)$	E	$E_{1728} \times E_{1728}$	id	i	1	4
3	$(1,5)$	E	$X(\mathbb{Q}(\xi_5))$	id	ξ_5	1	5
4	$(1,8)$	E	$X(\mathbb{Q}(\xi_8))$	id	ξ_8	1	8
5	$(2,4)$	E	$E_{1728} \times E_{1728}$	-id	i	2	4
6	$(2,6)$	E	$E_0 \times E_0$	$-$id	$-\rho$	2	6
7	$(2,8)$	E	$X(\mathbb{Q}(\xi_8))$	$-$id	ξ_8	2	8
8	$(2,10)$	E	$X(\mathbb{Q}(\xi_5))$	$-$id	$-\xi_5$	2	10
9	$(3,1)$	E_0	A	ρ	id	1	3
10	$(3,6)$	E_0	$E_0 \times E_0$	ρ	$-\rho$	3	6
11	$(3,12)$	E_0	$X(\mathbb{Q}(\xi_{12}))$	ρ	ξ_{12}	3	12
12	$(4,1)$	E_{1728}	A	i	id	1	4
13	$(4,2)$	E_{1728}	A	i	$-$id	2	4
14	$(4,8)$	E_{1728}	$X(\mathbb{Q}(\xi_8))$	i	ξ_8	4	8
15	$(4,12)$	E_{1728}	$X(\mathbb{Q}(\xi_{12}))$	i	ξ_{12}	4	12
16	$(6,2)$	E_0	A	$-\rho$	$-$id	2	6
17	$(6,3)$	E_0	$E_0 \times E_0$	$-\rho$	ρ	3	6
18	$(6,12)$	E_0	$X(\mathbb{Q}(\xi_{12}))$	$-\rho$	ξ_{12}	6	12

Below we give examples of an abelian threefolds $X = \mathbb{C}^3/(Z,D)\mathbb{Z}^6$ with an automorphism α for every type Nos 1 to 18 of Table 5.4. More examples are given in the thesis [**SCH**]. Moreover D. Schmidt shows that in several of the above cases every pair (X, α) is isomorphic to one of the examples given below. In the examples we describe the threefold X in terms of the period matrix, normalized in the form $\pi = (Z, D)$ with an element Z of the Siegel upper half space $\mathfrak{H}_3$ and an integer valued diagonal matrix D indicating the type of the polarization of X. The automorphism α of X is given in terms of its analytic representation $\rho_a(\alpha)$ and its rational representation $\rho_r(\alpha)$ with respect to the coordinates given by the period matrix. Hence the proof that α is an automorphism of X consists just in checking the equation

$$\rho_a(\alpha)(Z, D) = (Z, D)\rho_r(\alpha). \qquad (*)$$

Moreover we indicate the isomorphism type of the subgroup K in each case. With this notation the following list gives examples of abelian threefolds with automorphism (X, α) corresponding to Nos 1 to 18 of Table 5.4:

EXAMPLES 5.5.

No 1: Periodmatrix: $\pi = \begin{pmatrix} \frac{\tau-1+\rho}{3} & 0 & \rho-1 & 1 & 0 & 0 \\ 0 & 1+\rho & 0 & 0 & 1 & 0 \\ \rho-1 & 0 & -3\rho^2 & 0 & 0 & 3 \end{pmatrix}$ with $\tau \in \mathfrak{H}_1$.

Analytic and rational representation of α:

$$\rho_a(\alpha) = \begin{pmatrix} 1 & 0 & \frac{-1+\rho}{3} \\ 0 & \rho & 0 \\ 0 & 0 & \rho \end{pmatrix} \text{ and } \rho_r(\alpha) = \begin{pmatrix} 1 & 0 & 0 & 0 & 0 & 0 \\ 0 & 0 & 0 & 0 & 1 & 0 \\ -1 & 0 & 0 & 0 & 0 & 1 \\ -1 & 0 & -3 & 1 & 0 & 0 \\ 0 & -1 & 0 & 0 & -1 & 0 \\ 1 & 0 & -1 & 0 & 0 & -1 \end{pmatrix}$$

$K \simeq \mathbb{Z}/3\mathbb{Z}$.

No 2: Periodmatrix: $\pi = \begin{pmatrix} \frac{\tau-1+i}{2} & 0 & -1+i & 1 & 0 & 0 \\ 0 & i & 0 & 0 & 1 & 0 \\ -1+i & 0 & 2i & 0 & 0 & 2 \end{pmatrix}$ with $\tau \in \mathfrak{H}_1$.

Analytic and rational representation of α:

$$\rho_a(\alpha) = \begin{pmatrix} 1 & 0 & \frac{-1+i}{2} \\ 0 & i & 0 \\ 0 & 0 & i \end{pmatrix} \text{ and } \rho_r(\alpha) = \begin{pmatrix} 1 & 0 & 0 & 0 & 0 & 0 \\ 0 & 0 & 0 & 0 & 1 & 0 \\ -1 & 0 & 0 & 0 & 0 & 1 \\ -1 & 0 & -2 & 1 & 0 & 0 \\ 0 & -1 & 0 & 0 & 0 & 0 \\ 0 & 0 & -1 & 0 & 0 & 0 \end{pmatrix}$$

$K \simeq \mathbb{Z}/2\mathbb{Z}$

No 3: Periodmatrix: $\pi = \begin{pmatrix} \frac{\tau+16\xi_5+4\xi_5^2+12\xi_5^3}{5} & -6-2\xi_5^2-2\xi_5^3 & 5+4\xi_5+2\xi_5^2+4\xi_5^3 & 1 & 0 & 0 \\ -6-2\xi_5^2-2\xi_5^3 & 5\xi_5^2-5\xi_5^3 & -5-5\xi_5^2 & 0 & 5 & 0 \\ 5+4\xi_5+2\xi_5^2+4\xi_5^3 & -5-5\xi_5^2 & 5+5\xi_5+5\xi_5^2+5\xi_5^3 & 0 & 0 & 5 \end{pmatrix}$ with $\tau \in \mathfrak{H}_1$.

Analytic and rational representation of α:

$$\rho_a(\alpha) = \begin{pmatrix} 1 & \frac{-6-2\xi_5^2-2\xi_5^3}{5} & \frac{-6+4\xi_5+2\xi_5^3}{5} \\ 0 & \xi_5^2-\xi_5^3 & -1-\xi_5^3 \\ 0 & -1-\xi_5^2 & \xi_5+\xi_5^3 \end{pmatrix} \text{ and } \rho_r(\alpha) = \begin{pmatrix} 1 & 0 & 0 & 0 & 0 & 0 \\ -1 & 0 & -1 & 0 & 1 & 1 \\ -1 & 0 & -1 & 0 & 0 & 1 \\ -3 & 0 & 0 & 1 & 0 & -5 \\ 0 & -1 & 0 & 0 & 0 & 0 \\ -1 & 1 & -1 & 0 & 0 & 0 \end{pmatrix}$$

$K \simeq \mathbb{Z}/5\mathbb{Z}$

No 4: Periodmatrix: $\pi = \begin{pmatrix} \frac{\tau}{2}+1+i & -1-i & -1-i & 1 & 0 & 0 \\ -1-i & 2i & 0 & 0 & 2 & 0 \\ -1-i & 0 & 2i & 0 & 0 & 2 \end{pmatrix}$ with $\tau \in \mathfrak{H}_1$.

Analytic and rational representation of α:

$$\rho_a(\alpha) = \begin{pmatrix} 1 & 0 & \frac{1-i}{2} \\ 0 & 0 & i \\ 0 & 1 & 0 \end{pmatrix} \text{ and } \rho_r(\alpha) \begin{pmatrix} 1 & 0 & 0 & 0 & 0 & 0 \\ 0 & 0 & 0 & 0 & 0 & 1 \\ 0 & 1 & 0 & 0 & 0 & 0 \\ -1 & 0 & 0 & 1 & 0 & 2 \\ 1 & 0 & -1 & 0 & 0 & 0 \\ 0 & 0 & 0 & 0 & 1 & 0 \end{pmatrix}$$

$K \simeq \mathbb{Z}/2\mathbb{Z}$

No 5: Periodmatrix: $\pi = \begin{pmatrix} \frac{\tau}{2}-1+i & -1+i & -1+i & 1 & 0 & 0 \\ -1+i & 2i & 0 & 0 & 2 & 0 \\ -1+i & 0 & 2i & 0 & 0 & 2 \end{pmatrix}$ with $\tau \in \mathfrak{H}_1$.

Analytic and rational representation of α:

$$\rho_a(\alpha) = \begin{pmatrix} -1 & \frac{1+i}{2} & \frac{1+i}{2} \\ 0 & i & 0 \\ 0 & 0 & i \end{pmatrix} \text{ and } \rho_r(\alpha) = \begin{pmatrix} -1 & 0 & 0 & 0 & 0 & 0 \\ 0 & 0 & 0 & 0 & 1 & 0 \\ 0 & 0 & 0 & 0 & 0 & 1 \\ -2 & 0 & 0 & -1 & 2 & 2 \\ -1 & -1 & 0 & 0 & 0 & 0 \\ -1 & 0 & -1 & 0 & 0 & 0 \end{pmatrix}$$

$K \simeq \mathbb{Z}/2\mathbb{Z}$

No 6: Let (X, α) be as in No 1. Then $(X, -\alpha)$ is an example of for No 6 of Table 5.4.

No 7: Periodmatrix: $\pi = \begin{pmatrix} \frac{\tau}{2}+1+i & -1-i & -1-i & 1 & 0 & 0 \\ -1-i & 2i & 0 & 0 & 2 & 0 \\ -1-i & 0 & 2i & 0 & 0 & 2 \end{pmatrix}$ with $\tau \in \mathfrak{H}_1$.

Analytic and rational representation of α:

$$\rho_a(\alpha) = \begin{pmatrix} -1 & -1 & -\frac{1+i}{2} \\ 0 & 0 & i \\ 0 & 1 & 0 \end{pmatrix} \text{ and } \rho_r(\alpha) = \begin{pmatrix} -1 & 0 & 0 & 0 & 0 & 0 \\ -1 & 0 & 0 & 0 & 0 & 1 \\ -1 & 1 & 0 & 0 & 0 & 0 \\ -1 & 2 & 2 & -1 & -2 & 0 \\ 0 & 0 & -1 & 0 & 0 & 0 \\ -1 & 0 & 0 & 0 & 1 & 0 \end{pmatrix}$$

$K \simeq \mathbb{Z}/2\mathbb{Z}$

No 8: Let (X, α) be as in No 3. Then $(X, -\alpha)$ is an example for No 8 of Table 5.4.

No 9: Periodmatrix: $\pi = \begin{pmatrix} \frac{-1+\rho}{3} & 1 & 0 & 1 & 0 & 0 \\ 1 & 3z_1 & -3z_2 & 0 & 3 & 0 \\ 0 & -3z_2 & 3z_3 & 0 & 0 & 3 \end{pmatrix}$ with $Z = \begin{pmatrix} z_1 & z_2 \\ z_2 & z_3 \end{pmatrix} \in \mathfrak{H}_2$.

Analytic and rational representation of α:

$$\rho_a(\alpha) = \begin{pmatrix} \rho & 0 & 0 \\ 0 & 1 & 0 \\ 0 & 0 & 1 \end{pmatrix} \text{ and } \rho_r(\alpha) \begin{pmatrix} -2 & 3 & 0 & 3 & 0 & 0 \\ 0 & 1 & 0 & 0 & 0 & 0 \\ 0 & 0 & 1 & 0 & 0 & 0 \\ -1 & 0 & 0 & 1 & 0 & 0 \\ 1 & -1 & 0 & -1 & 1 & 0 \\ 0 & 0 & 0 & 0 & 0 & 1 \end{pmatrix}$$

$K \simeq \mathbb{Z}/3\mathbb{Z}$

No 10: Periodmatrix: $\pi = \begin{pmatrix} \frac{1}{2}+\rho & 0 & -\frac{1}{2} & 1 & 0 & 0 \\ 0 & \rho & 0 & 0 & 1 & 0 \\ -\frac{1}{2} & 0 & \frac{1}{2}+\rho & 0 & 0 & 1 \end{pmatrix}$.

Analytic and rational representation of α:

$$\rho_a(\alpha) = \begin{pmatrix} -\frac{1}{2}-\rho & 0 & \frac{1}{2} \\ 0 & -\rho & 0 \\ \frac{1}{2} & 0 & -\frac{1}{2}-\rho \end{pmatrix} \text{ and } \rho_r(\alpha) = \begin{pmatrix} 0 & 0 & 1 & -1 & 0 & 0 \\ 0 & 1 & 0 & 0 & -1 & 0 \\ 1 & 0 & 0 & 0 & 0 & -1 \\ 1 & 0 & 0 & 0 & 0 & 0 \\ 0 & 1 & 0 & 0 & 0 & 0 \\ 0 & 0 & 1 & 0 & 0 & 0 \end{pmatrix}$$

$K \simeq \mathbb{Z}/2\mathbb{Z} \times \mathbb{Z}/2\mathbb{Z}$

No 11: Periodmatrix: $\pi = \begin{pmatrix} \frac{-5-7\rho^2}{3} & -\frac{20+16\rho^2}{3} & 8\frac{2+\rho^2}{3} & 1 & 0 & 0 \\ -\frac{20+16\rho^2}{3} & -\frac{56+40\rho^2}{3} & 20\frac{1-\rho}{3} & 0 & 4 & 0 \\ 8\frac{2+\rho^2}{3} & 20\frac{1-\rho}{3} & -16\frac{2+\rho^2}{3} & 0 & 0 & 4 \end{pmatrix}$.

Analytic and rational representation of α:

$$\rho_a(\alpha) = \begin{pmatrix} -\frac{1+5\rho^2}{3} & \rho^2 & \frac{1+2\rho^2}{3} \\ -4\frac{1+2\rho^2}{3} & 2\rho^2 & \frac{1+5\rho^2}{3} \\ 4\frac{2+\rho^2}{3} & \rho & -2\frac{1+2\rho^2}{3} \end{pmatrix} \text{ and } \rho_r(\alpha) = \begin{pmatrix} 3 & 0 & 4 & 3 & -4 & 0 \\ -1 & 0 & -2 & -1 & 1 & 0 \\ 1 & 2 & -2 & 0 & 0 & 1 \\ -5 & -4 & 0 & -2 & 0 & -4 \\ -2 & -3 & 2 & 0 & -2 & -3 \\ 2 & 4 & -3 & 0 & 1 & 2 \end{pmatrix}$$

$K \simeq \mathbb{Z}/2\mathbb{Z} \times \mathbb{Z}/2\mathbb{Z}$

No 12: Periodmatrix: $\pi = \begin{pmatrix} \frac{i-1}{2} & 1 & 0 & 1 & 0 & 0 \\ 1 & 2z_1 & -2z_2 & 0 & 2 & 0 \\ 0 & -2z_2 & 2z_3 & 0 & 0 & 2 \end{pmatrix}$ with $Z = \begin{pmatrix} z_1 & z_2 \\ z_2 & z_3 \end{pmatrix} \in \mathfrak{H}_2$.

Analytic and rational representation of α:

$$\rho_a(\alpha) = \begin{pmatrix} i & 0 & 0 \\ 0 & 1 & 0 \\ 0 & 0 & 1 \end{pmatrix} \text{ and } \rho_r(\alpha) = \begin{pmatrix} -1 & 2 & 0 & 2 & 0 & 0 \\ 0 & 1 & 0 & 0 & 0 & 0 \\ 0 & 0 & 1 & 0 & 0 & 0 \\ -1 & 0 & 0 & 1 & 0 & 0 \\ 1 & -1 & 0 & -1 & 1 & 0 \\ 0 & 0 & 0 & 0 & 0 & 1 \end{pmatrix}$$

$K \simeq \mathbb{Z}/2\mathbb{Z}$

No 13: Let (X, α) be as in No 12. Then $(X, -\alpha^3)$ is an example for No 13 of Table 5.4.

No 14: Periodmatrix: $\pi = \begin{pmatrix} \frac{-1+3i}{5} & \frac{4-2i}{5} & \frac{4-2i}{5} & 1 & 0 & 0 \\ \frac{4-2i}{5} & \frac{-6+8i}{5} & \frac{-6-2i}{5} & 0 & 2 & 0 \\ \frac{4-2i}{5} & \frac{-6-2i}{5} & \frac{-6+8i}{5} & 0 & 0 & 2 \end{pmatrix}$.

Analytic and rational representation of α:

$$\rho_a(\alpha) = \begin{pmatrix} \frac{1+2i}{5} & \frac{-1-2i}{5} & 0 \\ \frac{-4+2i}{5} & \frac{-1+3i}{5} & i \\ \frac{6+2i}{5} & \frac{4+3i}{5} & 0 \end{pmatrix} \text{ and } \rho_r(\alpha) = \begin{pmatrix} 1 & 0 & 0 & 2 & 0 & 2 \\ 1 & -1 & -1 & 1 & 1 & 2 \\ 1 & 0 & -1 & 1 & 1 & 1 \\ -2 & 2 & 2 & -1 & -2 & -2 \\ 1 & -1 & -2 & 0 & 1 & 1 \\ 1 & -1 & -1 & 1 & 2 & 1 \end{pmatrix}$$

$K \simeq \mathbb{Z}/2\mathbb{Z} \times \mathbb{Z}/2\mathbb{Z}$

No 15: Periodmatrix: $\pi = \begin{pmatrix} \frac{i}{10} & \frac{9}{10} & -\frac{18}{5} & 1 & 0 & 0 \\ \frac{9}{10} & 99\frac{i}{10} & 27\frac{i}{5} & 0 & 9 & 0 \\ -\frac{18}{5} & 27\frac{i}{5} & 27\frac{i}{5} & 0 & 0 & 27 \end{pmatrix}$.

Analytic and rational representation of α:

$$\rho_a(\alpha) = \begin{pmatrix} 7\frac{i}{10} & \frac{1}{10} & -\frac{2}{15} \\ \frac{63}{10} & 11\frac{i}{10} & -4\frac{i}{5} \\ \frac{9}{5} & 3\frac{i}{5} & \frac{i}{5} \end{pmatrix} \text{ and } \rho_r(\alpha) = \begin{pmatrix} 0 & 9 & -27 & 7 & 0 & 0 \\ 1 & 0 & 0 & 0 & 1 & -6 \\ -1 & 0 & 0 & 0 & 0 & 7 \\ -4 & 0 & 0 & 0 & 0 & 27 \\ 0 & -1 & 0 & 0 & 0 & 0 \\ 0 & 1 & -4 & 1 & 0 & 0 \end{pmatrix}$$

$K \simeq \mathbb{Z}/3\mathbb{Z} \times \mathbb{Z}/3\mathbb{Z}$

No 16: Let (X, α) be as in No 9. Then $(X, -\alpha^2)$ is an example for No 16 of Table 5.4.

No 17: Let (X, α) be as in No 10. Then $(X, -\alpha)$ is an example for No 17 of Table 5.4.

No 18: Periodmatrix: $\pi = \begin{pmatrix} -\frac{5+7\rho^2}{3} & -\frac{14+16\rho^2}{3} & \frac{10+8\rho^2}{3} & 1 & 0 & 0 \\ -\frac{14+16\rho^2}{3} & -\frac{44+40\rho^2}{3} & \frac{28+20\rho^2}{3} & 0 & 4 & 0 \\ \frac{10+8\rho^2}{3} & \frac{28+20\rho^2}{3} & -\frac{20+16\rho^2}{3} & 0 & 0 & 4 \end{pmatrix}$.

Analytic and rational representation of α:

$$\rho_a(\alpha) = \begin{pmatrix} \frac{2-5\rho^2}{3} & -\frac{1}{2}+\rho^2 & \frac{-1+4\rho^2}{6} \\ \frac{2-8\rho^2}{3} & -1+2\rho^2 & \frac{-2+5\rho^2}{3} \\ \frac{2+4\rho^2}{3} & -\rho^2 & \frac{1-4\rho^2}{3} \end{pmatrix} \text{ and } \rho_r(\alpha) = \begin{pmatrix} 2 & 4 & 0 & 3 & -4 & 0 \\ -1 & -2 & 0 & -1 & 1 & 0 \\ 0 & 0 & 0 & 0 & 0 & 1 \\ -1 & 0 & 0 & 1 & -4 & -4 \\ -1 & -1 & 0 & 0 & -2 & -3 \\ 1 & 2 & -1 & 0 & 1 & 2 \end{pmatrix}$$

$K \simeq \mathbb{Z}/2\mathbb{Z} \times Z/2\mathbb{Z}$

Case II:

Let α be an automorphism of order $d \geq 3$ of an abelian threefold X with $\dim X_{\langle\alpha\rangle} \geq 1$ such that the eigenvalues of $\rho_a(\alpha)$ have orders $d_1 < d_2 < d_3$. Then there are α-stable abelian subvarieties E_1, E_2 and ρ of dimension 1 in X, such that $\alpha_{E_i} := \alpha|E_i$ is of order d_i for $i = 1, 2, 3$ and the addtion map

$$\mu : E_1 \times E_2 \times E_3 \to X$$

is an isogeny. Let K denote the kernel of the isogeny μ. According to Example 1.9 we have $\mathrm{ord}(\alpha_{E_i}) \in \{1, 2, 3, 4, 6\}$. Hence we have to study the following cases (as always d denotes the order of α, and let E and E' denote arbitrary elliptic curves):

TABLE 5.6.

No	(d_1, d_2, d_3)	E_1	E_2	E_3	d
a)	(1, 2, 3)	E	E'	E_0	6
b)	(1, 2, 4)	E	E'	E_{1728}	4
c)	(1, 2, 6)	E	E'	E_0	6
d)	(1, 3, 4)	E	E_0	E_{1728}	12
e)	(1, 3, 6)	E	E_0	E_0	6
f)	(1, 4, 6)	E	E_{1728}	E_0	12
g)	(2, 3, 4)	E	E_0	E_{1728}	12
h)	(2, 3, 6)	E	E_0	E_0	6
i)	(2, 4, 6)	E	E_{1728}	E_0	12
12)	(3, 4, 6)	E_0	E_{1728}	E_0	12

The construction of pairs (X, α) as in Table 5.6 is very similar to the construction of Examples 5.5 in Case I. In fact one uses a result analogous to Proposition 5.3 above. We omit the details here (see [**SCH**]). Of course there are always examples with trivial K, that is an isomorphism $\mu : E_1 \times E_2 \times E_3 \to X$. According to Proposition 1.4 we have $\dim X_{\langle\alpha\rangle} \geq 1$ in these cases. Below we give an example for each case No a) to k) with K nontrivial. As above we give a period matrix in normalized form, the corresponding analytic and rational representations and the isomorphism type of K. Let τ and τ' be elements of the upper half plane $\mathfrak{H}_1$ such that $E \sim \mathbb{C}/\mathbb{Z} + \tau\mathbb{Z}$ and $E' \sim \mathbb{C}/\mathbb{Z} + \tau'\mathbb{Z}$.

EXAMPLES 5.7.

No a): Periodmatrix: $\pi = \begin{pmatrix} \frac{2\tau-1+\rho}{3} & \tau & -2\tau & 1\ 0\ 0 \\ \tau & \frac{\tau'+3\tau}{2} & -3\tau & 0\ 1\ 0 \\ -2\tau & -3\tau & 6(\tau+1) & 0\ 0\ 3 \end{pmatrix}$

Analytic and rational representation of the automorphism α:

$$\rho_a(\alpha) \begin{pmatrix} \rho & 0 & \frac{\rho-1}{3} \\ 0 & -1 & -1 \\ 0 & 0 & 1 \end{pmatrix} \text{ and } \rho_r(\alpha) = \begin{pmatrix} -2 & 0 & 6 & 3 & 0 & 3 \\ 0 & -1 & 0 & 0 & 0 & 0 \\ -1 & -1 & 3 & 1 & 0 & 1 \\ -1 & 0 & 0 & 1 & 0 & 0 \\ 0 & 0 & -6 & 0 & -1 & -3 \\ 2 & 2 & -4 & -2 & 0 & -1 \end{pmatrix}$$

$K \simeq \mathbb{Z}/6\mathbb{Z}$

No b): Periodmatrix: $\pi = \begin{pmatrix} 2\tau' & -2\tau' & 0 & 1\ 0\ 0 \\ -2\tau' & 2\tau'+2\tau & -2\tau & 0\ 1\ 0 \\ 0 & -2\tau & 1+2\tau+i & 0\ 0\ 2 \end{pmatrix}$.

Analytic and rational representation of the automorphism α:

$$\rho_a(\alpha) = \begin{pmatrix} -1 & 0 & 0 \\ 2 & 1 & 0 \\ -1+i & -1+i & i \end{pmatrix} \text{ and } \rho_r(\alpha) = \begin{pmatrix} -1 & 2 & 0 & 1 & 1 & 2 \\ 0 & 1 & 0 & 1 & 1 & 2 \\ 0 & 0 & 1 & 1 & 1 & 2 \\ 0 & 0 & 0 & -1 & 0 & 0 \\ 0 & 0 & 0 & 2 & 1 & 0 \\ 0 & 0 & -1 & -1 & -1 & -1 \end{pmatrix}$$

$K \simeq (\mathbb{Z}/2\mathbb{Z})^2$

No c): Let (X, α) be the pair of No a). Then $(X, -\alpha)$ is an example for No c) of Table 5.6.

No d): Periodmatrix: $\pi = \begin{pmatrix} \frac{\tau-1+\rho}{3} & 0 & -1+\rho & 1 & 0 & 0 \\ 0 & i & 0 & 0 & 1 & 0 \\ -1+\rho & 0 & -3\rho^2 & 0 & 0 & 3 \end{pmatrix}$.

Analytic and rational representation of the automorphism α:

$$\rho_a(\alpha) = \begin{pmatrix} 1 & 0 & \frac{\rho-1}{3} \\ 0 & i & 0 \\ 0 & 0 & \rho \end{pmatrix} \text{ and } \rho_r(\alpha) \begin{pmatrix} 1 & 0 & 0 & 0 & 0 & 0 \\ 0 & 0 & 0 & 0 & 1 & 0 \\ -1 & 0 & 0 & 0 & 0 & 1 \\ -1 & 0 & -3 & 1 & 0 & 0 \\ 0 & -1 & 0 & 0 & 0 & 0 \\ 1 & 0 & -1 & 0 & 0 & -1 \end{pmatrix}$$

$K \simeq \mathbb{Z}/6\mathbb{Z}$

No e): Periodmatrix: $\pi = \begin{pmatrix} \frac{\tau-4\rho^2}{3} & 0 & 2-2\rho & 1 & 0 & 0 \\ 0 & 4\rho & 6\rho & 0 & 2 & 0 \\ 2-2\rho & 6\rho & 3+12\rho & 0 & 0 & 6 \end{pmatrix}$.

Analytic and rational representation of the automorphism α:

$$\rho_a(\alpha) = \begin{pmatrix} 1 & \rho-1 & 2\frac{1-\rho}{3} \\ 0 & -\rho^2 & 0 \\ 0 & \frac{3}{2} & \rho \end{pmatrix} \text{ and } \rho_r(\alpha) = \begin{pmatrix} 1 & 0 & 0 & 0 & 0 & 0 \\ -3 & 0 & 0 & 0 & 2 & -3 \\ 2 & 0 & 0 & 0 & -1 & 2 \\ -4 & 0 & 6 & 1 & 0 & 0 \\ 0 & -2 & -3 & 0 & 1 & 0 \\ -1 & -1 & -2 & 0 & 1 & -1 \end{pmatrix}$$

$K \sim \mathbb{Z}/6\mathbb{Z} \times \mathbb{Z}/6\mathbb{Z}$

No f): Let (X, α) be the pair of No g) below. Then $(X, -\alpha^{-1})$ is an example for No f) of Table 5.6.

No g): Periodmatrix: $\pi = \begin{pmatrix} \frac{\tau-5+3i+2\rho}{6} & -2+i+\rho & -5+3i+2\rho & 1 & 0 & 0 \\ -2+i+\rho & 3+2i+3\rho & 6+6i+6\rho & 0 & 1 & 0 \\ -5+3i+2\rho & 6+6i+6\rho & 12+18i+12\rho & 0 & 0 & 6 \end{pmatrix}$.

Analytic and rational representation of the automorphism α:

$$\rho_a(\alpha) = \begin{pmatrix} 1 & -i+\rho & \frac{1+3i+2\rho^2}{6} \\ 0 & -2i+3\rho & i-\rho \\ 0 & -6i+6\rho & 3i-2\rho \end{pmatrix} \text{ and } \rho_r(\alpha) = \begin{pmatrix} 1 & 0 & 0 & 0 & 0 & 0 \\ -1 & 0 & 0 & 0 & 5 & -12 \\ 0 & 0 & 0 & 0 & -2 & 5 \\ -2 & -5 & -12 & 1 & 0 & 0 \\ 3 & -5 & -12 & 0 & -3 & 6 \\ 1 & -2 & -5 & 0 & -1 & 2 \end{pmatrix}$$

$K \simeq \mathbb{Z}/2\mathbb{Z}$

No h): Let (X, α) be the pair of No e). Then $(X, -\alpha)$ is an example for No h) of Table 5.6.

No i): Let (X, α) be a pair of No d). Then $(X, -\alpha)$ is an example for No i) of Table 5.6 .

No k): One can show (see [**SCH**], Sec 5.35) that if (X, α) is as in No k) of Table 5.6, then necessarily K is trivial, that is $X \in E_0 \times E_{1728} \times E_0$ and $\alpha = \rho \times i \times (-\rho)$.

References

[CAV] Lange, H., Birkenhake, Ch.: *Complex Abelian Varieties*, Grundlehren 302, Springer Verlag (1992).

[FU] Fujiki, A.: *Finite Automorphism Groups of Complex Tori of Dimension Two*, Publ. RIMS, Kyoto Univ. 24, 1-97 (1988).

[BGL] Birkenhake, Ch., González-Aguilera, V., Lange, H.: *Automorphism Groups of 3-dimensional Complex Tori*, Preprint (1998).

[BL] Birkenhake, Ch., Lange, H.: *Cubic theta relations*, Journal fur die reine angew. Math 407, 167-177 (1990)

[MUM] Mumford, D.: *Abelian Varieties*, Bombay (1974).

[RO] Roan, S.: *Complex 3-Tori with an Order 7 Special Automorphism*, Preprint (1990).

[RAU] Rauch, H.E., Lewittes, J.: *The Riemann surface of Klein with 168 automorphisms*, Problems in Analysis, a symposium in honor of Solomon Bochner,

[RIE] Ries, X.: *The splitting of some Jacobi varieties using Automorphism Groups*, Contemp. Math. 102, 81-124 (1997)

[SCH] Schmidt,D.: *Automorphismen 2 und 3-dimensionaler abelscher Varietäten*, Dissertation, Universität Erlangen (1997)

Ch. Birkenhake, Mathematisches Institut, Bismarckstr. 1 1/2, D 91054 Erlangen
E-mail address: birken@mi.uni-erlangen.de

V. González-Aguilera, Departamento de Matemática. Universidad Santa María, Casilla 110-V. Valparaiso
E-mail address: vgonzale@mat.utfsm.cl

H. Lange, Mathematisches Institut, Bismarckstr. 1 1/2, D 91054 Erlangen
E-mail address: lange@mi.uni-erlangen.de

Contemporary Mathematics
Volume **240**, 1999

On restrictions of the Picard bundle

L. Brambila-Paz, L. Hidalgo-Solís, and J. Muciño-Raymundo

Abstract. The Picard sheaves have been studied from differents aspects. Li proves the stability of the Picard bundle $\mathcal{W}$ over the moduli space $\mathcal{M}(n,d)$ of stable bundles of rank n and degree d. In general the restrictions of stable bundles need not be stable. In this paper we study the restiction $\mathcal{W}_\xi$ of the Picard bundle $\mathcal{W}$ to the subvariety $\mathcal{M}(n,\xi)$ of stable bundles with fixed determinant ξ. We give a condition to get polystability. If such condition is satisfied for rank 2 then $\mathcal{W}_\xi$ is stable and the connected component of the moduli space of stable bundles over $\mathcal{M}(2,\xi)$ with the same Hilbert polynomial as $\mathcal{W}_\xi$ containing $\mathcal{W}_\xi$ is isomorphic to the Jacobian J of the curve.

Let X be a non-singular projective curve of genus $g \geq 2$ over **C**. Let $\mathcal{M}(n,d)$ be the moduli space of stable bundles of rank n and degree d over X. If $(n,d)=1$ then $\mathcal{M}(n,d)$ is a fine moduli space, i.e. there exists a universal family, $\mathcal{U}$, called a Poincaré bundle, over $X \times \mathcal{M}(n,d)$. The direct images of $\mathcal{U}$ are called Picard sheaves over $\mathcal{M}(n,d)$.

The Picard sheaves have been studied from differents aspects (see [**13**], [**4**], [**7**], [**11**], [**1**]). If $d >> 0$, the Picard sheaf, $\pi_*\mathcal{U} = \mathcal{W}$, is actually a vector bundle. For $n=1$, the stability of $\mathcal{W}$ was proved by Kempf [**7**] for $d = 2g-1$ and by Ein and Lazarfeld [**4**] for $d \geq 2g$. For $n \geq 2$ and $d > 2gn$, Y. Li proved the stability of $\mathcal{W}$, in [**9**].

The moduli space $\mathcal{M}(n,d)$ has natural subvarieties, namely the Jacobian of the curve and the moduli space $\mathcal{M}(n,\xi)$ of stable bundles with fixed determinant ξ.

Li in [**9**], Theorem 2.5, proved that the restriction $\mathcal{W}_J$ of the Picard bundle $\mathcal{W}$ to the Jacobian is stable with respect to any theta divisor. By the results of Uhlenbeck and Yau (see [**18**]) there are Hermitian Einstein metrics h and h_1 in $\mathcal{W}$ and $\mathcal{W}_J$ respectively. If such metrics coincide then we prove (see Theorem 3.1)

The restriction $\mathcal{W}_\xi$ of the Picard bundle $\mathcal{W}$ to $\mathcal{M}(n,\xi)$ is polystable with respect to any polarization of $\mathcal{M}(n,\xi)$.

If Theorem 3.1 holds and $n=2$ then

$\mathcal{W}_\xi$ is stable.

1991 *Mathematics Subject Classification.* Primary: 14D20; Secondary: 14, 53C07.

The first author is member of the VBAC Research group of Europroj. The work forms part of the project Moduli of bundles in algebraic geometry, funded by the EU International Scientific Cooperation Initiative (Contract No. CI1*-CT93-0031). The first and third author also acknowledges support from CONACYT (Project No. 3231-E9307) and DGAPA-UNAM.

Moreover, let H be any ample divisor on $\mathcal{M}(2,\xi)$ and $\mathcal{M}(\mathcal{W}_\xi)$ the moduli space of H-stable bundles with same Hilbert polynomial as $\mathcal{W}_\xi$ on $\mathcal{M}(2,\xi)$ then

The connected component $\mathcal{M}(\mathcal{W}_\xi)^0$ of $\mathcal{M}(\mathcal{W}_\xi)$ containing $\mathcal{W}_\xi$ is isomorphic to the Jacobian $J(X)$.

In section 1 we shall recall the relation between Hermitian-Einstein bundles and stable bundles. We prove that given a Hermitian-Einstein vector bundle E over the product of two Kähler manifolds, $M \times N$, such that the restriction in one factor is Hermitian-Einstein then the restriction to the other factor is Hermitian-Einstein, with respect to the respective forms (see Lemma 1.2).

In section 2 we prove that there is a Hermitian-Einstein bundle over $J_0(X) \times \mathcal{M}(n,\xi)$ such that the restriction to each factor is $\mathcal{W}_J$ and $\mathcal{W}_\xi$ respectively. In section 3 we prove that $\mathcal{W}_\xi$ is polystable if certain condition in the metrics is satisfied. If such condition is satisfied for rank 2, we use Balaji and Vishwanath results in [**1**], to see that in this case $\mathcal{W}_\xi$ is stable for any polarisation and that the connected component $\mathcal{M}(\mathcal{W}_\xi)^0$ of $\mathcal{M}(\mathcal{W}_\xi)$ containing $\mathcal{W}_\xi$ is isomorphic to the Jacobian $J(X)$.

Acknowledgments: We would like to thank the referee for his useful comments. The first author thanks N. Raghavendra for previous discussions in Hermitian-Einstein bundles.

1. Hermitian-Einstein bundles

In this section we recall the definition of Hermitian-Einstein bundles and prove two Lemmas that we will use.

Let (M, g) be a compact Kähler manifold of complex dimension m with Kähler form Φ. A Hermitian-Einstein vector bundle (E, h) with respect to g is a holomorphic vector bundle E, together with a Hermitian metric h in E such that the curvature $R(\nabla)$ of the canonical connection ∇ is of bidegree $(1,1)$, and satisfies the condition

$$R(\nabla) \wedge \Phi^{m-1} = \lambda \Phi^m \otimes \mathbf{1}_E$$

for some constant $\lambda \in \mathbf{C}$.

Let $\mathcal{E}$ be a coherent sheaf over M. The Φ-slope of $\mathcal{E}$, is by definition the quotient

$$\mu_\Phi(\mathcal{E}) := \frac{d_\Phi(\mathcal{E})}{rk(\mathcal{E})}$$

where $rk(\mathcal{E})$ is the rank and $d_\Phi(\mathcal{E})$ is the degree of $\mathcal{E}$ with respect to Φ, i.e.

$$d_\Phi(\mathcal{E}) := \frac{1}{(m-1)!}\int_M c_1(\mathcal{E}) \wedge \Phi^{m-1}.$$

A vector bundle E is called Φ-stable if for any coherent subsheaf F with $rk(F) < rk(E)$,

$$\mu_\Phi(F) < \mu_\Phi(E).$$

A bundle E is Φ-polystable if it is the direct sum of Φ-stable bundles of the same slope. It is clear that Φ-polystable and simple implies Φ-stable.

REMARK 1.1. Kobayashi in [**8**] and Lübke in [**10**] prove that every Hermitian-Einstein vector bundle (E, h) with respect to g is Φ-polystable. Moreover, Uhlenbeck and Yau in [**18**] proved that every Φ-stable bundle over (M, g) admits a Hermitian-Einstein metric with respect to g, which is unique up to multiplication by a positive scalar. When M is a compact Riemann surface, such results were obtained by Narasimhan and Seshadri in [**15**] (see also [**5**]).

In general the restriction of a Hermitian-Einstein vector bundle does not need to be Hermitian-Einstein. However, when the base is a product of two Kähler manifolds we can say something.

Let (M, g_M) and (N, g_N) be compact Kähler manifolds, of complex dimension m and n respectively. Let g be the Kähler metric on $M \times N$ given by $g = g_M + g_N$ and (E, h) be a Hermitian holomorphic vector bundle over $M \times N$. Let Φ_M, Φ_N and Φ denote the Kähler forms of (M, g_M), (N, g_N) and $(M \times N, g)$, respectively, so $\Phi = \Phi_M + \Phi_N$.

LEMMA 1.2. *Suppose that (E, h) is Hermitian-Einstein with respect to g and that for every point $x \in M$, the restriction (E_x, h_x) of (E, h) to $\{x\} \times N$ is Hermitian-Einstein with respect to g_N. Then, for every point $y \in N$, the restriction (E^y, h^y) of (E, h) to $M \times \{y\}$ is Hermitian-Einstein with respect to g_M.*

Proof. Let U and V be holomorphic coordinate domains on M and N, respectively, and fix a point $(x_0, y_0) \in U \times V$. Denote by Ω, Ω_{x_0} and Ω^{y_0} the curvature matrices of (E, h), (E_{x_0}, h_{x_0}) and (E^{y_0}, h^{y_0}) with respect to the frames S, S_{x_0} and S^{y_0}, respectively.

We know that

$$\Phi^{m+n-1} = (a\Phi_M^m \Phi_N^{n-1} + b\Phi_M^{m-1}\Phi_N^n)$$

and

$$\Phi^{m+n} = c\Phi_M^m \Phi_N^n.$$

Since U and V are coordinate domains, on the restrictions we have

$$\Omega = \Omega_{x_0} + \sum_{i=1}^{r} \beta_{U,i} \wedge \gamma_{V,i} + \Omega^{y_0},$$

where $\beta_{U,i} \in A^1(U)$, $\gamma_{V,i} \in A^1(V)$.

Thus,

$$\begin{aligned} \Omega\Phi^{m+n-1} &= (\Omega_{x_0} + \textstyle\sum_{i=1}^r \beta_{U,i} \wedge \gamma_{V,i} + \Omega^{y_0})(a\Phi_M^m \Phi_N^{n-1} + b\Phi_M^{m-1}\Phi_N^n) \\ &= a'\Omega_{x_o}\Phi_M^m\Phi_N^{n-1} + b'\Omega^{y_0}\Phi_M^{m-1}\Phi_N^n \end{aligned} \tag{1}$$

where a', b' are constants.

By hypothesis (E_x, h_x) and (E, h) are Hermitian-Einstein,

$$i.e. \;\; \Omega_{x_0}\Phi_N^{n-1} = \lambda'\Phi_N^n I \quad \text{and} \quad \Omega\Phi^{m+n-1} = \lambda\Phi^{m+n} I \tag{2}$$

where I is the identity matrix.

From (1) and (2) we have

$$\begin{aligned} c\lambda\Phi_M^m\Phi_N^n I &= \lambda\Phi^{m+n} I \\ &= \Omega\Phi^{m+n-1} \\ &= a'\Omega_{x_o}\Phi_M^m\Phi_N^{n-1} + b'\Omega^{y_0}\Phi_M^{m-1}\Phi_N^n \\ &= a''\Phi_M^m\Phi_N^n I + b''\Omega^{y_0}\Phi_M^{m-1}\Phi_N^n. \end{aligned} \tag{3}$$

Thus, $\Omega^{y_0}\Phi_M^{m-1} = \lambda'\Phi_M^m I$, i.e. (E^{y_0}, h^{y_0}) is Hermitian-Einstein with respect to g_M. □

We shall quote the following result since we will need it later.

LEMMA 1.3. Let $f : (A, g_A) \rightarrow (B, g_B)$ be a local biholomorphism between two compact Kähler manifolds. Let g_B and g_A be the Kähler metrics of A and B, respectively, and $f^*(g_B) = g_A$. If (E, h) is Hermitian-Einstein over B with respect to Φ_B then $(f^*(E), f^*(h))$ is Hermitian-Einstein with respect to Φ_A, where Φ_B and Φ_A are the respectively Kähler forms.

Proof: The proof follows from [**8**], IV, Prop. 1.8, since it is a local biholomorphism.

2. Stability

Let A be a smooth complex projective variety of dimension n with an ample divisor H on A and E a vector bundle over A. The H-slope of E, denoted by $\mu_H(E)$, is the quotient

$$\mu_H(E) = c_1(E) \cdot [H]^{n-1}/rk(E)$$

where $rk(E)$ is the rank of E. The vector bundle E is call H-stable if for every coherent subsheaf F with $rk(F) < rk(E)$ one has

$$\mu_H(F) < \mu_H(E). \tag{4}$$

The bundle E is H-polystable if it is direct sum of H-stable bundles of the same slope.

REMARK 2.1. If H is an ample line bundle and Φ is a closed $(1,1)$-form representing the first Chern class $c_1(H)$, then we say either H-stable or Φ-stable.

Note that if A is a curve then the stability of the vector bundle is independent of the line bundle H.

In general it is not easy to see when a vector bundle is stable. Mehta and Ramanathan in [**12**] proved that if $[\omega]$ is the hyperplane class of $A \subset \mathbf{P}^N$ and X the intersection of A with a hypersurface of degree $d >> 0$, then a bundle E over A is ω-stable if and only if the restriction of E to X is $\omega_{|X}$-stable. In the case that A is a principal polarized abelian variety we have the following result.

THEOREM 2.2. Let (A, θ_A) be a principally polarized abelian variety of dimension g and E a vector bundle over A. There exists a non-singular irreducible projective curve X and an embedding $\phi : X \rightarrow A$ such that the θ_A-stability of E is implied from the stability of $\phi^*(E)$.

Proof: Welters proved in [**19**] that for any principally polarized abelian variety (A, θ_A) there is non-singular irreducible projective curve X, (not unique) such that (A, θ_A) is a Prym-Tyurin variety for the curve X of exponent $e = 3^{g-1}(g-1)!$. By Welters' Criterion (see [**19**]) there is a embedding $\phi : X \rightarrow A$ such that the algebraic 1-cycles $\phi_*(X)$ and $\frac{e}{(g-1)!}[\theta_A]^{g-1}$ are numerically equivalent.

Now let $F \subset E$ be a torsion free subsheaf of E such that E/F is also a torsion free sheaf. Since Codim $(Sing(F)) \geq 2$ and X generates A as an abelian group we may assume $\phi(X) \subset A - Sing(F)$, thus $\phi^*(F)$ is a subbundle of $\phi^*(E)$. If $\phi^*(E)$ is stable then

$$\mu(\phi^*(F)) < \mu(\phi^*(E))$$

i.e.

$$\frac{c_1(\phi^*(F))}{rk(\phi^*(F))} < \frac{c_1(\phi^*(E))}{rk(\phi^*(E))}. \tag{5}$$

Since

$$c_1(\phi^*(-)) = \phi_*(X) \cdot c_1(-) \equiv c_1(-) \cdot \frac{e}{(g-1)!}[\theta_A]^{g-1} \tag{6}$$

and $\mu_{\theta_A}(-) = c_1(-) \cdot [\theta_A]^{g-1}/rk(-)$, we have from (5) and (6) that

$$\mu_{\theta_A}(F) < \mu_{\theta_A}(E).$$

Hence, E is θ_A-stable.

REMARK 2.3. When A is the Jacobian of a non-singular curve, Theorem 2.2 was proved by Li in [**9**], Lemma 2.8. Actually, we use similar arguments for the proof.

Some projective varieties, like abelian varieties or moduli spaces, carries natural bundles different from the trivial or the induced from the tangent bundle, namely the Picard bundles. Mukai in [**13**] use the Fourier functor to study Picard sheaves over abelian varieties. However, we consider Picard sheaves in the following sense.

Let $\mathcal{M}(n,d) = \mathcal{M}$ be the moduli space of stable bundles of rank n and degree d over a non-singular algebraic curve X of genus $g \geq 2$. If $(n,d) = 1$ then there is a universal family called the Poincaré bundle. Denote by $\mathcal{U}$ a Poincaré bundle over $X \times \mathcal{M}$. Note that $\mathcal{U}$ is unique up to tensoring by a line bundle from $\mathcal{M}$.

The direct image $\pi_*(\mathcal{U}) = \mathcal{W}$ over $\mathcal{M}(n,d)$ is a sheaf that is called the *Picard sheaf*. If $d > n(2g-2)$ then the direct image $\pi_*(\mathcal{U}) = \mathcal{W}$ over $\mathcal{M}(n,d)$ is a vector bundle that is called the *Picard bundle*. By Riemann-Roch's theorem the rank of $\mathcal{W}$ is $d + n(1-g)$.

Let θ be the generalized theta divisor over $\mathcal{M}(n,d)$. Li in [**9**] proved that if $d > 2ng$ then $\mathcal{W}$ is θ-stable.

REMARK 2.4. There is a natural Kähler metric on $\mathcal{M}(n,d)$ (see [**14**]) called the *theta metric*, which we denote by g_θ. If Φ_θ denotes the Kähler form of this metric, then by [**3**] Theorem 3.27, the cohomology class of $n\Phi_\theta$ equals θ. Therefore, from the results of Uhlembeck and Yau (see [**18**]) and the θ-stability of $\mathcal{W}$, there exists a Hermitian metric k in $\mathcal{W}$, which is Hermitian-Einstein with respect to g_θ.

We shall consider the restriction of $\mathcal{W}$ to natural subvarieties of $\mathcal{M}(n,d)$.

If $\mathcal{M}(n,\xi) = \mathcal{M}_\xi$ denote the moduli space of stable bundles in $\mathcal{M}(n,d)$ with fixed determinat ξ and $J_0(X)$ the Jacobian of line bundles over X of degree 0 then we have the map

$$f : J_0(X) \times \mathcal{M}(n,\xi) \to \mathcal{M}(n,d)$$

defined as $(L,E) \mapsto L^{-1} \otimes E$.

For generic $L \in J_0(X)$ and $E \in \mathcal{M}(n,\xi)$ the restrictions of

$$f : J_0(X) \times \mathcal{M}(n,\xi) \to \mathcal{M}(n,d)$$

to $J_0(X) \times \{E\}$ and $\{L\} \times \mathcal{M}(n,\xi)$ induce embeddings. Actually, if J^n denote the group of n-torsion points in $J_0(X)$ then

$$\mathcal{M}(n,\xi) \times J_0(X) \to \mathcal{M}(n,d).$$

is a fibration with fibre J^n (see [**9**], 4.8).

Denote by θ_J the theta divisor in $J_0(X)$ and by θ_ξ the restriction of the generalized theta divisor θ in $\mathcal{M}(n,d)$ to $\mathcal{M}(n,\xi)$. As in Remark 2.4, (see also [**3**]) we have Kähler metrics $g_{n\theta_J}$ and g_{θ_ξ} in $J_0(X)$ and $\mathcal{M}(n,\xi)$, respectively, such that the associated Kähler forms are in the class of θ_J and θ_ξ, respectively.

From [**2**] we have that

$$f^*(\theta) = n\theta_J + \theta_\xi := \Theta \tag{7}$$

and

$$g_\Theta = g_{n\theta_J} + g_{\theta_\xi}.$$

Hence, since $\mathcal{W}$ is θ-stable we have

PROPOSITION 2.5. *$f^*(\mathcal{W})$ is semistable. Moreover, there exists a metric h in $f^*(\mathcal{W})$ such that $(f^*(\mathcal{W}), h)$ is Hermitian-Einstein with respect to g_Θ.*

Proof: The semistability follows from [**6**], lemma 3.2.2. To prove that is Hermitian-Einstein, we have from Remark 2.4, that there exists a metric k in $\mathcal{W}$ such that $(\mathcal{W}, k)$ is Hermitian-Einstein with respect to g_θ. Denote by h the pull-back $f^*(k)$ under the morphism $f : J_0(X) \times \mathcal{M}(n,\xi) \to \mathcal{M}(n,d)$. Hence, from Lemma 1.1 $(f^*(\mathcal{W}), h)$ is Hermitian-Einstein with respect to g_Θ since f is a local biholomorphism (see [**9**], 4.8).

3. Restrictions

From Proposition 2.5 we have a Hermitian-Einstein bundle

$$p : (f^*(\mathcal{W}), h) \to J_0(X) \times \mathcal{M}(n,\xi)$$

with respect to $g_\Theta = g_{n\theta_J} + g_{\theta_\xi}$. For generic $L \in J_0(X)$ and $E \in \mathcal{M}(n,\xi)$, denote by $\mathcal{W}_{J,E}$ and $\mathcal{W}_{\xi,L}$ the restriction of the Picard bundle $\mathcal{W}$ over $\mathcal{M}(n,d)$ to the embeddings of $J_0(X)$ and $\mathcal{M}(n,\xi)$ given by the restriction of $f : J_0(X) \times \mathcal{M}(n,\xi) \to \mathcal{M}(n,d)$. That is,

$$f^*(\mathcal{W})_{|_{J_0(X)\times\{E\}}} = \mathcal{W}_{J,E} \quad \text{and} \quad f^*(\mathcal{W})_{|_{\{L\}\times\mathcal{M}_\xi}} = \mathcal{W}_{\xi,L}.$$

Li in [**9**], Theorem 2.5, proved that the restriction $\mathcal{W}_{J,E}$ is stable with respect to any theta divisor in $J(X)$. In this case, the stability of $\mathcal{W}_{J,E}$ follows from the stability of the restriction of $\mathcal{W}_{J,E}$ to the curve, which is embedded in $J_0(X)$ by the Abel-Jacobi map (see [**9**]).

Since $\mathcal{W}_{J,E}$ is θ_J-stable and therefore $n\theta_J$-stable, there exists a metric h_E such that $(\mathcal{W}_{J,E}, h_E)$ is Hermitian-Einstein with respect to $g_{n\theta_J}$. Let h' be the metric on $f^*(\mathcal{W})$ such that $h'_{|_{J\times\{E\}}} = h_E$, for a generic E.

THEOREM 3.1. *If the metrics h' and h on $f^*(\mathcal{W})$ are gauge equivalent then the restriction $\mathcal{W}_{\xi,L}$ of the Picard bundle $\mathcal{W}$ to $\mathcal{M}(n,\xi)$ is θ_ξ-polystable.*

Proof: From Proposition 2.5 $(f^*(\mathcal{W}), h)$ is Hermitian-Einstein with respect to g_Θ. By assumption the restriction of h to $J_0(X) \times \{E\}$ is h_E. Hence, for a generic E the restriction of $(f^*(\mathcal{W}), h)$ to $J_0(X) \times \{E\}$ (which is $(\mathcal{W}_{J,E}, h_E)$), is Hermitian-Einstein with respect to $g_{n\theta_J}$. Therefore, from Lemma 1.1 we have that for a generic $L \in J_0(X)$, $(\mathcal{W}_{\xi,L}, h_|) = (f^*(\mathcal{W}), h)_{|\{L\}\times\mathcal{M}_\xi}$, is Hermitian-Einstein with respect to g_{θ_ξ}. Therefore, from Remark 1.1, $\mathcal{W}_{\xi,L}$ is θ_ξ-polystable.

For rank 2, Balaji and Vishwanath in [**1**] proved that

$$H^0(\mathcal{M}_\xi, \mathrm{ad}(\mathcal{W}_\xi)) = 0 \tag{8}$$

and

$$H^1(\mathcal{M}_\xi, \mathrm{ad}(\mathcal{W}_\xi)) = g. \tag{9}$$

They use Thaddeus description of the moduli space of stable pairs (see [**17**]). The equality (8) implies that $\mathcal{W}_\xi$ is simple. Therefore, if the hypothesis of Theorem 3.1 holds then for rank 2 we have that $\mathcal{W}_\xi$ *is* θ_ξ*-stable.*

Denote by $\mathcal{M}(\mathcal{W}_\xi)$ the moduli space of stable vector bundles over $\mathcal{M}(2,\xi)$ that have the same Hilbert polynomial as $\mathcal{W}_\xi$. Denote by $\mathcal{M}^0(\mathcal{W}_\xi)$ the connected component of $\mathcal{M}(\mathcal{W}_\xi)$ that contains $\mathcal{W}_\xi$. By deformation theory we have that

$$\dim H^1(\mathcal{M}_\xi, \mathrm{End}(\mathcal{W}_\xi)) = \dim T\mathcal{M}^0(\mathcal{W}_\xi)_{|\mathcal{W}_\xi}, \tag{10}$$

where $T\mathcal{M}^0(\mathcal{W}_\xi)_{|\mathcal{W}_\xi}$ is the tangent space of $\mathcal{M}^0(\mathcal{W}_\xi)$ at $\mathcal{W}_\xi$.

Since $H^1(\mathcal{M}_\xi, \mathrm{End}(\mathcal{W}_\xi)) = H^1(\mathcal{M}_\xi, \mathcal{O}) \oplus H^1(\mathcal{M}_\xi, \mathrm{ad}(\mathcal{W}_\xi))$ and $H^1(\mathcal{M}_\xi, \mathcal{O}) = 0$ (see [**16**]), we have from equality (9) that $\dim H^1(\mathcal{M}_\xi, \mathrm{End}(\mathcal{W}_\xi)) = g$. Hence, from (10) we have that

$$\dim \mathcal{M}^0(\mathcal{W}_\xi) = g.$$

Let $\mathcal{U}_\xi$ be a universal bundle over $X \times \mathcal{M}_\xi$ and p_i the projection in the i-factor, for $i = 1, 2$. The natural map $\phi : J(X) \to \mathcal{M}^0(\mathcal{W}_\xi)$ defined as $L \mapsto p_{2*}(\mathcal{U}_\xi \otimes p_1^* L)$ is injective (see [**1**]). Since $J(X)$ is complete, ϕ is injective and has the same dimension, thus ϕ is an isomorphism. That is,

The connected component $\mathcal{M}(\mathcal{W}_\xi)^0$ *of* $\mathcal{M}(\mathcal{W}_\xi)$ *containing* $\mathcal{W}_\xi$ *is isomorphic to the Jacobian* $J(X)$. i.e.

$$\mathcal{M}^0(\mathcal{W}_\xi) \cong J(X).$$

References

[1] V. Balaji, P. A. Vishwanath, *Deformations of Picard sheaves and moduli of pairs*, Duke Math. Journal. Vol. **76**, No. 3, (1994), 773-792 .

[2] A. Beauville, M. S. Narasimhan, S. Ramanan, *Spectral curves and the generalised theta divisor*, Journal reine angew. Math. **398** (1989), 169-179.

[3] I. Biswas, N. Raghavendra, *Determinants of parabolic bundles on Riemann surfaces*, Proc. Indian Acad. Sci. (Math. Sci.) **103** (1993) 41-71.

[4] L. Ein, R. Lazarfeld, *Stability and restrictions of the Picard Bundles with an application to the normal bundle*, Lecture Notes **176** (Trieste 1989/Bergen 1989) 149-156, London Math Soc.

[5] S. K. Donaldson, *A new proof of a theorem of Narasimhan and Seshadri*, Jour. of Diff. Geom. **18** (1983), 269-278.

[6] D. Huybrechts, M. Lehn *The Geometry of the Moduli Spaces of Sheaves* Aspects of Mathematics, Vol **31**, Vieweg 1997.

[7] G. Kempf, *Rank g Picard bundles are stable*, American Journal of Mathematics **112** (1990), 397-401.

[8] S. Kobayashi, *Differential Geometry of Complex Vector Bundles*, Iwanami-Shoten and Princeton University Press, 1987.

[9] Y. Li, *Spectral curves, theta divisors and Picard bundles*, Inter. J. Math. **2** (1991), 525-550.

[10] M. Lübke, *Stability of Einstein-Hermitian vector bundles*, Manuscripta Mathematica **42** (1983), 245-257.

[11] A. Mattuck, *Picard bundles*, Illinois J. Math. **5** (1961), 550-564.

[12] V.B. Mehta, A. Ramanathan, *Semistable sheaves on projective varieties and the rstriction to curves*, Math. Ann. **258** (1982), 213-226.

[13] S. Mukai, *Duality between* $D(X)$ *and* $D(\widehat{X})$ *with applications to Picard sheaves*, Nag. Math. J. **81**, (1981), 153-175.

[14] M.S. Narasimhan, *Elliptic operators and differential geometry of moduli spaces of vector bundles on compact Riemann surfaces*, Proc. Inter.Conf. on Functional Analysis and related topics (Tokyo, 1969), University of Tokyo Press, 1970, 68-71.

[15] M. S. Narasimhan, C. S. Seshadri, *Stable and unitary vector bundles on a compact Riemann surface*, Ann. Math. **82** (1965), 540-567.

[16] M. S. Narasimhan, S.Ramanan, *Deformations of the moduli space of vector bundles over an algebraic curve*, Ann. of Math. **2** 101, (1975), 391-417.

[17] M. Thaddeus *Stable pairs, linear systems and the Verlinde formula*, Invent. Math **117** No. 2, 317-353, (1994)

[18] K. Uhlenbeck, S. T. Yau, *On the existence of Hermitian-Yang-Mills connections in stable vector bundles*, Comm. Pure Appl. Math. **39** (1986), 5257-5293.

[19] G. Welters, *Curves of twice the minimal class on principally polarizad abelian varieties*, Indig. Math. **94**, 87-109 (1987).

CIMAT, Apdo. Postal 402, C.P. 36240 Guanajuato, México
E-mail address: `lebp@fractal.cimat.mx`

Departamento de Matemáticas, UAM - Iztapalapa México, 09340, D.F.
E-mail address: `lhso@xanum.uam.mx`

Instituto de Matemáticas, Unidad Morelia, UNAM Morelia Michoacán.
E-mail address: `muciray@servidor.unam.mx`

Contemporary Mathematics
Volume **240**, 1999

Representations of triangle groups in the Heisenberg group

Alberto M. Castro

ABSTRACT. In this paper we classify the discrete representations of triangle groups in the Heisenberg group of dimension five ($\mathcal{H}_5$). For the case where the triangle group is of spherical or Euclidean type these representations are understood, so we concentrate in the description of hyperbolic representations. We prove that representations of triangle groups into $\mathcal{H}_5$ are in one to one correspondence with finite subgroups of $SU(2)$, and that the triangle group is hyperbolic iff the corresponding group in $SU(2)$ is non-cyclic. For the cyclic non-trivial case we show that the triangle group must be of Euclidean type. The trivial group in $SU(2)$ corresponds to spherical triangle groups.

1. Introduction

The purpose of this paper is to classify discrete representations of triangle groups in the Heisenberg group of dimension five ($\mathcal{H}_5$). The representations we will study are of a special type, they are characterized by the fact that the generating reflections of the group Δ will be mapped to reflections on vertical planes of $\mathcal{H}_5$. These representations arise as the stabilizers of cusps of complex ideal tetrahedral groups.

Complex ideal tetrahedral groups are defined in [**Cas98**] as subgroups of the group of holomorphic isometries ($PU(3,1)$) of complex hyperbolic space of dimension three, denoted $\mathbf{H}^3_{\mathbf{C}}$. The generators of these groups are complex reflections of order two about linear two-dimensional complex subspaces of $\mathbf{H}^3_{\mathbf{C}}$, and they are determined by a set of points $Q = \{q_0, q_1, q_2, q_3\}$ in the boundary of $\mathbf{H}^3_{\mathbf{C}}$. These are the *vertices* of the tetrahedron and we will ask that no three vertices are $\mathbf{C}$-collinear. Therefore for any three points in Q there exists a unique complex two dimensional subspace through these points. Denote by P_k the plane that does not contain the point q_k. The planes $P_0, \dots, P_3$ form the *faces* of the tetrahedron. Associated to each plane P_k there exists a unique complex reflection (for definitions see [**Gol99**]). The group T generated by these reflections is the *complex ideal hyperbolic group associated to* Q.

Given a vertex q call G_q the group generated by the three generators of T that stabilize q. In [**Cas98**] we prove that G_q is the image of a representation of a triangle group Δ (for a definition see below) into the stabilizer of q in $PU(3,1)$

1991 *Mathematics Subject Classification.* 22E40 (32H20, 32M05, 20H10).

The author was partially suported by a Faculty Grant from Mount Holyoke College.

(call it H):

$$\phi : \Delta \to H \subset PU(3,1).$$

The group Δ can be of any type and H can be identified with the group of similarities of the Heisenberg group of dimension five. The Heisenberg group of dimension five is isomorphic to the boundary of complex hyperbolic space of (complex) dimension three, minus a point. In this case the point is the vertex q.

We expect to use the results of this paper to complete the classification of complex ideal tetrahedral groups initiated in [**Cas98**]. Groups of this type may provide interesting examples of manifolds or orbifolds with nice properties (e.g. minimal volume, etc.).

A triangle group Δ is a Coxeter group with presentation

$$\Delta = \langle \alpha, \beta, \gamma \,|\, \alpha^2, \beta^2, \gamma^2, (\alpha\beta)^p, (\beta\gamma)^q, (\gamma\beta)^r \rangle, \tag{1.1}$$

where the group is called spherical, Euclidean or hyperbolic, provided that the number $\frac{1}{p} + \frac{1}{q} + \frac{1}{r}$ is bigger, equal to or less than one, respectively. The problem of describing representations of triangle groups of spherical and Euclidean type was completely solved in [**Cas98**]. In this paper we will study representations of hyperbolic triangle groups and we give a complete classification of the discrete representations of triangle groups in the Heisenberg group.

We will frequently use the index two subgroup Δ^+ generated by $a = \alpha\beta$, $b = \beta\gamma$ and $c = \gamma\alpha$ that will be called the "orientation preserving" triangle group associated to Δ. The presentation of this group is

$$\Delta^+ = \langle a, b, c \,|\, a^p, b^q, c^r, abc \rangle. \tag{1.2}$$

The Heisenberg group of (real) dimension five (denoted $\mathcal{H}_5$) is a nilpotent group of degree two and is topologically equivalent to the Cartesian product of $\mathbf{C}^2$ with $\mathbf{R}$: $\mathcal{H}_5 \cong \mathbf{C}^2 \times \mathbf{R}$. The group law is defined by

$$(\zeta, t) \cdot (\vartheta, s) = (\zeta + \vartheta, t + s + 2\,\mathrm{Im}\langle\langle \zeta, \vartheta \rangle\rangle),$$

where (ζ, t) and (ϑ, s) belong to $\mathbf{C}^2 \times \mathbf{R}$ and $\langle\langle \cdot, \cdot \rangle\rangle$ is the standard positive definite Hermitian product in $\mathbf{C}^2$. The center of $\mathcal{H}_5$ is $Z = \{(0,t) | t \in \mathbf{R}\} \cong \mathbf{R}$. The *vertical projection* Π_V is defined to be the natural map

$$\begin{aligned} \Pi_V : &\mathcal{H}_5 \to \mathbf{C}^2 \\ &(\zeta, v) \mapsto \zeta. \end{aligned}$$

The group $\mathcal{H}_5$ acts on itself by left translations:

$$T_{(\zeta,v)}(\vartheta, u) = (\zeta, v) \cdot (\vartheta, u).$$

for $(\zeta, v), (\vartheta, u) \in \mathcal{H}_5$. And the group $U(2)$ acts on $\mathcal{H}_5$ by:

$$A \cdot (\zeta, v) = (A\zeta, v),$$

for $A \in U(2)$ and $(\zeta, v) \in \mathcal{H}_5$. The semidirect products of these two groups is called the group of isometries of $\mathcal{H}_5$

$$\mathrm{Isom}(\mathcal{H}_5) = \mathcal{H}_5 \rtimes U(2).$$

The Heisenberg translations T of the form $T_{(0,v)}$ are called *vertical translations,* the number $|v|$ is the *translation length* of T. The group $\mathrm{Isom}(\mathcal{H}_5)$ is a subgroup of the group of similarities of $\mathcal{H}_5$ (see Section 2.1 and [**Gol99**]).

We now describe the representations that we will consider in this work. For further details we refer the reader to [**Cas98**]. Call the representation ϕ and G its image in $\mathrm{Isom}(\mathcal{H}_5)$. We need to define the image of each generator of the group Δ (equation (1.1)) and verify that each relation of the group Δ is satisfied by the elements of G. Associate to each generator ν of Δ a reflection r_ν in the group of isometries of $\mathcal{H}_5$ ($\mathrm{Isom}(\mathcal{H}_5)$) along a vertical two chain C_ν. (For definitions see section 2.) The vertical projection of C_ν is a complex line l_ν in $\mathbf{C}^2$. If the representation ϕ is discrete the product of any two generators of Δ must be of finite order, the therefore the angle between any of two of these lines (say l_α and l_β) must be a rational multiple of 2π:

$$\theta_{\alpha\beta} = \angle\{l_\alpha, l_\beta\} = p\pi/q, \quad p, q \in \mathbf{Z}.$$

Using these properties it easily follows that the generators α, β and γ map to elliptic isometries of $\mathrm{Isom}(\mathcal{H}_5)$ of orders p, q and r, respectively. These reflections are given in terms of the angles $\theta_{\alpha\beta}$, $\theta_{\beta\gamma}$, $\theta_{\gamma\alpha}$ and other geometric information of the chains C_α, C_β and C_γ. (For details see Theorem 2.1.)

We have then a representation

$$\phi : \Delta \to G \subset \mathrm{Isom}(\mathcal{H}_5) \cong \mathcal{H}_5 \rtimes U(2).$$

And it is easy to check that

$$\phi|_{\Delta^+} : \Delta^+ \to G^+ \subset \mathrm{Isom}^+(\mathcal{H}_5) \cong \mathcal{H}_5 \rtimes SU(2).$$

where $\mathrm{Isom}^+(\mathcal{H}_5)$ is the subgroup of what we call "orientation preserving" isometries and that is isomorphic to $\mathcal{H}_5 \rtimes SU(2)$. We shall also call ϕ the restriction map $\phi|_{\Delta^+}$. Representations of spherical groups are trivial, as it is shown in the following theorem that follows as a direct consequence of Theorem 4.4 of [**Cas98**].

THEOREM 1.1 (Spherical Groups). *Discrete representations of spherical orientation preserving triangle groups are always trivial (i.e. the image is the identity). Note also that in this case the full triangle group Δ maps to a group isomorphic to $\mathbf{Z}/2\mathbf{Z}$.*

Therefore we only need to concentrate on the cases of Euclidean or hyperbolic triangle groups. We will assume from now on, unless it is stated otherwise, that all triangle groups are of Euclidean or hyperbolic type. In this case the intersection of any two vertical cochains from the set $\{C_\alpha, C_\beta, C_\gamma\}$ is a one cochain through the vertex at infinity and another vertex of the tetrahedron. Define $\mathcal{L}$ to be the projection in $\mathbf{C}^2 = \Pi_V(\mathcal{H}_5)$ of these three chains. (Note that $\mathcal{L}$ could have been defined solely in terms of the vertices of the tetrahedron, the definition given here has the advantage of eliminating any reference to the ideal tetrahedron in $\mathbf{H}^3_{\mathbf{C}}$.)

THEOREM 1.2. *Let $\Delta \to G \subset \mathrm{Isom}(\mathcal{H}_5)$ be a discrete representation. The group Δ is Euclidean if and only if the set $\mathcal{L}$ is contained in a totally real subspace of $\mathbf{C}^2$.*

A totally real subset V of $\mathbf{C}^2$ lifts under $\Pi_V : \mathcal{H}_5 \to \mathbf{C}^2$ to an infinite $\mathbf{R}$-sphere (Section 2), which is the boundary of a totally real subspace of $\mathbf{H}^3_{\mathbf{C}}$. It is called infinite because it contains the point at infinity of $\mathcal{H}_5$. The subspace V is kept invariant under the induced group of transformations G in $PU(3,1)$.

This is indeed a rigidity theorem, since it can be concluded from it that a discrete representation of an Euclidean triangle group Δ is always obtained as an extension to $\mathcal{H}_5$ of a representation of Δ in a totally real subspace V of $\Pi_V(\mathcal{H}_5) \cong \mathbf{C}^2$.

Non trivial representations of hyperbolic triangle groups are possible as we show in [**Cas98**, Theorem 5.6], although the kernel is big. In the results of this paper we will present a systematic way of finding discrete representations of hyperbolic triangle groups, and prove that the list is complete. The main ingredient for the classification theorem is the following result.

THEOREM 1.3. *Let Δ be a Euclidean or hyperbolic triangle group. If the representation $\phi : \Delta \to \mathrm{Isom}(\mathcal{H}_5) \cong \mathcal{H}_5 \rtimes U(2)$ is discrete, the projection of $G^+ = \phi(\Delta^+)$ by the natural map $\pi : \mathcal{H}_5 \rtimes SU(2) \to SU(2)$ is a non trivial finite group $F = \pi(G^+)$. Moreover, the group Δ is Euclidean iff F is cyclic.*

After proving this theorem we will show that associated to every finite non-cyclic group $F \subset SU(2)$ there exists a unique representation of a hyperbolic triangle group, up to isometry. We will call the image of the representation the *lifted group associated to F* or simply *lifted group.*

By putting together the results of [**Cas98**] and of this paper we have the following theorem that describes representations of triangle groups in the Heisenberg group:

THEOREM 1.4. *Let $\phi : \Delta \to \mathbf{Sim}(\mathcal{H}_5)$ be a discrete representation of a triangle group Δ. And denote also by $\phi : \Delta^+ \to \mathbf{Sim}(\mathcal{H}_5)$ the associated representation of the orientation preserving subgroup Δ^+ of Δ. Then $G \subset \mathrm{Isom}(\mathcal{H}_5)$ and $G^+ \subset \mathrm{Isom}^+(\mathcal{H}_5)$. Moreover, if the triangle group is of*

- **Spherical type:** *The representation is trivial, in the sense that $\phi(\Delta) \cong \mathbf{Z}/2\mathbf{Z}$, or, equivalently, $\phi(\Delta^+) \cong id$.*
- **Euclidean type:** *The representation ϕ is faithful and G preserves an infinite $\mathbf{R}$-sphere in $\mathcal{H}_5$. Moreover, if $\pi : \mathrm{Isom}^+(\mathcal{H}_5) \to SU(2)$ denotes the canonical projection, $\pi(G^+)$ is a cyclic group.*
- **Hyperbolic type:** *$\pi(G^+)$ is a finite non-cyclic subgroup of $SU(2)$ and the generators of the lifted subgroups are listed in Table 2.*

This result will be proved in Section 4.

The groups obtained in the Euclidean and spherical cases are discrete as is shown in [**Cas98**]. Table 2 lists the possible discrete representations of hyperbolic triangle groups, we finish this paper by showing that all these groups are discrete.

THEOREM 1.5. *All the representations of hyperbolic triangle groups in $\mathcal{H}_5$ with generators given in Table 2 are discrete.*

This paper is organized as follows: In Section 2 we introduce the notation and present some of the background and preliminary material; Section 3 is devoted to the proof of Theorem 1.3 which is based on three technical lemmas that are proved in Section 5; In Section 4 we will present the classification of representations of triangle groups in $\mathcal{H}_5$ and prove Theorem 1.4; In Section 5 we prove Theorem 1.5; Finally Section 6 presents all the (more technical) proofs of the lemmas of Section 3.

2. Preliminaries

2.1. The Heisenberg Group. The *Heisenberg group* is a degree two nilpotent group, which can be constructed in several different ways. In this paper we will use the conventions and notation of [**Gol99**]. Thus for us the Heisenberg group is a central extension of $\mathbf{C}^{n-1}$ by $\mathbf{R}$

$$1 \to \mathbf{C}^{n-1} \to \mathcal{H}_{2n-1} \to \mathbf{R} \to 1.$$

If (ζ, v), $(\vartheta, u) \in \mathcal{H}_{2n-1} \cong \mathbf{C}^{n-1} \times \mathbf{R}$ their product is defined by:

$$(\zeta, v) \cdot (\vartheta, u) = (\zeta + \vartheta, u + v + 2\,\mathrm{Im}\langle\langle \zeta, \vartheta \rangle\rangle)$$

where $\langle\langle \cdot, \cdot \rangle\rangle$ denotes the standard positive definite hermitian product in $\mathbf{C}^{n-1}$. The Heisenberg group is important in group theory and in geometry: the boundary of complex hyperbolic space ($\mathbf{H}^n_{\mathbf{C}}$) minus one point is naturally isomorphic to the Heisenberg group ($\mathcal{H}_{2n-1}$). Several metrics can be defined in this group, of these the most important from our perspective are the *Carnot-Caratheodory* and the *Cygan* metrics. The latter is easier to define, the distance d_C is defined by:

$$d_C((\zeta, v), (\vartheta, u)) = |(\zeta, v)^{-1} \cdot (\vartheta, u)|$$

where $|\cdot|$ is the gauge

$$|(\zeta, v)| = \left(\|\zeta\|^4 + v^2\right)^{1/4}.$$

(See for example **[Kor83]** and **[Cyg78]**.)

In the Carnot-Caratheodory metric the distance between two points p and q is defined as the infimum of the lengths over all the paths connecting p and q that are horizontal with respect to the contact structure in $\mathcal{H}_{2n-1}$. This canonical contact structure comes from the CR structure on $\mathcal{H}_{2n-1}$ when regarded as the boundary of $\mathbf{H}^n_{\mathbf{C}}$ minus one point. (See **[Gol99]** for details.)

The Carnot-Caratheodory and the Cygan metrics turn out to be equivalent **[BM98]**.

The subgroup of the similarity group ($\mathbf{Sim}(\mathcal{H}_{2n-1})$) that leaves invariant the Cygan metric is called the *Heisenberg isometries* and it is denoted $\mathrm{Isom}(\mathcal{H}_{2n-1})$. This group is isomorphic to $\mathcal{H}_{2n-1} \rtimes U(n-1)$. Where $\mathcal{H}_{2n-1}$ acts by left translations as follows

$$T_{(\zeta,v)}(\vartheta, u) = (\zeta, v) \cdot (\vartheta, u)$$

with $(\zeta, v) \in \mathcal{H}_{2n-1}$. And $U(n-1)$ acts by

$$A \cdot (\zeta, v) = (A\zeta, v), \quad \text{for } A \in U(n-1).$$

The subgroup $\mathcal{H}_{2n-1} \rtimes SU(n-1) \subset \mathrm{Isom}(\mathcal{H}_{2n-1})$ of "orientation preserving" isometries will be denoted $\mathrm{Isom}^+(\mathcal{H}_{2n-1})$.

2.2. Representations of triangle groups. In **[Cas98]** it is proved that, up to a Heisenberg similarity, the complex lines l_1, l_2 and l_3 in $\mathbf{C}^2 = \Pi_V(\mathcal{H}_5)$ that determine the two-chains P_1, P_2 and P_3, respectively, are

$$\begin{aligned} l_1 &= \mathrm{Span}_{\mathbf{C}}\, e_1 \\ l_2 &= \mathrm{Span}_{\mathbf{C}}\, q \\ l_1 &= e_1 + \mathrm{Span}_{\mathbf{C}}(q - e_1) \end{aligned}$$

where $q = (q_1, q_2)$ represents the point of intersection of l_2 and l_3 and $q_1 \in \mathbf{C}$ and $q_2 \in \mathbf{R}$. With this "normal" form for the configuration of two-chains we can find a normal or standard form for the generators of the group G.

THEOREM 2.1. *Suppose $\phi : \Delta \to \mathbf{Sim}(\mathcal{H}_5)$ is a representation of a triangle group. The group G is generated by the following Heisenberg rotations:*

$$s_1(\zeta, v) = (R_1\zeta, v), \qquad s_2(\zeta, v) = (R_2\zeta, v), \tag{2.1}$$
$$s_3(\zeta, v) = (R_3 \circ T_{(w,0)})(\zeta, v)$$

with

$$R_1 = \begin{pmatrix} 1 & 0 \\ 0 & -1 \end{pmatrix}, \qquad R_2 = \begin{pmatrix} \cos 2\theta & e^{i\alpha}\sin 2\theta \\ e^{-i\alpha}\sin 2\theta & -\cos 2\theta \end{pmatrix}, \tag{2.2}$$

(2.3)

$$R_3 = \begin{pmatrix} \cos 2\psi & e^{i\beta}\sin 2\psi \\ e^{-i\beta}\sin 2\psi & -\cos 2\psi \end{pmatrix} \quad \text{and} \quad w = R_3 e_1 - e_1 = \begin{pmatrix} \cos 2\psi - 1 \\ e^{-i\beta}\sin 2\psi \end{pmatrix}$$

Where the angles are given by

$$\theta = \angle\{\mathbf{C}\cdot q, \mathbf{C}\cdot e_1\}, \quad \alpha = \arg\langle\langle e_1, q\rangle\rangle$$
$$\psi = \angle\{\mathbf{C}\cdot (q-e_1), \mathbf{C}\cdot e_1\}, \quad \beta = \arg\langle\langle e_1, q-e_1\rangle\rangle.$$

In particular, it is clear then that the image of the representation ϕ can be restricted to $\mathrm{Isom}(\mathcal{H}_5)$.

We will call the representation ϕ of this theorem the *standard representation* of the triangle group Δ in $\mathrm{Isom}(\mathcal{H}_5)$.

2.3. Quaternions and $SU(2)$. Throughout the paper S^1 will denote the set of unit complex numbers and $\mathbb{H}$ the set of quaternions:

$$\mathbb{H} = \{a + jb \,|\, a, b \in \mathbf{C},\ j^2 = -1,\ xj = j\bar{x},\ \forall x \in \mathbf{C}\}.$$

The norm is defined in $\mathbb{H}$ by $\|a+jb\| = \sqrt{|a|^2+|b|^2}$. The set of quaternions of unit norm will be denoted $S^1\mathbb{H}$. Remember that $SU(2)$ and $S^1\mathbb{H}$ are isomorphic via the isomorphism:

$$u + jv \to \begin{pmatrix} u & v \\ -\bar{v} & \bar{u} \end{pmatrix}. \tag{2.4}$$

The following simple lemma will be particularly useful, and is a well known property of the Weil group of $SU(2)$.

LEMMA 2.2. [**Cas98**, Lemma 2.9] *Let $\mathbf{T} \subset SU(2)$ be a maximal torus. If g is an element of $SU(2)$ such that $g^{-1}\mathbf{T}g \subset \mathbf{T}$ and $g \notin \mathbf{T}$ then $g^{-1}tg = t^{-1}$ for all $t \in \mathbf{T}$. If we identify $SU(2)$ with the unit quaternions $S^1\mathbb{H}$ and let $\mathbf{T} \cong S^1 \subset S^1\mathbb{H}$, then $g = vj$, for some $v \in S^1$.*

2.4. Finite subgroups of $SU(2)$. The classification of finite subgroups of $SU(2)$ can be found in [**Wol84**] and [**DV64**] (among other places). The map

$$\rho : SU(2) \to SO(3)$$

defines a two to one covering of $SO(3)$. This is the universal cover of $SO(3)$, with kernel $\{\pm I\}$.

It is well known that the finite subgroups of $SO(3)$ are *cyclic groups*, *dihedral groups* ($\mathcal{D}_m \cong \Delta_{(2,2,m)}$), and *polyhedral groups*. The polyhedral groups are the groups of symmetries of the Platonic solids: the tetrahedral group ($\mathcal{T} = \Delta_{(2,3,3)}$); the octahedral group ($\mathcal{O} = \Delta_{(2,3,4)}$); and the icosahedral group ($\mathcal{I} = \Delta_{(2,3,5)}$).

Following [**Wol84**] define the *binary dihedral* groups by

$$\mathcal{D}^*_m = \rho^{-1}(\mathcal{D}_m), \tag{2.5}$$

and the *binary polyhedral* groups by

$$\mathcal{T}^* = \rho^{-1}(\mathcal{T}),\ \ \mathcal{O}^* = \rho^{-1}(\mathcal{O}),\ \ \mathcal{I}^* = \rho^{-1}(\mathcal{I}). \tag{2.6}$$

THEOREM 2.3. [**Wol84**, Theorem 2.6.7] *Every finite subgroup of $SU(2)$ is either cyclic, binary dihedral or binary polyhedral. A finite subgroup of $SU(2) \cong S^1\mathbb{H}$ is contained in a complex subfield of $\mathbb{H}$ if and only if it is cyclic and it is contained in a real subfield if and only if it is cyclic of order 1 or 2.*

3. Representations in the Heisenberg Group

In this section we will assume that the representation ϕ is discrete. Our main goal here is to prove the following theorem.

THEOREM 3.1. *Suppose that Δ is a hyperbolic triangle group, and let G be the group $\phi(\Delta) \subset \mathrm{Isom}(\mathcal{H}_5)$. If G is discrete the projection of G into $U(2)$ is a finite group.*

We will break the proof of this theorem in a number of lemmas and propositions. This theorem, together with Theorem 2.1 form the backbone of the proof of the classification theorems that we will prove in the next section (Theorems 4.1 and 4.2).

If Δ is a Euclidean or hyperbolic triangle group the normal closure of the group generated by the element $\eta = acb \in \Delta^+$ has infinite order in Δ^+— see for example [**Cas98**]. Let $\pi : \mathrm{Isom}^+(\mathcal{H}_5) \cong \mathcal{H}_5 \rtimes SU(2) \to SU(2)$ be the natural projection. Define the map ψ by $\psi = \phi \circ \pi$. This map defines a representation $\psi : \Delta \to SU(2)$ which factors through the representation $\phi : \Delta \to \mathrm{Isom}(\mathcal{H}_5)$. Let A be the element $\psi(\eta) \in SU(2)$, and let $N(\eta)$ denote the normal closure in $\Delta^+$of $\langle\eta\rangle$ — the subgroup generated by η. Denote by $N(A) = \psi(N(\eta))$, the normal closure in $\psi(\Delta^+)$ of $\langle A\rangle$. Finally call G^+ the image of Δ^+ under ϕ; this image lies in $\mathrm{Isom}^+(\mathcal{H}_5)$.

Since the element η has infinite order it is natural to ask whether the same is true for its image into $SU(2)$, or if the discreteness condition forces this image to be of finite order. The following result follows from Lemma 5.1 and Theorem 5.4 of [**Cas98**] and answers this question in the Euclidean case.

THEOREM 3.2. *Let $\phi : \Delta \to G \subset \mathrm{Isom}(\mathcal{H}_5)$ be a discrete representation then $A = \pi\phi(\eta) = I$ if and only if Δ^+ is Euclidean. In this case $N(A)$ is also trivial.*

To answer the same question for the hyperbolic case we need to do some work. Let us begin by defining x_a, x_b and x_c to be the linear parts of the images under the representation ϕ of the generators a, b and c of Δ^+, respectively. The following two lemmas, whose proof will be delayed to Section 6, will be used to prove the main result of this section.

LEMMA 3.3. *If the subgroup $N(A)$ of $SU(2)$ is infinite, then the group $N(A)$ is abelian and dense in the maximal torus $\mathbf{T} \subset SU(2)$ that contains it.*

LEMMA 3.4. *Under the hypothesis of the previous lemma the elements x_a, x_b and x_c are not contained in $\mathbf{T} = \overline{N(A)}$.*

PROPOSITION 3.5. *If the representation ϕ is discrete, then the projection A of the element $\eta \in \Delta$ (of infinite order in Δ) to $SU(2)$ is of finite order. In particular Δ is Euclidean if and only if $A = I$.*

PROOF. We begin by proving that A must have finite order. Suppose the contrary and identify $SU(2)$ with the group of unit quaternions. Normalize in such a way that $\mathbf{T} = \overline{N(A)}$ is the torus consisting of the unit complex numbers S^1. By Lemma 3.4 the elements x_a, x_b and x_c do not belong to $\mathbf{T}$, then by

Lemma 2.2 $x_\kappa = d_\kappa j$, with $d_\kappa \in S^1$ ($\kappa = a, b, c$). But this is impossible because $A = d_a j\, d_c j\, d_b j = -d_a \overline{d_c} d_b j \notin \mathbf{T} = \overline{N(A)}$. Therefore A must have finite order. The second part of this theorem follows directly from Theorem 3.2. □

PROPOSITION 3.6. *Suppose Δ is hyperbolic and $A = \psi(\eta)$ has finite order, then $N(A)$ is a finite group.*

PROOF. Let k denote the order of the element A of $SU(2)$, and suppose that the group $N(A)$ is infinite. The group $N(A)$ can not be dense in $SU(2)$, since, by Lemma 3.3, $N(A)$ is abelian. Since $N(A)$ is infinite, it has to be dense in a maximal torus $\mathbf{T} \cong S^1$ of $SU(2)$.

However, all the elements of $N(A)$ have order at most k, and there is only a finite number of elements of order k in S^1, contradicting the assumption that $N(A)$ is infinite. □

Theorem 3.1 follows immediately from the last two propositions and the fact that, since $N(\eta)$ is a finite index subgroup of Δ, $N(A)$ is a finite index subgroup of the group $K = \psi(G)$.

4. The classification

We will now classify all discrete representations of triangle groups in the Heisenberg group. Theorems 1.1 in this paper and Theorem 1.4 in [**Cas98**] take care of the cases when Δ is Euclidean or spherical. For the remaining case —hyperbolic triangle groups— Theorem 3.1 together with Theorem 2.1 will be used to produce lifts of finite subgroups of $SU(2)$ into $\mathrm{Isom}^+(\mathcal{H}_5)$. We will prove:

THEOREM 4.1. *All discrete representations of hyperbolic triangle groups in $\mathcal{H}_5$ can be obtained by lifting non trivial finite groups of $SU(2)$ via the natural projection*

$$\psi : \mathrm{Isom}^+(\mathcal{H}_5) \to SU(2).$$

4.1. Statements of results. Call $\mathcal{G}$ the set of equivalence classes of non-cyclic finite subgroups of $SU(2)$. Two groups are equivalent if there is an automorphism of $SU(2)$ taking one to the other.

THEOREM 4.2. *For every element of $\mathcal{G}$ with representative F there exists a unique group G_F —up to conjugation by an isometry of $\mathcal{H}_5$— acting by isometries in $\mathcal{H}_5$ and such that $F = \pi(G_F)$. The group G_F is a representation of a hyperbolic triangle group in $\mathrm{Isom}(\mathcal{H}_5)$. Conversely, every representation of a hyperbolic triangle group Δ in $\mathrm{Isom}(\mathcal{H}_5)$ is conjugate in $\mathrm{Isom}(\mathcal{H}_5)$ to a lift of a finite group of $SU(2)$. A complete list of generators of representatives of classes in $\mathcal{G}$ is given in Table 1. The list of generators for the corresponding groups G_F is given in Table 2.*

Table 1 lists generators of the binary dihedral and polyhedral groups. These generators are chosen so that they will be the linear parts of the lifts G_F to $\mathcal{H}_5$ listed in Table 2. The proof of Theorem 4.2 will be broken into lemmas that will be proven in Section 4.3. Section 4.2 presents some technical results that are used in this proof as well. Clearly Theorem 3.1 and Theorem 4.2 imply Theorem 4.1.

COROLLARY 4.3. *If $\phi : \Delta^+ \to G^+ \subset \mathrm{Isom}(\mathcal{H}_5)$ is a discrete representation of a hyperbolic triangle group and if $K = \pi(G^+)$, then the class of the group K modulo automorphisms of $SU(2)$ is an element of $\mathcal{G}$.*

Group	x_b, x_c
$\mathcal{D}_m^*$	$\begin{pmatrix} i\sin\theta_m & -\cos\theta_m \\ -\cos\theta_m & -i\sin\theta_m \end{pmatrix}$ $\begin{pmatrix} \cos\theta_m & i\sin\theta_m \\ i\sin\theta_m & \cos\theta_m \end{pmatrix}$
$\mathcal{T}^*$	$\begin{pmatrix} (-1+i\sqrt{2})/2 & 1/2 \\ -1/2 & -(1+i\sqrt{2})/2 \end{pmatrix}$ $\begin{pmatrix} -1/2 & -(1+i\sqrt{2})/2 \\ (-1+i\sqrt{2})/2 & -1/2 \end{pmatrix}$
$\mathcal{O}^*$	$\begin{pmatrix} (\sqrt{2}+i)/2 & 1/2 \\ -1/2 & (\sqrt{2}-i)/2 \end{pmatrix}$ $\begin{pmatrix} -1/2 & (-\sqrt{2}+i)/2 \\ (\sqrt{2}+i)/2 & -1/2 \end{pmatrix}$
$\mathcal{I}^*$	$\begin{pmatrix} (\tau+i\tau^{-1})/2 & 1/2 \\ -1/2 & (\tau-i\tau^{-1})/2 \end{pmatrix}$ $\begin{pmatrix} -1/2 & (-\tau+i\tau^{-1})/2 \\ (\tau+i\tau^{-1})/2 & -1/2 \end{pmatrix}$

$x_a = \begin{pmatrix} 0 & 1 \\ -1 & 0 \end{pmatrix}$ for all groups, $\tau = (\sqrt{5}-1)/2$,
$\theta_m = 2\pi/m$ if m is odd and $\theta_m = \pi/m$ if m is even.

TABLE 1. List of generators of representatives of classes in $\mathcal{G}$

A question that is not answered by these theorems is what happens if the image of G in $SU(2)$ is cyclic. The following theorem shows that this case is different from the rest.

THEOREM 4.4. *Suppose that the group Δ is Euclidean or hyperbolic, and that the image of the group G^+ in $SU(2)$ is cyclic. Then Δ is a Euclidean triangle group.*

PROOF. If the group K is cyclic in $SU(2)$ then by Theorem 2.3 K is contained in a complex subfield of $SU(2) \cong S^1\mathbb{H}$. In (4.1) through (4.3) below, we have the generators of K associated to the standard representation of Δ in $\mathrm{Isom}(\mathcal{H}_5)$.

By the isomorphism of equation (2.4) we have that x_a, x_b and x_c correspond to the unit quaternions:

$$\eta_a = \cos 2\theta + je^{i\alpha}\sin 2\theta$$
$$\eta_b = u + jv$$
$$\eta_c = \cos 2\psi - je^{i\beta}\sin 2\psi$$

with u and v as in (4.4). Conjugating these numbers by $\rho = e^{-i\alpha/2}$ we obtain

$$\eta_a' = \rho^{-1}\eta_a\rho = \cos 2\theta + j\sin 2\theta$$
$$\eta_b' = \rho^{-1}\eta_b\rho = u + je^{-i\alpha}v$$
$$\eta_c' = \rho^{-1}\eta_c\rho = \cos 2\psi - je^{i(\beta-\alpha)}\sin 2\psi$$

Since K is cyclic η_a', η_b' and η_c' commute. This implies that $e^{i(\beta-\alpha)} \in \mathbf{R}$ or, equivalently, that $\beta - \alpha$ equals 0 or π. This means that (with the notation of 2.1)

$$\arg(q_1 - 1) - \arg q_1$$

equals 0 or π. This can only happen if q_1 is a real number, in which case $q = (q_1, q_2)$ is a real vector (see 2.2). Then, by Lemma 5.1 and Theorem 5.3 of [**Cas98**], the group Δ must be Euclidean, and the group $G = \phi(\Delta)$ must leave an infinite **R**-sphere in $\mathcal{H}_5$ invariant. $\square$

4.2. Overview of the proof of the main theorem. In this and next subsections we will prove Theorem 4.2. The proof is divided in two parts. We first show that all binary dihedral and polyhedral groups of $SU(2)$ can be represented as the linear part of the standard representation of an orientation preserving hyperbolic triangle group (see 2.2). We then compute all the lifts of groups from $SU(2)$ to $\mathrm{Isom}(\mathcal{H}_5)$. To find the lifts is enough to observe that the linear part of the group Δ^+ conveys all the information necessary to reconstruct the affine translations of the group Δ. This is a simple consequence of the form of the standard representation of these groups.

To prepare for the proofs of the results stated in the previous subsection we need to set some notation and perform a few elementary computations.

As in Section 3 denote by x_a, x_b and x_c the linear parts of the images under the representation ϕ of the elements $a, b, c \in \Delta^+$. Using the notation of Theorem 2.1 $x_a = R_1R_2$, $x_b = R_2R_3$ and $x_c = R_3R_1$. More explicitly:

$$x_a = \begin{pmatrix} \cos 2\theta & e^{i\alpha}\sin 2\theta \\ -e^{-i\alpha}\sin 2\theta & \cos 2\theta \end{pmatrix} \tag{4.1}$$

$$x_b = \begin{pmatrix} u & v \\ -\overline{v} & \overline{u} \end{pmatrix} \tag{4.2}$$

$$x_c = \begin{pmatrix} \cos 2\psi & -e^{i\beta}\sin 2\psi \\ e^{-i\beta}\sin 2\psi & \cos 2\psi \end{pmatrix} \tag{4.3}$$

with

$$\begin{aligned} u &= \cos 2\theta \cos 2\psi + e^{i(\alpha-\beta)} \sin 2\theta \sin 2\psi \\ v &= e^{i\beta}\cos 2\theta \sin 2\psi - e^{i\alpha}\cos 2\psi \sin 2\theta \end{aligned} \tag{4.4}$$

Note that by construction

$$x_a\, x_b\, x_c = 1. \tag{4.5}$$

This means that by choosing x_a, x_b and x_c of finite orders (say p, q and r, respectively), we will have a representation of an oriented triangle group Δ^+ of type (p, q, r). (Compare with the presentation of Δ^+ in equation (1.2).)

REMARK 4.5. The image G^+ of the group Δ^+ under the representation ϕ is generated by the Heisenberg isometries

$$f(\zeta, v) = (x_a\zeta, v), \tag{4.6}$$

and

$$g(\zeta, v) = (x_b \circ T_{(w,0)})(\zeta, v) = (x_b(w + \zeta), v + 2\,\mathrm{Im}\langle\!\langle w, \zeta\rangle\!\rangle). \tag{4.7}$$

With x_a and x_b given in Table 1 and w as in Table 2.

Part of what we claim in Theorem 4.2 can be rephrased as follows. For any finite subgroup F of $SU(2)$ there exist angles α, β, θ and ψ (as in Theorem 2.1) so that the group generated by $\{x_a, x_b, x_c\}$ is isomorphic to the group F. Since any two such groups are conjugate under an automorphism of $SU(2)$, which induces an automorphism of Isom($\mathcal{H}_5$), finding the groups generated by $\{x_a, x_b, x_c\}$ is equivalent to finding all finite groups of $SU(2)$, up to automorphism. In [**Cas98**] we relate these angles to geometric information of the configuration of 2-chains associated to the standard generators of G. We will not repeat this here.

Before starting note that

$$x_a^j = \begin{pmatrix} \cos 2j\theta & e^{i\alpha}\sin 2j\theta \\ -e^{-i\alpha}\sin 2j\theta & \cos 2j\theta \end{pmatrix} \tag{4.8}$$

for all integers j, and similarly for x_c. This implies that the order of x_a (respectively, x_c) is determined by the angle θ (respectively ψ).

By knowing these three matrices one can reconstruct the group $G = \phi(\Delta)$ in $\mathcal{H}_5$ as follows: by equation (2.2) we know R_1 and therefore s_1 (2.1), the other generators of G are given by

$$R_2 = R_1 x_a, \quad R_3 = x_c R_1 \quad \text{and} \quad w = R_3 e_1 - e_1 = x_c e_1 - e_1. \tag{4.9}$$

(See equations (2.1) and (2.3).)

4.3. Construction of lifts to Isom($\mathcal{H}_5$). In view of the work done in Section 4.2 the construction of the lifts of finite groups of $SU(2)$ to Isom($\mathcal{H}_5$) can now be achieved with little effort.

Theorem 2.3 guarantees the uniqueness of finite subgroups in $SU(2)$, so any representation of a binary group will be equivalent to any other. As noted in Section 2.4 all finite subgroups of $SO(3)$ are triangle groups. Their lifts to $SU(2)$ will not be triangle groups, but by (4.5) they will be *representations* of hyperbolic (and

Finite group	Generator R_3, w of G
$\mathcal{D}_m^*$	$\begin{pmatrix} \cos\theta_m & -i\sin\theta_m \\ i\sin\theta_m & -\cos\theta_m \end{pmatrix}$ $\begin{pmatrix} \cos\theta_m - 1 \\ i\sin\theta_m \end{pmatrix}$
$\mathcal{T}^*$	$\begin{pmatrix} -1/2 & (-1-i\sqrt{2})/2 \\ (-1+i\sqrt{2})/2 & 1/2 \end{pmatrix}$ $\begin{pmatrix} -3/2 \\ (-1+i\sqrt{2})/2 \end{pmatrix}$
$\mathcal{O}^*$	$\begin{pmatrix} -1/2 & (\sqrt{2}-i)/2 \\ (\sqrt{2}+i)/2 & 1/2 \end{pmatrix}$ $\begin{pmatrix} -3/2 \\ (\sqrt{2}+i)/2 \end{pmatrix}$
$\mathcal{I}^*$	$\begin{pmatrix} -1/2 & (\tau-i\tau^{-1})/2 \\ (\tau+i\tau^{-1})/2 & 1/2 \end{pmatrix}$ $\begin{pmatrix} -3/2 \\ (\tau+i\tau^{-1})/2 \end{pmatrix}$

$R_1 = \begin{pmatrix} 1 & 0 \\ 0 & -1 \end{pmatrix}$ and $R_2 = \begin{pmatrix} 0 & 1 \\ 1 & 0 \end{pmatrix}$ for all groups; also $\tau = (\sqrt{5}-1)/2$.
$\theta_m = 2\pi/m$ if m is odd and $\theta_m = \pi/m$ if m is even.
The generators of each group are $\sigma_1(\zeta, v) = (R_1\zeta, v)$, $\sigma_2(\zeta, v) = (R_2\zeta, v)$ and $\sigma_3(\zeta, v) = (R_3 \circ T_{(w,0)})(\zeta, v) = (R_3(\zeta + w), v + 2\,\mathrm{Im}\langle\langle w, \zeta\rangle\rangle)$.

TABLE 2. List of generators in Isom($\mathcal{H}_5$) of the lifts of groups corresponding to classes in $\mathcal{G}$.

hence infinite) triangle groups. This follows from the fact that every element of $SO(3)$ of even order lifts to an element in $SU(2)$ of twice its order. We then have, for example, that the dihedral group $\mathcal{D}_m \cong \Delta_{(2,2,m)}$ lifts to the group $\mathcal{D}_m^*$ which is a representation of $\Delta_{(4,4,m)}$ if m is odd and of $\Delta_{(4,4,2m)}$ if m is even. But $\Delta_{(4,4,k)}$ is hyperbolic for all $k \geq 3$. The same conclusions hold for the polyhedral subgroups of $SO(3)$ and their lifts to $SU(2)$. Following this principle Table 3 shows the list of finite subgroups of $SU(2)$, and the associated hyperbolic triangle group, so that the finite group is a representation of that triangle group.

To construct the representatives of the classes of groups in $\mathcal{G}$ we will use the results of the previous subsection and equations 4.1 through 4.8.

Group	Hyperbolic triangle group
$\mathcal{D}_m^*$	$\begin{cases} \Delta_{(4,4,m)} & \text{if } m \text{ is odd,} \\ \Delta_{(4,4,2m)} & \text{if } m \text{ is even} \end{cases}$
$\mathcal{T}^*$	$\Delta_{(4,3,3)}$
$\mathcal{O}^*$	$\Delta_{(4,3,8)}$
$\mathcal{I}^*$	$\Delta_{(4,3,5)}$

TABLE 3. Binary dihedral, and polyhedral groups, and their associated triangle groups

LEMMA 4.6. *Let $\mathcal{D}_m^*$ be a binary dihedral group of $SU(2)$ then up to an automorphism of $SU(2)$ $\mathcal{D}_m^*$ is generated by the following matrices*

$$x_a = \begin{pmatrix} 0 & 1 \\ -1 & 0 \end{pmatrix}, \qquad x_b = \begin{pmatrix} i\sin\theta_m & -\cos\theta_m \\ \cos\theta_m & -i\sin\theta_m \end{pmatrix}, \tag{4.10}$$

$$x_c = \begin{pmatrix} \cos\theta_m & i\sin\theta_m \\ i\sin\theta_m & \cos\theta_m \end{pmatrix} \tag{4.11}$$

where ξ is a unit complex number and

$$\theta_m = \begin{cases} 2\pi/m & \text{if } m \text{ is odd,} \\ \pi/m & \text{if } m \text{ is even.} \end{cases}$$

Moreover the group $F = \langle x_a, x_b \rangle \subset SU(2)$ is a representation of $\Delta_{(4,4,m)}$ if m is odd and of $\Delta_{(4,4,2m)}$ if m is even.

PROOF. First suppose that m is odd, then let x_a be of order 4, and x_c of order m. We can do this by setting $\theta = \pi/4$ and $\psi = 2\pi/m$. Replacing these values in equations (4.1), (4.2) and (4.3), we obtain

$$x_a = \begin{pmatrix} 0 & e^{i\alpha} \\ -e^{-i\alpha} & 0 \end{pmatrix} \qquad x_b = \begin{pmatrix} e^{i(\alpha-\beta)}\sin\theta_m & -e^{i\alpha}\cos\theta_m \\ e^{-i\alpha}\cos\theta_m & e^{i(\beta-\alpha)}\sin\theta_m \end{pmatrix}$$

$$x_c = \begin{pmatrix} \cos\theta_m & -e^{i\beta}\sin\theta_m \\ e^{-i\beta}\sin\theta_m & \cos\theta_m \end{pmatrix}$$

Conjugate these matrices by the matrix

$$U = \begin{pmatrix} 1 & 0 \\ 0 & e^{-i\alpha} \end{pmatrix} \in U(2),$$

and write $\xi = e^{i(\alpha-\beta)}$. We should choose for ξ to ensure that x_b has order 4 and then that the group is in fact a representation of the corresponding triangle group (see Table 3). Without loss of generality let $\operatorname{Im}\xi > 0$. The characteristic polynomial of x_b is:

$$\chi_b(\lambda) = \lambda^2 - 2\sin(2\pi/m)\operatorname{Re}\xi\,\lambda + 1.$$

The roots of this polynomial are of order four if and only if $\operatorname{Re}\xi = 0$. Therefore $\xi = i$. With this choice we obtain the matrices of equations (4.10) and (4.11). □

LEMMA 4.7. *Let $\mathcal{B}$ be a binary polyhedral group of $SU(2)$. Then, up to an automorphism of $SU(2)$, $\mathcal{B}$ is generated by the matrices*

$$x_a = \begin{pmatrix} 0 & 1 \\ -1 & 0 \end{pmatrix}, \qquad x_b = \begin{pmatrix} \mu\sqrt{3}/2 & 1/2 \\ -1/2 & \bar{\mu}\sqrt{3}/2 \end{pmatrix}, \tag{4.12}$$

$$x_c = \begin{pmatrix} -1/2 & -\bar{\mu}\sqrt{3}/2 \\ \mu\sqrt{3}/2 & -1/2 \end{pmatrix} \tag{4.13}$$

where μ is a unit complex number. In particular the matrix x_a has order four and the matrix x_c has order three.

PROOF. All binary polyhedral groups can be treated in a similar way. First observe that these groups are all representations of hyperbolic triangle groups of type $(4, k, 3)$, with k equal to 3, 8 or 5. To find the polyhedral group associated to $\Delta_{(4,k,3)}$ We will choose matrices x_a, x_b and x_c of orders 4, k, and 3, respectively. From equations (4.1), (4.2) and (4.3) we see that the conditions on x_a and x_c are met if $\theta = \pi/4$ and $\psi = \pi/3$. Therefore

$$x_a = \begin{pmatrix} 0 & e^{i\alpha} \\ -e^{-i\alpha} & 0 \end{pmatrix}, \qquad x_b = \begin{pmatrix} \sqrt{3}e^{i(\alpha-\beta)}/2 & e^{i\alpha}/2 \\ -e^{-i\alpha}/2 & \sqrt{3}e^{i(\beta-\alpha)}/2 \end{pmatrix}, \tag{4.14}$$

$$x_c = \begin{pmatrix} -1/2 & -e^{i\beta}\sqrt{3}/2 \\ \sqrt{3}e^{-i\beta}/2 & -1/2 \end{pmatrix} \tag{4.15}$$

Conjugating by

$$U = \begin{pmatrix} 1 & 0 \\ 0 & e^{-i\alpha} \end{pmatrix}$$

defines an automorphism of $SU(2)$ that maps the matrices x_a, x_b and x_c to the matrices in equations (4.12) and (4.13), with $\mu = e^{i(\alpha-\beta)}$. □

By Table 3 all binary polyhedral groups are representations of a triangle hyperbolic group of type $(4, k, 3)$, with k equal to 3, 8 or 5. To obtain the representation of $\mathcal{B}$ we just need to find the value of μ for which x_b will have order k. In the next lemma we do that. To avoid repetitions we will choose μ so that $\operatorname{Im}\mu > 0$.

LEMMA 4.8. *If $k = 3$ then $\mu = (-1+i\sqrt{2})/\sqrt{3}$, if $k = 8$ then $\mu = (\sqrt{2}+i)/\sqrt{3}$ and if $k = 5$ then $\mu = (\tau + i\tau^{-1})/\sqrt{3}$, where $\tau = (\sqrt{5}-1)/2$ is the golden ratio.*

PROOF. The characteristic polynomial of the matrix x_b is

$$\chi_b(\lambda) = \lambda^2 - \sqrt{3}y\lambda + 1$$

where $y = \operatorname{Re}\mu$. The roots of χ_b are

$$\sqrt{3}y/2 \pm i\sqrt{1 - \left(\sqrt{3}y/2\right)^2}.$$

Therefore choosing $y = 2\cos(2\pi/k)/\sqrt{3}$, for $k = 3, 8, 5$ we will have that x_b has the desired order. Since $\operatorname{Im}\mu > 0$, $\mu = y + i\sqrt{1-y^2}$. The rest of the proof is a simple computation if we remember that $\cos(2\pi/5) = \tau/2$. □

The proof of Theorem 4.2 is now straightforward.

PROOF OF THEOREM 4.2. Let G be any representation of a triangle group, by Theorem 2.1 the projection $G^+ \to K$ is a subgroup of $SU(2)$ and the group K is finite (Theorem 3.1).

By Theorem 2.3 the matrices of Table 1 are the generators for representatives of all the classes of groups in $\mathcal{G}$. Call the groups generated by these generators *standard representatives.* By lemmas 4.6, 4.7 and 4.8 the lifts of the standard representatives are the groups whose generators are listed in Table 2. □

5. Discreteness

The question that remains to be answered is to whether all the lifts G_F of finite groups F of $SU(2)$ constructed in the previous section are discrete in $\mathcal{H}_5$. This section consists of the proof the the following theorem.

THEOREM 5.1. *The groups generated by the Heisenberg isometries of Table 2 are discrete in* $\operatorname{Isom}(\mathcal{H}_5)$.

As before $\phi : \Delta \to \operatorname{Isom}(\mathcal{H}_5)$ is the representation of Δ in the isometries of the $\mathcal{H}_5$, defined in previous sections. To prove the discreteness theorem we need a representation of $\operatorname{Isom}(\mathcal{H}_5)$ in the group $PU(3,1)$. For convenience we will define the group $PU(3,1)$ using the Hermitian pairing of vectors $x, y \in \mathbf{C}^4$ defined by $(x|y) = x^*Qy$. Where $\cdot^*$ represents the transpose conjugate matrix and Q is the matrix

$$Q = \begin{pmatrix} 0 & 0 & 0 & 1 \\ 0 & 1 & 0 & 0 \\ 0 & 0 & 1 & 0 \\ 1 & 0 & 0 & 0 \end{pmatrix}.$$

(Other choices are possible and for further reference see [**Eps87**].) We will call $PU(Q)$ the unitary group associated to Q to distinguish it from the standard group which we denote $PU(3,1)$. Following [**Par98**] the Heisenberg isometries can be put in correspondence with elements of $PU(Q)$ by:

$$T_{(\zeta,v)} \mapsto \begin{pmatrix} 1 & -\zeta^* & -|\zeta|^2/2 + iv/2 \\ 0 & I & \zeta \\ 0 & 0 & 1 \end{pmatrix}, \quad (\zeta, v) \in \mathcal{H}_5 \tag{5.1}$$

$$U \mapsto \frac{1}{(\det U)^{1/4}} \begin{pmatrix} 1 & 0 & 0 \\ 0 & U & 0 \\ 0 & 0 & 1 \end{pmatrix}, \quad U \in U(2). \tag{5.2}$$

PROOF OF THEOREM 5.1. Let Δ^+ be an orientation preserving triangle group of hyperbolic type with presentation given by (1.2). Since Δ^+ has index two in Δ to prove that G is discrete it suffices to prove that $G^+ = \phi(\Delta^+)$ is discrete. We can write a presentation for G^+ with generators a and b as follows:

$$\Delta^+ = \langle a, b | a^p, b^q, (ab)^r \rangle.$$

Any element of Δ^+ has then the form

$$a^{n_1}b^{m_1}\dots a^{n_k}b^{m_k}. \tag{5.3}$$

With

$$0 \le n_1 \le p-1, \quad 0 \le m_1 \le q-1,$$

and

$$1 \le n_j \le p-1, \quad 1 \le m_j \le q-1, \quad \text{for all } 1 < j \le k.$$

By Remark 4.5 the group G^+ is generated by the Heisenberg isometries f and g of equations (4.6) and (4.7). We have that

$$A \circ T_{(\vartheta,u)} = T_{(A\vartheta,v)} \circ A. \tag{5.4}$$

Hence we can write the image under ϕ of any element of Δ^+ of the form (5.3) as:

$$x_a^{n_1}x_b^{m_1}\dots x_a^{n_k}x_b^{m_k}T_{(M_1w,0)}\dots T_{(M_{2k}w,0)}.$$

Where the integers $(n_1, m_1, \dots, n_k, m_k)$ are the same as in (5.3) and $M_j \in K$ are the corresponding matrices obtained by applying (5.4) to the image under ϕ of (5.3).

By (5.1) and (5.2) and the previous remarks we can see that any matrix in G^+ has the form

$$\begin{pmatrix} 1 & -\tau^* & -|\tau|^2/2 + it/2 \\ 0 & U & U\tau \\ 0 & 0 & 1 \end{pmatrix}, \tag{5.5}$$

where $U \in K$ and τ is a vector in $\mathbf{C}^2$ given by

$$\tau = M_1w + \dots + M_kw, \tag{5.6}$$

with $M_j \in K$, for all j, and $w = R_3e_1 - e_1$ as in Theorem 2.1. A simple computation shows that for all the groups G of Table 2, the corresponding matrices x_a and x_b (Table 1) belong to the set of two by two matrices with coefficients in $\mathcal{O}/2$. Were $\mathcal{O}$ is the ring of integers of a cyclotomic field $\mathcal{F}$, and the notation $\mathcal{O}/2$ represents elements of $\mathcal{O}$ divided by two. The field $\mathcal{F}$ is given by:

$$\mathcal{F} = \begin{cases} \mathcal{C}_8 & \text{for } \mathcal{T}^* \\ \mathcal{C}_8 & \text{for } \mathcal{O}^* \\ \mathcal{C}_{20} & \text{for } \mathcal{I}^* \\ \mathcal{C}_n & \text{for } \mathcal{D}_n^*,\ n \text{ odd} \\ \mathcal{C}_{2n} & \text{for } \mathcal{D}_{2n}^*,\ n \text{ even.} \end{cases}$$

Where $\mathcal{C}_n$ denotes the cyclotomic field $\mathbf{Q}(e^{2i\pi/n})$. Moreover,

$$w \in \mathcal{O}/2 \times \mathcal{O}/2.$$

Since K is generated by x_a and x_b, and since K is finite (Theorem 3.1) we have that there exists an integer $N \ge 2$ such that $K \subset M(2, \mathcal{O}/N)$.

By (5.6) we can conclude that the vector τ will be a vector in $\mathcal{O}/2N \times \mathcal{O}/2N$. The entries τ and U in the generic element of G^+ given by (5.5) are elements of $\mathcal{O}/2N$. To prove that G^+ is discrete we need to understand the behavior of the term $|\tau|^2/2 + it/2$ in (5.5).

We have already that the group of isometries of $\mathbf{C}^2$ defined by $\Pi_V(G)$ is discrete. By Selberg's lemma [**Sel60**] there exists a finite index torsion free subgroup Ω of $\Pi_V(G)$, and by a theorem of Bieberbach (see [**Bib**] and also [**Wol84**, Theorem

3.2.1]) the group Ω has a finite index subgroup Λ of affine translations. The group Λ has a lift Γ to $\mathrm{Isom}(\mathcal{H}_5)$ which is a finite index subgroup of Heisenberg translations in G. More explicitly by [**Aus77**], there exists a basis $v_1, \ldots, v_4$ of $\mathbf{C}^2$ as an $\mathbf{R}$ vector space such that $\mathrm{Im}\langle\langle v_3, v_4\rangle\rangle = k\,\mathrm{Im}\langle\langle v_1, v_2\rangle\rangle$, where k is a positive integer. The group Γ has the property that

$$\Gamma = \langle {\Pi_V}^{-1}(\Pi_V(\Gamma))\rangle,$$

(where ${\Pi_V}^{-1}(X)$ represents lifts of the elements of X in G). In the notation of [**Aus77**] Γ is said to be a *lifted subgroup* of $\mathcal{H}_5$. This implies that the intersection of Γ with the center of $\mathcal{H}_5$ is generated by a vertical translation of translation length $4\,\mathrm{Im}\langle\langle v_1, v_2\rangle\rangle$. Since Γ is discrete in $\mathrm{Isom}(\mathcal{H}_5)$ it follows that the group G is discrete in $\mathrm{Isom}(\mathcal{H}_5)$. □

REMARK 5.2.
1. It is interesting to note that the lift of the dihedral group $\mathcal{D}_2^*$ to $\mathrm{Isom}(\mathcal{H}_5)$, is a representation of $\Delta_{(4,4,4)}$, the unit quaternions. This example was already found in [**Cas98**], with slightly different methods.
2. Since all these groups are discrete subgroups of a linear group they have a torsion free subgroup of finite index. Moreover they admit a finite index subgroup of Heisenberg translations $\Gamma \subset G$. Then Γ is arithmetic [**Rag72**, pag. 77], and therefore so is G.
3. Using the computer program GAP [**S**$^+$**95**] we were able to compute the entries of some of the matrices of the groups G in their representation in $PU(Q)$. For all the polyhedral groups we were able to show that $G \subset PU(Q, \mathcal{O}/2)$, where $\mathcal{O}$ is the appropriate ring of integers. We conjecture that the same is true for dihedral groups. For our purposes in this paper a more refined proof of this fact was not necessary.

6. Proof of Lemmas

We will now prove Lemmas 3.3 and 3.4. Remember that $\phi : \Delta \to \mathrm{Isom}(\mathcal{H}_5)$ is a discrete representation of a triangle group of Euclidean or hyperbolic type. To prove Lemma 3.3 we will need a simple modification of a result from Auslander:

THEOREM 6.1. *Let G be a Lie group and A a closed connected nilpotent normal subgroup. Let $\pi : G \to G/A$ be the natural map. Let H be a closed subgroup of G such that H° the identity component of H, is trivial ($H^\circ = \{e\}$). Let $U = \overline{\pi(H)}$, the closure of $\pi(H)$. Then the identity component U° of U is nilpotent.*

This theorem is a modification of Theorem 8.24 [**Rag72**, page 24] (which states the same conclusion for solvable groups). Its short proof will be included here for the sake of completeness, and it goes along the same lines as the proof of the cited theorem.

PROOF. For a subset K of a Lie group G define $K^{(n)}$ inductively by:

$$K^{(0)} = K, \qquad K^{(n)} = \{aba^{-1}b^{-1} | a \in K,\ b \in K^{(n-1)}\}.$$

By [**Rag72**, Proposition 8.8, page 143] there exists Ω, a neighborhood of $\pi(e)$ in G/A such that for all compact subsets $K \subset \pi^{-1}(\Omega)$, $K^{(n)}$ converges to e as n goes to ∞, in the sense that for any neighborhood W of e, there exist m large enough such that $K^{(m)} \subset W$.

Since $\pi(H)$ is dense in U it follows that the closed subgroup generated by $\pi(H)\cap\Omega$ contains U°. Let K be a compact subset of $H\cap\pi^{-1}(\Omega)$. Then $K^{(n)}\to e$ as $n\to\infty$. Thus for n large $K^{(n)}=\{e\}$.

The fact that $K^{(m)}=\{e\}$ for m large enough implies, by Lemma 8.11 in [**Rag72**, page 143], that the subgroup generated by K is nilpotent. Hence every finitely generated subgroup of the group generated by $H\cap\pi^{-1}(\Omega)$ is nilpotent, therefore $\pi(H)\cap\Omega$ is nilpotent as well. Since U° is generated by any neighborhood of $\{e\}$ in U°, it follows that U° is nilpotent. □

PROOF OF LEMMA 3.3. If Δ^+ is a Euclidean group then $N(\eta)$ is abelian. therefore $N(A)=\psi(N(\eta))$ is abelian, and it is contained in a maximal torus $\mathbf{T}\subset SU(2)$. Moreover, since A is of infinite order $N(A)$ is dense in $\mathbf{T}$.

Let us suppose that Δ^+ is hyperbolic, and assume that the representation Δ^+ is discrete in $\mathcal{H}_5$. The group $N(\eta)$ has finite index in Δ^+. We claim that its image $N(A)$ is an abelian subgroup of $SU(2)$. Apply Theorem 6.1 with $G=\mathrm{Isom}(\mathcal{H}_5)$, $R=\mathcal{H}_5$ and $H=N(\phi(\eta))$, which is the normalizer of the image of the group generated by η in $\mathrm{Isom}^+(\mathcal{H}_5)$, and a subgroup of G^+. Since G^+ is discrete, its identity component is trivial. Therefore the theorem asserts that $\overline{\pi(H)}^\circ=\overline{N(A)}^\circ$ is nilpotent in $SU(2)$, since it is also connected it must be abelian.

To finish we should show that $\overline{N(A)}=\overline{N(A)}^\circ$. The group $N(A)$ is generated by the union of the cyclic subgroups:

$$\langle g^{-1}Ag\rangle=\{g^{-1}A^kg\mid k\in\mathbf{Z}\},\quad g\in G^+.$$

But $\overline{N(g^{-1}Ag)}^\circ=\overline{N(A)}^\circ$, so $\overline{N(A)}^\circ$ contains a set of generators for $N(A)$ and therefore $\overline{N(A)}^\circ=\overline{N(A)}$. □

PROOF OF LEMMA 3.4. Let $\mathbf{T}=\overline{N(A)}$. Suppose that x_a, then $A^{x_a}=A=x_ax_cx_b$ which implies that $(x_cx_b)x_a=x_a(x_cx_b)$, and $x_cx_b=x_a^{-1}A\in\mathbf{T}$.

If $x_c\in\mathbf{T}$, then $A=x_ax_cx_b=x_cx_ax_b=x_c(x_ax_bx_c)x_c^{-1}=1$. So $x_c\notin\mathbf{T}$, then Lemma 2.2 implies that $\varsigma^{x_c}=\varsigma$ for all $\varsigma\in\mathbf{T}$. In particular, choosing $\varsigma=x_cx_b$,

$$x_b^2x_c^2=1 \tag{6.1}$$

Conjugate $x_cx_bx_a=A^{(x_cx_b)^{-1}}$ by x_c, by Lemma 2.2 one obtains $(x_cx_bx_a)^{x_c}=(x_cx_bx_a)^{-1}=x_a^{-1}x_b^{-1}x_c^{-1}$. On the other hand $(x_cx_bx_a)^{x_c}=x_bx_ax_c$. Therefore $x_bx_ax_c=x_a^{-1}x_b^{-1}x_c^{-1}$. Simplifying this last equation we obtain finally:

$$(x_bx_a)^{-1}=x_c^2.$$

By replacing this in equation (6.1) and simplifying we obtain:

$$x_b=x_ax_bx_a. \tag{6.2}$$

From the relations in the presentation of Δ^+ we can also write $x_ax_bx_c=1$. Substituting in this last equation x_b from equation (6.2) and simplifying one obtains:

$$x_a^{-2}=A=x_ax_cx_b.$$

Which contradicts the hypothesis that A has infinite order. Therefore $x_a\notin\mathbf{T}$.

Analogous arguments show that x_b and x_c are not in $\mathbf{T}$ either. □

REMARK 6.2. Note that in Lemma 3.3 is the only place where the discreteness condition is used.

References

[Aus77] Louis Auslander. *Lecture Notes on nil-theta functions.* Regional conference series in mathematics, no. 34. AMS, Providence, RI, 1977.

[Bib] L. Bieberbach. Uber die Bewegungsgruppen der Euklidischen Raume I, *Mathematische Annalen,* 70:297–336, 1911; II, *ibid* 72:400–412, 1912.

[BM98] Ara Basmajian and Robert Miner. Discrete subgroups of complex hyperbolic motions, *Inventiones Mathematicæ*, 131, no. 1, 85–163, 1998.

[Cas98] Alberto M. Castro. Complex hyperbolic ideal tetrahedral groups, preprint, 1998.

[Cyg78] Jacek Cygan. Wiener's test for the brownian motion on the Heisenberg group. *Colloquium Mathematicum,* 39:367–373, 1978.

[DV64] Patrik Du Val. *Homographies quaternions and rotations.* Oxford Mathematical Monographs. Oxford University Press, 1964.

[Eps87] D. B. A. Epstein. Complex hyperbolic geometry. In *Analytic and geometric aspects of hyperbolic space*, ed. D.B.A. Epstein, London Math. Soc. Lecture Notes, Camb. U. Press 1987.

[Gol99] William M. Goldman. *Complex hyperbolic geometry.* To appear in Oxford Mathematical Monographs. Oxford University Press, Oxford, U. K., 1999.

[Kor83] A. Korányi. Geometric aspect of analysis on the Heisenberg group. In *Topics in Modern Harmonic Analysis.* Roma: Instituto Nazionale di Alta Matematica, 1983.

[Par98] John R. Parker. On the volumes of cusped complex hyperbolic manifolds and orbifolds, *Duke Math. J.*, 94, no. 3, 433–464, 1998.

[Rag72] M. S. Raghunathan. *Discrete subgroups of Lie groups*, volume 68 of *Ergebnisse der Mathematik und ihrer Grenzgebeite.* Springer, Berlin, 1972.

[Sel60] A. Selberg. On discontinuous groups in higher dimensional symmetric spaces. In *Contributions to function theory.* Bombay: Thata institute of Fundamental Research 1960.

[S+95] Martin Schönert et al. *GAP - Groups, Algorithms, and Programming.* Lehrstuhl D für Mathematik, Rheinisch Westfälische Technische Hochschule, Aachen, Germany, fifth edition, 1995.

[Wol84] Joseph A. Wolf. *Spaces of constant curvature.* Publish or Perish, Wilmington, Delaware, fifth edition, 1984.

DEPARTMENT OF MATHEMATICS, MOUNT HOLYOKE COLLEGE, SOUTH HADLEY, MA 01075, USA

Current address: Department of Mathematics, Amherst College, Amherst MA 01002

E-mail address: amcastro@amherst.edu

Contemporary Mathematics
Volume **240**, 1999

Recent Progress on the Geometry of Univalence Criteria

Martin Chuaqui and Brad Osgood

1. Introduction

This paper is a survey of some new techniques and new results on sufficient conditions in terms of the Schwarzian derivative for analytic functions defined in the unit disk to be univalent. Along with univalence we consider the questions of quasiconformal and homeomorphic extensions of the mapping.

Let f be analytic and locally univalent. Its Schwarzian derivative is

$$Sf = \left(\frac{f''}{f'}\right)' - \frac{1}{2}\left(\frac{f''}{f'}\right)^2.$$

If $u = (f')^{-1/2}$ then

$$u'' + \frac{1}{2}(Sf)u = 0.$$

Conversely if u is a solution of

$$u'' + pu = 0 \tag{1.1}$$

and

$$f(z) = \int_{z_0}^{z} u^{-2}(\zeta)\, d\zeta \tag{1.2}$$

then $Sf = 2p$. We recall the chain rule

$$S(f \circ g) = ((Sf) \circ g))(g')^2 + Sg \tag{1.3}$$

and that the Schwarzian is identically zero exactly for Möbius transformations. Let $\mathbf{D}$ denote the unit disk.

There has been progress in several areas, but the innovations we treat here come primarily from an injectivity criterion for conformal, local diffeomorphisms of an n-dimensional Riemannian manifold into the n-sphere. The criterion involves a generalization of the Schwarzian derivative which depends both on the conformal factor of the mapping and on the underlying Riemannian metric. The scalar curvature of the metric and the metric diameter of the manifold enter as bounds for

1991 *Mathematics Subject Classification.* Primary: 30C45, 30F45; Secondary: 34C11.
This work was partially supported by FONDECYT Grant 1971055.

the Schwarzian. A majority of the known classical univalence criteria follow from this general result.

The proof of the general criterion synthesizes several key ingredients that are present in the proofs of many classical criteria, most particularly the Sturm comparison theorem for second order ordinary differential equations, applied to (1.1). Comparison theorems can be considered as statements about convexity, and this turns out to be the essential notion through which we would like to understand and explain the phenomena of univalence and extensions.

2. The Nehari Class and the Ahlfors-Weill Extension

Nehari's theorem of 1949 in [**Neh49**] is perhaps the best known univalence criterion, and its connection to quasiconformal mappings discovered in 1962 by Ahlfors and Weill [**AW62**] are together the model cases for other classical crtiteria and their generalizations. Instead of beginning with the very general formulations now available we would therefore like to highlight the new ideas in the context of these two fundamental results.

THEOREM 2.1 (Nehari, Ahlfors-Weill). *If f is analytic in $\mathbf{D}$ and*

$$|Sf(z)| \leq \frac{2}{(1-|z|^2)^2}, \quad z \in \mathbf{D}, \tag{2.1}$$

then f is univalent. If

$$|Sf(z)| \leq \frac{2t}{(1-|z|^2)^2}, \quad z \in \mathbf{D}, \tag{2.2}$$

for some $0 \leq t < 1$ then f has a $\frac{1+t}{1-t}$-quasiconformal extension to $\widehat{\mathbf{C}} = \mathbf{C} \cup \infty$.

We let N be the set of all (univalent) functions that satisfy (2.1) and we call it the Nehari Class. N is a large class of functions, containing, for example, all convex conformal mappings, *i.e.*, those mappings with a (euclidean) convex image, [**Paa31**], [**Neh76**].

There is an interpretation of (2.1) in terms of convexity in hyperbolic geometry. Let $f \in N, \Omega = f(\mathbf{D})$ and let $\lambda_\Omega|dw|$ be the Poincaré metric of Ω. From calculations to be discussed in the next section – based essentially on a generalization of the differential equation (1.1) – it follows that $\lambda_\Omega^{1/2}$ is a hyperbolically convex function on Ω. This means that $\lambda_\Omega^{1/2}$ is convex, in the usual sense, along all hyperbolic geodesics in Ω, or equivalently that the Hessian of $\lambda_\Omega^{1/2}$, computed with respect to the Poincaré metric, is positive definite. Since (2.1) is invariant under Möbius transformations of the image, this convexity property is true for all Möbius shifts, $M(\Omega)$. This Möbius invariant hyperbolic convexity turns out to be equivalent to (2.1).

Nehari's proof related the univalence of f to the disconjugacy of solutions of (1.1). An upper bound on the Schwarzian allows one to invoke variants of the Sturm comparison theorem. Normalizing via Möbius transformations of the range and domain are possible because of the chain rule (1.3) and Schwarz's lemma, according to which

$$(1-|w|^2)^2|Sf(w)| = (1-|z|^2)^2|S(f \circ g)(z)|, \quad w = g(z) = e^{i\theta}\frac{z-\zeta}{1-\bar{\zeta}z}.$$

There were three elements in the proof of the Ahlfors-Weill result. First, one can consider $f(rz)$, $r < 1$, and assume that a function satisfying (2.2) is analytic on $\overline{\mathbf{D}}$. This is then later removed by taking a limit as $r \to 1$. Second, and most remarkable, is an explicit formula for the extension:

$$E_f(z) = \begin{cases} f(z) & \text{for } |z| \le 1, \\ f(\zeta) + \dfrac{(1-|\zeta|^2)f'(\zeta)}{\bar{\zeta} - \frac{1}{2}(1-|\zeta|^2)\frac{f''}{f'}(\zeta)}, & \zeta = 1/\bar{z}, \quad \text{for } |z| > 1. \end{cases} \tag{2.3}$$

The complex dilatation of E_f at a point z outside the disk is

$$\mu_{E_f}(z) = -\frac{1}{2}\left(\frac{\zeta}{\bar{\zeta}}\right)^2 (1-|\zeta|^2)^2 Sf(\zeta), \quad \zeta = 1/\bar{z}. \tag{2.4}$$

From this and (2.2) the complex dilatation is bounded by $t < 1$, and E_f will be proved to be quasiconformal once we know that it is a homeomorphism. This is then the third point in the argument, for (2.2) also imples that E_f is a local homeomorphism. Its range is $\widehat{\mathbf{C}}$ and hence by the Monodromy Theorem it is a global homeomorphism. Note that regularity on $\overline{\mathbf{D}}$ is essential in analyzing the continuity of the extension at the boundary. Note also that the proof gives no information on what happens to the extension in the limiting case as $t \to 1$.

The steps in the Ahlfors-Weill proof have been superceded, and the extension E_f does more than advertized. To explain, we first recall some results from an important paper of Gehring and Pommerenke in 1985, [**GP84**].

We adopt the terminology from [**CO98a**] and say that f is an *extremal function* for (2.1) if $f(\mathbf{D})$ is not a Jordan domain. This is the case for the mapping

$$L(z) = \frac{1}{2}\log\frac{1+z}{1-z}$$

onto a parallel strip, for which $SL(z) = 2/(1-z^2)^2$. The image is not a Jordan domain, and neither is $(A \circ L \circ B)(\mathbf{D})$ for any Möbius transformation A and any Möbius automorphism B of $\mathbf{D}$. Such Möbius conjugates of L are the only extremals in N:

THEOREM 2.2 (Gehring-Pommerenke). *If $f \in N$ then f has a continuous extension to $\overline{\mathbf{D}}$. The function L is the unique extremal function in N up to Möbius transformations of the range and of $\mathbf{D}$ onto itself.*

By suitable normalizations and use of comparison theorems one can establish a logarithmic modulus of continuity for f, and this suffices to prove the continuity up to the boundary. This replaces the first part of the Ahlfors-Weill argument, and we will add some further remarks in Section 3.

We let N^* be the set N minus all Möbius conjugations of the extremal L. Thus if $f \in N^*$ then $f(\mathbf{D})$ is a Jordan domain, and so, for topological reasons, has a homeomorphic extension to $\widehat{\mathbf{C}}$. In fact, (2.3) already gives a homeomorphic extension when $f \in N^*$, that is for $t = 1$ in (2.2). This is the main new result in this section and we explain now where it comes from.

Let $f \in N$ and let $\Omega = f(\mathbf{D})$. Again we let $\lambda_\Omega |dw|$ be the Poincaré metric of Ω. Define

$$\Lambda_\Omega(w) = w + \frac{1}{\partial_w \log \lambda_\Omega(w)}, \tag{2.5}$$

where

$$\partial_w = \frac{1}{2}\left(\frac{\partial}{\partial u} - i\frac{\partial}{\partial v}\right).$$

Thus $\Lambda_\Omega(w)$ moves away from w by a vector in the direction $\nabla \log \lambda_\Omega(w)$ of magnitude $2/|\nabla \log \lambda_\Omega(w)|$. When $f \in N^*$ we will show that Λ_Ω is a reflection across $\partial\Omega$; we think of it as a *gradient reflection.* In terms of Λ_Ω the Ahlfors-Weill extension is

$$E_f(z) = \begin{cases} f(z) & \text{for } |z| \le 1, \\ \Lambda_\Omega(f(\zeta)), \quad \zeta = 1/\bar{z}, & \text{for } |z| > 1. \end{cases} \tag{2.6}$$

Whatever mapping properties (2.6) is to enjoy must already be present in Λ_Ω. For example, if Λ_Ω is to be injective, then the Poincaré density λ_Ω must have at most one critical point, because such a point will be mapped under Λ_Ω to the point at infinity.

Λ_Ω has the important property of being *conformally natural.* This means that

$$\Lambda_{M(\Omega)}(M(w)) = M(\Lambda_\Omega(w)), \quad w \in \Omega \tag{2.7}$$

for any Möbius transformation M. We conclude from this that for Λ_Ω to be injective the density of the Poincaré metric must have at most one critical point for every shift of the image by a Möbius transformation. This goes the other way too, for if $\Lambda_\Omega(w_1) = \Lambda_\Omega(w_2)$ for two distinct points then some Möbius transformation takes this common value to infinity, showing that the density of the Poincaré metric of the shifted domain has at least two critical points.

Next, suppose that λ_Ω has a unique critical point. We claim that Ω is bounded. The proof of this uses the fact that the hyperbolic convexity of $\lambda_\Omega^{1/2}$ in Ω is equivalent to the function

$$u_f(z) = \frac{1}{\sqrt{|f'(z)|(1-|z|^2)}} \tag{2.8}$$

being hyperbolically convex in $\mathbf{D}$. The corresponding critical point in $\mathbf{D}$ may be assumed to be the origin. We thus have a convex function with a unique critical point, whose growth is therefore at least linear. This gives a lower bound in (2.8), and the resulting upper bound for $|f'|$ is of the form

$$|f'(z)| \le \frac{(1-|z|^2)}{(a+bL(|z|))^2} = -\frac{d}{d|z|}\left(\frac{1}{b(a+bL(|z|))}\right),$$

where a and b are positive constants depending on f. This shows that f is bounded. So Ω is bounded and the unique critical point is sent by Λ_Ω to infinity. This is an interesting global conclusion to draw. The Möbius invariant formulation of this result is that, if for every Möbius shift of Ω the density of the Poincaré metric has at most one critical point, then Λ_Ω takes values in the complement of $\overline{\Omega}$.

What about the boundary behavior of Λ_Ω? Again, the assumption is the absence of multiple critical points in every Möbius shift. A convex function with a unique critical point not only has at least linear growth, but its radial derivative moving away from the critical point must be bounded away from zero outside some small disk centered at the point. Since all is computed in the hyperbolic metric, this means that

$$\frac{1}{\lambda}|\partial_w \lambda_\Omega^{1/2}| \geq c > 0$$

away from the critical point. Thus

$$|\partial_w \log \lambda_\Omega| \geq 2c\lambda_\Omega^{1/2}, \tag{2.9}$$

which proves that Λ_Ω is the identity at the boundary because $\lambda_\Omega \to \infty$ there when the domain is bounded. We mention in passing that the exponent $1/2$ in the estimate above is sharp for functions in N^*.

To summarize, starting with $f \in N$, $\Omega = f(\mathbf{D})$, if the density of Poincaré metric of every Möbius shift of Ω has at most one critical point, then Λ_Ω is a reflection across $\partial\Omega$. When does this happen? Suppose that for some Möbius shift of Ω there are two critical points. By convexity, the hyperbolic geodesic joining them must consist entirely of critical points. Back in $\mathbf{D}$ that geodesic can be taken to lie along the real axis, and it is then easy to see that the original function f must be Möbius conjugate to the logarithm L. We can therefore now state, [**CO94**]:

THEOREM 2.3. *If $f \in N^*$ then the Ahlfors-Weill extension E_f is a homeomorphism of the sphere to itself.*

The Ahlfors-Weill theorem itself follows immediately by virtue of the formula (2.4) for the complex dilatation.

REMARK 2.4. Using comparison theorems requires not just the differential equation $u'' + qu = 0$ but also initial conditions. Typically one sets up the problem, in the disk, with $u(0) = 1$ and $u'(0) = 0$. With f as in (1.2) one then has $f(0) = 0$, $f'(0) = 1$ and $f''(0) = 0$. In terms of the Poincaré metric the condition $f''(0) = 0$ translates exactly to $\log \lambda_\Omega$ having a critical point at $f(0)$.

See [**CO93**] for some uses of comparison theorems to prove distortion theorems for the class N. One useful fact that comes out of that work is that it is possible to normalize a function $f \in N$, by a Möbius transformation, to have $f''(0) = 0$ without introducing a pole. We also refer to [**COP96**] and [**CO98b**] for further mapping properties of functions in the Nehrai class and its generalizations, especially in relation to John domains. We will not review the particular results here because we want to stay close to the theme of using convexity, and concentrate on the general geometric methods. See also [**Osg98**] for more background on the Schwarzian.

3. A General Univalence Criterion: Extensions and Extremals

The title of this section comes from the paper [**CO98a**]. Our intention is now to show how univalence, homeomorphic and quasiconformal extensions and convexity play out in a more general setting. We work with conformal metrics on $\mathbf{D}$ and with a generalized Schwarzian derivative.

Let g be a metric (tensor) on $\mathbf{D}$ and let $g_0 = |dz|^2$ denote the euclidean metric. For a smooth, real-valued function ψ on $\mathbf{D}$ we define a symmetric, traceless 2-tensor

$$B_g(\psi) = \mathrm{Hess}_g\, \psi - d\psi \otimes d\psi - \frac{1}{2}\left\{\Delta_g \psi - \|\,\mathrm{grad}_g\, \psi\|_g{}^2\right\} g.$$

The Hessian, gradient, Laplacian, and norm are computed with respect to g. If $f\colon (\mathbf{D}, g) \to (\mathbf{C}, g_0)$ is a conformal, local diffeomorphism with $f^* g_0 = e^{2\psi} g$, its

Schwarzian tensor is defined by

$$\mathcal{S}_g f = B_g(\psi).$$

This definition is in [**OS92**], see [**OSar**] for another approach. For the present discussion it is less important to know all the aspects of the generalization than simply to keep in mind that the Schwarzian tensor is computed with respect to a background metric g, and that it changes when g changes. When there is conformal change in g the Schwarzian tensor changes in a simple way, governed by a generalization of the chain rule, (1.3). When g is the euclidean metric $\mathcal{S}_g f$ can be written as the matrix

$$\mathcal{S}_g f = \begin{pmatrix} \operatorname{Re} Sf & -\operatorname{Im} Sf \\ -\operatorname{Im} Sf & -\operatorname{Re} Sf \end{pmatrix}.$$

In [**OS90**] the authors obtained a general univalence criterion in terms of $\mathcal{S}_g f$ that involves both the curvature of the metric and a diameter term. Let $K(g)$ denote the Gaussian curvature of the metric g. In the two-dimensional case the result can be stated as:

THEOREM 3.1. *Let f be analytic or meromorphic in $(\mathbf{D}, g)$ and locally univalent. Suppose that any two points in $\mathbf{D}$ can be joined by a geodesic of length $< \delta$, for some $0 < \delta \le \infty$. If*

$$||\mathcal{S}_g f||_g \le \frac{2\pi^2}{\delta^2} - \frac{1}{2}K(g) \tag{3.1}$$

then f is univalent.

Many known criteria for univalence follow from Theorem 3.1 by choosing different conformal background metrics g. Nehari's criterion (2.1) results by taking g to be $|dz|^2/(1-|z|^2)^2$, the Poincaré metric, with $K = -4$ and $\delta = \infty$. As it turns out, the Schwarzian of f is the same in the hyperbolic and euclidean metrics, and in computing the hyperbolic norm $||\mathcal{S}_g f||_g$ one gets two factors of $(1-|z|^2)$ because $\mathcal{S}_g f$ is a 2-tensor.

There is a second order differential equation that goes along with $\mathcal{S}_g f$ involving the Hessian of a function associated to the conformal factor of f. The use of comparison theorems is then still very much central to the analysis, though the geometric set-up has changed.

We consider metrics of the form

$$g = e^{2\sigma} g_0.$$

Let f be a conformal, local diffeomorphism of $(\mathbf{D}, g)$ into $(\mathbf{C}, g_0)$. If we write $\varphi = \log|f'|$ then $f^* g_0 = e^{2(\varphi - \sigma)} g$, and hence $\mathcal{S}_g f = B_g(\varphi - \sigma)$. We define

$$u_f = e^{(\sigma - \varphi)/2}. \tag{3.2}$$

We refer to u_f as the *associated function.* Note that if we use the euclidean metric in both the domain and the range of f then $u_f = |f'|^{-1/2}$.

A calculation, straightforward but somewhat involved, gives the following:

THEOREM 3.2. *If* $||\mathcal{S}_g f||_g \le \frac{2\pi^2}{\delta^2} - \frac{1}{2}K(g)$ *then* $\operatorname{Hess}_g u_f + \frac{\pi^2}{\delta^2} u_f\, g \ge 0$.

Analyzing the Hessian equation along geodesics is the main part of the proof of Theorem 4.

If the metric g is complete then $\delta = \infty$ and we only consider metrics of non-positive curvature. We write the condition (3.1) in the form

$$||\mathcal{S}f_g||_g \leq \frac{1}{2}|K(g)|. \tag{3.3}$$

The conclusion of Theorem 3.2 is then that

$$\operatorname{Hess}_g u_f \geq 0,$$

that is, u_f is a convex function on $\mathbf{D}$ with respect to the metric g. It may not look like it, but in the case of the Nehari class this is precisely the statement that the function u_f in (2.8) is hyperbolically convex in $\mathbf{D}$, or that $\lambda_\Omega^{1/2}$ is hyperbolically convex on $\Omega = f(\mathbf{D})$, $f \in N$.

As with the Nehari class, the convexity is a characteristic property for functions to satisfy (3.3). And again the nature of the critical points of u_f have consequences that are global:

THEOREM 3.3. *Let g be complete and f a meromorphic function in $(\mathbf{D}, g)$. Then $||\mathcal{S}_g f||_g \leq \frac{1}{2}|K(g)|$ if and only if $u_{M\circ f}$ is convex for all Möbius transformations M. If u_f has a critical point at which u_f is positive, then f is analytic. If the critical point is unique then f is bounded.*

Extension to the Boundary. We assume that g is complete. A typical way to use the convexity of u_f is the following. Let γ be a geodesic in the metric g in the disk, parametrized by arclength $\tau = d_g(z, z_0)$, where $z_0 \in \gamma$ is a fixed point and $d_g(z, z_0)$ is distance. Since the metric is complete, starting at z_0 we can follow γ in both directions indefinitely. Let γ_+, γ_- denote these two halves. Suppose that the derivative of u_f at z_0 in the direction of γ_+ is positive, say $b > 0$, Convexity implies that u_f will grow at least linearly along γ_+, that is,

$$u_f(z) \geq a + bd_g(z, z_0),$$

or

$$|f'(z)| \leq \frac{e^\sigma}{(a + bd_g(z, z_0))^2} = -\frac{d}{ds}\frac{1}{b(a + bd_g(z, z_0)}, \tag{3.4}$$

where s is the euclidean arclength. This implies that the integral

$$\int_{\gamma_+} |f'(z)||dz| \tag{3.5}$$

is finite, which implies that $f(z)$ has a limit as z tends to the boundary ∂D along γ_+.

The finiteness of the integral (3.5), together with a very natural geometric condition on the behavior of the geodesics near the boundary, allow us to generalize to this setting the result of Gehring and Pommerenke regarding continuous extension to the closed disk. The necessary geometric condition, which we must take as an assumption on the metric, appears frequently in areas of differential geometry concerned with the 'visibility' of a boundary at infinity. We formulate it as follows:

DEFINITION 3.4. The complete metric g on $\mathbf{D}$ has the *Unique Limit Point* property (ULP) if:
(a) Let $z_0 \in \mathbf{D}$. If $\gamma(t)$, $0 \leq t < T \leq \infty$ is a maximally extended geodesic starting at z_0 then $\lim_{t\to T}\gamma(t)$ exists (in the euclidean sense). We denote it by $\gamma(T) \in \partial\mathbf{D}$.
(b) The limit point is a continuous function of the initial direction at z_0.
(c) Let $\zeta \in \partial\mathbf{D}$. Then there is a geodesic starting at z_0 whose limit point on $\partial\mathbf{D}$ is ζ.

The problem of extending the map to $\overline{\mathbf{D}}$ is local, and we work on sectors defined by a vertex z_0 near the boundary and all geodesics emanating from z_0 toward the boundary. For small enough angle of the sector at z_0, and with a proper normalization of the mapping f, we can achieve that the derivative of u_f at z_0 in the direction of any of the geodesics is (uniformly) positive. This ensures at least a linear growth of u_f along each geodesic. By (3.4) and (3.5) all integrals

$$\int_{\gamma_+} |f'(z)||dz|$$

for γ_+ in the sector will be finite. The limit $\lim_{z\in\gamma_+} f(z)$ must lie on the boundary of the image, and it is not difficult to show that it depends continuously on the direction of the geodesic γ_+ at z_0. This shows that (locally) the boundary of the image can be continuously parametrized, which ensures the continuous extension of the mapping f.

This result also holds when the metric is not complete, but the argument is more involved. We mention also that in specific cases, the bound for $|f'|$ obtained from (3.4) can be used to show directly that the mapping f is Hölder continuous, or even Lipschitz.

Extremal Functions. The fact that a function satisfying (3.1), in both the complete or incomplete case, admits a continuous extension to the closed disk motivates the definition of extremal function for the criterion. This is a function satisfying (3.1) which is not injective on $\partial\mathbf{D}$. These are the analogues for the general criterion of the logarithm function L. An analysis of the geometry of extremal functions depends on being able to join by a geodesic in $\mathbf{D}$ the points on $\partial\mathbf{D}$ where injectivity is lost. We call this an *extremal geodesic.* Again, we need to assume that such joining of boundary points is always possible, and we adopt the following definition:

DEFINITION 3.5. The metric g on $\mathbf{D}$ has the *Boundary Points Joined* property (BPJ) if any two points on $\partial\mathbf{D}$ can be joined by a geodesic which lies in $\mathbf{D}$ except for its endpoints.

We assume again that the metric g is complete. Suppose f is extremal, with $f(\zeta_1) = f(\zeta_2)$, $|\zeta_1| = |\zeta_2| = 1$. Via a Möbius transformation we may assume that the common value of f is the point at infinity. But then the associated function u_f must be constant along the geodesic γ joining ζ_1, ζ_2, for otherwise, by the finiteness of, say, (3.5), one of $f(\zeta_1)$, $f(\zeta_2)$ would be finite. Some technical considerations imply that the constant value of u_f along the extremal geodesic is its absolute minimum. Therefore, the gradient of u_f vanishes identically along γ, which provides enough information to characterize the extremal function along the curve. In particular, one shows that equality must hold in (3.3) along γ. Thus, for example, if strict inequality holds in (3.3), then the image is a Jordan domain.

We want to emphasize this last point in connection with our use of the term 'extremal'. An extremal function satisfies (3.3) (or (3.1) in the incomplete case) with equality along an extremal geodesic. However, equality can hold in the criterion along some geodesic without the function being an extremal. For example, for Nehari's criterion (2.1) the interval $(-1,1)$ is an extremal geodesic for the (extremal) function

$$L(z) = \frac{1}{2}\log\frac{1+z}{1-z}.$$

But we also have $|Sf(z)| = 2/(1-|z|^2)^2$ along $(-1,1)$ for the function

$$f(z) = \frac{1}{\sqrt{2}}\frac{(1+z)^{\sqrt{2}} - (1-z)^{\sqrt{2}}}{(1+z)^{\sqrt{2}} + (1-z)^{\sqrt{2}}}, \quad Sf(z) = \frac{-2}{(1-z^2)^2},$$

and $f(\mathbf{D})$ is a Jordan domain, in fact a quasidisk. Hence, in our sense, f is not an extremal function for Nehari's criterion.

Finally, a geometrically striking fact is that the image of an extremal geodesic under an extremal function is a *euclidean* circle. It lies in the image domain except for one point, the point on the boundary of the image which is the image of two points on $\partial\mathbf{D}$ where the function is not one-to-one.

Reflections, Homeomorphic, and Quasiconformal Extensions. We now construct an extension modeled on the Ahlfors-Weill formula (2.5), (2.6). We can do this in general only when the metric g is complete, which we again assume. Let f satisfy (3.3) and let $\Omega = f(D)$. We define a function ρ on Ω by the equation

$$f^*(\rho^2|dw|^2) = e^{2\sigma}|dz|^2. \tag{3.6}$$

In other words,

$$\rho(f(z))|f'(z)| = e^{\sigma},$$

or

$$\rho(f(z)) = u_f^2(z). \tag{3.7}$$

By (3.6), f is an isometry between the two given metrics, and by (3.7), $\rho^{1/2}$ is convex in the metric $\rho^2|dw|^2$. Now set

$$\Lambda_\Omega(w) = w + \frac{1}{\partial_w \log\rho}, \tag{3.8}$$

and define an extension, as before, via

$$E_f(z) = \begin{cases} f(z) & \text{for } |z| \le 1, \\ \Lambda_\Omega(f(\zeta)), \quad \zeta = 1/\bar{z}, & \text{for } |z| > 1. \end{cases} \tag{3.9}$$

Λ_Ω is again conformally natural, which, together with the convexity of $\rho^{1/2}$, allows us to proceed with the analysis of the mapping properties of the relfection Λ_Ω as in the case of the Nehari class. The working hypothesis is that for all Möbius shifts of f, the induced convex function $\rho^{1/2}$ has at most one critical point. This implies that Λ_Ω is one-to-one, and that it assumes values in the complement of the closure of Ω. An estimate analogous to (2.9), together with completeness of the metric, implies that Λ_Ω is the identity on $\partial\Omega$. One thus obtains:

THEOREM 3.6. *The mapping E_f in (3.9) is a homeomorphic extension of f if and only if for every Möbius shift the convex function $\rho^{1/2}$ has at most one critical point.*

If for some Möbius shift the convex function possesses two critical points, then the entire geodesic segment joining them must consist of critical points. As in the analysis of extremal functions, this implies equality in (3.3) along some geodesic segment in $\mathbf{D}$. In particular, we can conclude:

COROLLARY 3.7. *If strict inequality holds in (3.3) then the extension E_f is a homeomorphism.*

It is natural to ask whether the failure of the extension to be a homeomorphism occurs exactly when the image domain is not Jordan, *i.e.*, when f is an extremal. Such was the case for the Nehari class. The answer is yes, when the metric g is real analytic. But if the metric is just C^∞ there are simple examples showing that Ω can even be equal to $\mathbf{D}$ but with the associated function u_f having a geodesic segment of critical points. In the absence of real analyticity, the vanishing of the gradient along a geodesic *segment* does not force its vanishing along the entire geodesic, *i.e.*, out to the boundary, and this is the necessary condition for the failure of univalence at the endpoints of the geodesic.

Finally we consider the quasiconformality of the extension. As before, it is not difficult to compute its Beltrami coefficient outside $\mathbf{D}$. In absolute value it is

$$\frac{2}{|K(g)|}||\mathcal{S}_g f||_g.$$

This means that the extension E_f is quasiconformal exactly when (3.3) is replaced by the stronger inequality $||\mathcal{S}_g f||_g \le (t/2)|K(g)|$ for some $t < 1$, just as for the Ahlfors-Weill theorem.

There are several interesting questions still to address here, especially in connection with extremal functions. For example, we know something, but not much, about situations where there is a unique extremal function, or a unique extremal geodesic for an extremal function. This is an interesting question. In addition, for a given metric and a given geodesic γ, there is, as mentioned above, a defining equation that characterizes when an f is extremal along γ. This equation provides enough information so that, in principle, one can solve for f along the curve. We would like to understand better when the map f defined on γ extends to a univalent map of $\mathbf{D}$ which satisfies (3.3). By construction it will produce equality in (3.3) along γ.

Techniques for Incomplete Metrics. Recall that the full statement (3.1) of the general injectivity criterion allows for incomplete metrics, with the diameter term appearing as part of the bound for the Schwarzian tensor. Several of the results in the previous paragraphs do work for incomplete metrics, for example continuous extension to $\overline{\mathbf{D}}$ and the analysis of properties of extremal functions. What does not go through without modification is the work on homeomorphic and quasiconformal extensions.

This is a genuine issue. For example, in Nehari's original paper with (2.1) he also proved the very simple univalence criterion

$$|Sf(z)| \le \frac{\pi^2}{2}, \tag{3.10}$$

which, in turn, is actually the easiest case of the general criterion (3.1); take g to be the euclidean metric on $\mathbf{D}$ – curvature 0, diameter 2. Furthermore, it follows from another result of Gehring and Pommerenke, in the same paper cited earlier, that the image of $\mathbf{D}$ by a function satisfying (3.10) will, in fact, be *quasidisk* as long as it is a Jordan domain. Such an f does then have a quasiconformal extension, but it is not the Ahlfors-Weill extension. However, a modified version of Ahlfors-Weill works. The full details are in [**CO**], but we sketch the main points here.

To begin with, the function

$$F(z) = \tan\left(\frac{\pi}{2}z\right)$$

is, up to rotations of $\mathbf{D}$ and Möbius transformations of the image, the only extremal function for (3.10). It also determines a complete, radial, metric on $\mathbf{D}$ by

$$F'(|z|)|dz| = \cos^{-2}\left(\frac{\pi}{2}|z|\right)|dz|. \tag{3.11}$$

The usefulness of considering this metric comes from following lemma, which can be considered as a sharpening of the Sturm comparison theorem.

LEMMA 3.8 (Relative Convexity). *Let u, v, and q be defined in $[0,1)$ and suppose that*

$$u'' + qu \geq 0, \tag{3.12}$$

and that

$$v'' + qv = 0. \tag{3.13}$$

Then the function

$$w = \left(\frac{u}{v}\right) \circ F^{-1}$$

is convex, where F is defined by $F' = 1/v^2$.

Recall from the introduction that a function f with $Sf = 2p$ can be written as the integral

$$f(z) = \int_{z_0}^{z} u^{-2}(\zeta)\, d\zeta,$$

where $u'' + pu = 0$. In the case $|2p| \leq \pi^2/2 = SF = 2q$, the lemma implies that

$$\sqrt{\frac{F'(|z|)}{|f'(z)|}}$$

is convex along rays from the origin, but when the rays are parametrized by arclength of the metric (3.11). In other words, the associated function u_f of f relative to (3.11) is radially convex. A little more effort gives the full convexity on $\mathbf{D}$, and that a function satisfying (3.10) must also satisfy the general criterion (3.3) with the complete metric (3.11). The condition takes the form

$$\left|\zeta^2 Sf(z) + \frac{\pi^2}{4}\tan^2\left(\frac{\pi}{2}|z|\right) + \frac{\pi}{2}\frac{1}{|z|}\tan\left(\frac{\pi}{2}|z|\right) - \frac{\pi^2}{4}\right| \leq$$
$$\frac{\pi^2}{4}\tan^2\left(\frac{\pi}{2}|z|\right) + \frac{\pi}{2}\frac{1}{|z|}\tan\left(\frac{\pi}{2}|z|\right) + \frac{\pi^2}{4}.$$

The maximum principle implies that for nonextremals it must be that

$$|Sf(z)| < \frac{\pi^2}{2},$$

which in turn implies a strict bound in the big inequality, above. Hence by Corollary 1, the gradient extension defined by (3.8) is a homeomorphism. Unfortunately, the reflection, and hence the corresponding extension E_f, will in general fail to be quasiconformal in a neighborhood of $\partial\mathbf{D}$; the magnitude of the complex dilatation can tend to 1.

Here is where one can modify the construction. To overcome the failure of quasiconformality at the boundary, we perturb the metric (3.11), to one of the form

$$F'(|z|)^{\alpha}|dz|\,,$$

where $\alpha < 1$ must be chosen sufficiently close to 1, and depending on the given nonextremal function f. In terms of f the reflection takes the form

$$\Lambda(z) = f(z) + \frac{2|z|f'(z)}{\dfrac{\alpha\pi}{2}\tan\left(\dfrac{\pi}{2}|z|\right) - |z|\dfrac{f''}{f'}(z)}.$$

The parameter α can be chosen independent of f if the nonextremal function satisfies the stronger inequality $|Sf(z)| \leq t\pi^2/2$ for some $t < 1$. Then Λ will give a quasiconformal reflection in the image for any α with $\max\{1/2, t\} < \alpha < 1$.

The full arguments supporting these facts, which apply to more general criteria than (3.10), are rather technical. We refer to [**CO**].

REMARK 3.9. The conditions (ULP) and (BPJ) must be hypotheses in many of our results, but they are not restrictive conditions on a metric. Several sufficient conditions are given in [**CO98a**].

It would take us too far afield to discuss the important work of C. Epstein. Though in a different direction, it has been very influential on our own work. We refer to his papers [**Eps86**] and [**Eps87**].

References

[AW62] L. Ahlfors and G. Weill, *A uniqueness theorem for Beltrami equations*, Proc. Amer. Math. Soc. **13** (1962), 975–978.

[CO] M. Chuaqui and B. Osgood, *Finding complete conformal metrics to extend conformal mappings*, To appear in Indiana Math. J.

[CO93] M. Chuaqui and B. Osgood, *Sharp distortion theorems associated with the Schwarzian derivative*, Jour. London Math. Soc.(2) **48** (1993), 289–298.

[CO94] M. Chuaqui and B. Osgood, *Ahlfors-Weill extensions and critical points of the Poincaré metric*, Comment. Math. Helv. **69** (1994), 659–668.

[CO98a] M. Chuaqui and B. Osgood, *General univalence criteria in the disk: extensions and extremal functions*, Ann. Acad. Sci. Fenn. **23** (1998), 101–132.

[CO98b] M. Chuaqui and B. Osgood, *John domains and a univalence criterion of Ahlfors*, Results in Math. **33** (1998), 203–207.

[COP96] M. Chuaqui, B. Osgood, and Ch. Pommerenke, *John domains, quasidisks, and the Nehari class*, Jour. für die Reine u. Angew. Math. **471** (1996), 77–114.

[Eps86] C. Epstein, *The hyperbolic Gauss map and quasiconformal reflections*, Jour. für die Reine u. Angew. Math. **372** (1986), 96–135.

[Eps87] C. Epstein, *Univalence criteria in hyperbolic space*, Jour. für die Reine u. Angew. Math. **380** (1987), 196–214.

[GP84] F.W. Gehring and Ch. Pommerenke, *On the Nehari univalence criterion and quasicircles*, Comment. Math. Helv. **59** (1984), 226–242.

[Neh49] Z. Nehari, *The Schwarzian derivative and schlicht functions*, Bull. Amer. Math. Soc. **55** (1949), 545–551.

[Neh76] Z. Nehari, *A property of convex conformal maps*, Jour. d'Analyse Math. **30** (1976), 390–393.

[OS90] B. Osgood and D. Stowe, *A generalization of Nehari's univalence criterion*, Comment. Math. Helv. **65** (1990), 234–242.

[OS92] B. Osgood and D. Stowe, *The Schwarzian derivative and conformal mapping of Riemannian mainfolds*, Duke Math. Journal **67** (1992), 57–97.

[Osg98] B. Osgood, *Old and new on the Schwarzian derivative*, Quasiconformal mappings in analysis, 1998, pp. 275–308.

[OSar] B. Osgood and D. Stowe, *The schwarzian derivative, conformal connections, and Möbius structures*, Jour. d'Analyse Math. (to appear).

[Paa31] V. Paatero, *Über die konforme Abbildung von Gebieten deren Ränder von beschränkter Drehung sind*, Ann. Acad. Sci. Fenn. Ser. A XXXIII **9** (1931), 1–79.

P. UNIVERSIDAD CATOLICA DE CHILE
E-mail address: `mchuaqui@mat.puc.cl`

STANFORD UNIVERSITY
E-mail address: `osgood@unitdisk.stanford.edu`

Contemporary Mathematics
Volume **240**, 1999

On anticonformal automorphisms of order >2 of Riemann surfaces

Antonio F. Costa and Ana M. Porto

Abstract. Let f be an anticonformal automorphism of order $2q$. Then f^q is an involution of S and f induces an automorphism $\overline{f}$ on the Klein surface $S/<f^q>$. We establish that when q is odd then the topological type of f it is determined by the topological type of $S/<f^q>$ and by the rotation indices of the action of $\overline{f}^{\ell}$ on the fixed points and invariant components of $\partial(S/<f^q>)$, $\ell = 1, ..., q-1$. When q is even and $\overline{f}^{q/2}$ has fixed points points then the topological type of f is determined by the rotation indices of $\overline{f}^q$ acting on the nonorientable closed surface $S/<f^q>$. We present two topologically non-equivalent anticonformal automorphisms f_1 and f_2 of order four with $\overline{f}_i$, $i = 1, 2$, fixed point free. In section 2 we give a characterization of the two topological types of anticonformal automorphisms of order 4 with square fixed point free in terms of the action of such automorphisms on the homology of the surface. Finally in section 3 we give a proof of the fact that if the squares of two anticonformal automorphisms of order 4 are conformally conjugate and fixed point free then the automorphisms are topologically conjugate. This last result has been proved in [3] and we give here a geometrical alternative proof.

Let S be a compact Riemann surface with genus > 1 and f be an automorphism of S. We denote by $<f>$ the group generated by f and by $Fix(f)$ the set of fixed points of f. If p is a fixed point of f then we can find a local coordinate system (U, φ) such that $\varphi(p) = 0$ and $\varphi \circ f \circ \varphi^{-1}(z) = z.e^{i\alpha}$, for $z \in \varphi(U)$. For each $p \in Fix(f)$, the angle $\alpha = \alpha(f, p)$ is uniquely determined up multiple of 2π, independent of the local coordinated system and will be called rotation index for f in p.

Let $f_1$and f_2 be two conformal or anticonformal automorphisms of S. The automorphisms f_1 and f_2 are topologically conjugate or have the same topological type if there exists a homeomorphism h of S such that $f_1 = h \circ f_2 \circ h^{-1}$.

If f_1 and f_2 are conformal automorphisms of order n, J. Nielsen proved, [4], that f_1 and f_2 are topologically conjugate if $\{\alpha(f_1^m, p), p \in Fix(f_1^m)\} = \{\alpha(f_2^m, p), p \in Fix(f_2^m)\}$, $m = 1, ..., n-1$.

If f is an anticonformal involution then the topological type of f is given by the number of closed curves fixed by f and by orientability character of $S/<f>$.

1991 *Mathematics Subject Classification.* Primary 30F10; Secondary 57M60.

Key words and phrases. Riemann surface, anticonformal automorphism

The first author were partially supported by DGICYT PB 95-0017.

If f_1 and f_2 are conformal automorphisms of order $2q$, $q > 2$, it is possible to determine wether f_1 and f_2 are topologically conjugate by means of computing some invariants related to $S/ < f_i >$, $i = 1, 2$ and the monodromy epimorphism $\omega_i : \pi_1 O(S/ < f_i >) \to \pi_1 O(S/ < f_i >)/\pi_1(S) = \{f_i^\alpha \pi_1(S) : \alpha = 1, ..., 2q\} = C_{2q}$, $i = 1, 2$, (see [5] and [1]).

We present here a different approach. Let f be an anticonformal automorphism of order $2q$. Then f^q is an involution of S and f induces an automorphism $\overline{f}$ on the Klein surface (possibly nonorientable and with boundary) $S/ < f^q >$. The rotation indices for fixed points of $\overline{f}^\ell$, $\ell = 1, ..., q-1$ can also be defined, but if $S/ < f^q >$ is nonorientable then $\alpha = \alpha(\overline{f}^\ell, p)$ is uniquely determined up change of sign and multiple of 2π. For each boundary component C of $S/ < f^q >$ invariant by $\overline{f}^\ell$ we define the rotation index of $\overline{f}^\ell$ in C by $\alpha = \alpha(\overline{f}^\ell, C)$ if the action of $\overline{f}^\ell$ restricted to C is topologically conjugate to the rotation of angle α on a circle, α is uniquely determined up multiple of 2π, if $S/ < f^q >$ is orientable, and up multiple of 2π and sign, if $S/ < f^q >$ is nonorientable. We establish that when q is odd then the topological type of f it is determined by the rotation indices of the action of $\overline{f}^\ell$ on the fixed points and invariant components of $\partial(S/ < f^q >)$, $\ell = 1, ..., q-1$. When q is even the situation it is more complex. If $\overline{f}^{q/2}$ has fixed points then the topological type of f is determined by the rotation indices of fixed points of $\overline{f}^\ell$ acting on the nonorientable closed surface $S/ < f^q >$. But if $\overline{f}^{q/2}$ is fixed point free then it is not enough to know the rotation indices to determine the topological type of f. In fact we present two topologically non-equivalent anticonformal automorphisms f_1 and f_2 of order four with $\overline{f}_i$, $i = 1, 2$, fixed point free.

In section 2 we give a characterization of the two topological types of anticonformal automorphisms of order 4 with square fixed point free in terms of the action of such automorphisms on the homology of the surface.

Finally in section 3 we give a proof of the fact that if the squares of two anticonformal automorphisms of order 4 are conformally conjugated and fixed point free then the automorphisms are topologically conjugate. This result has been proved in [3] and we give here a geometrical alternative proof.

1. Anticonformal automorphisms and rotation indices.

Let S be a compact Riemann surface with genus > 1 and f be an anticonformal automorphism. Then f^q is an anticonformal involution and $S/ < f^q >= \overline{S}$ is a Klein surface (possibly with boundary and non-orientable). The action of f on S induces an automorphism $\overline{f}$ of $\overline{S}$.

PROPOSITION 1. *Assume that the order of f is $2q$ with q odd. The topological type of f is determined by the topological type of $\overline{S}$ and by the rotation indices in the fixed points and in the invariant connected components of $\partial\overline{S}$ by $\overline{f}$.*

PROOF. The action of f on S defines an orbifold structure on $S/ < f >$ and an orbifold covering $p : S \to S/ < f >$. The covering p is determined by the monodromy representation $\omega : \pi_1 O(S/ < f >) \to Z_{2q}$, where $\pi_1 O(S/ < f >)$ is the fundamental group of the orbifold $S/ < f >$. Let $H_1 O(S/ < f >)$ be the abelianization of $\pi_1 O(S/ < f >)$, then p is determined by the epimorphism $T : H_1 O(S/ < f >) \to Z_{2q}$. For each conic point z in $S/ < f >$ let D_z be a neigbourhood disc of z and $[\partial D_z]$ be the element of $H_1O(S/<f>)$ determined by the curve ∂D_z. For each boundary component C in $S/<f>$ let $[C]$ be the element of $H_1O(S/<f>)$ determined by the curve C. The isotropy invariants for f are the numbers $T[\partial D_z]$ and $T[C]$ for all conic points and boundary components of $S/<f>$ (or $\pm T[\partial D_z]$ and $\pm T[C]$ for all conic points and boundary components if $S/<f>$ is nonorientable). The isotropy invariant $T[\partial D_z]$ is determined by the rotation indices in $p^{-1}(z)$ for $\overline{f}^{\frac{q}{order(T[\partial D_z])}}$ and $T[C]$ by the rotation indices in C for $\overline{f}^{\frac{q}{order(T[C])}}$. Since q is odd by the Theorems 0.1 and 0.2 of [1] then the topological type of f is determined by the orientability character of $S/<f>$ and the isotropy invariants of f. The Klein surface $\overline{S}$ is orientable if and only if S is orientable and the isotropy invariants are determined by the rotation indices of $\overline{f}^r$. ∎

PROPOSITION 2. *Assume that f has order $2q$ with q even and that f^q has fixed points. Then the topological type of f is determined by the rotation indices in the fixed points for the action of $\overline{f}^r$, $r = 1, ..., q$, on $\overline{S}$.*

PROOF. Since q is even then $\overline{S}$ has no boundary and so there is no rotation indices for the boundary components of $\overline{S}$. The Proposition follows from Theorem 0.2 in [1] taking consideration that the condition on the existence of fixed points of f^q avoids the use of conditions (i) and (iii) of such theorem. ∎

Let us assume that f has order $2q$ and q is even. The automorphism f induces an automorphism f_4 of order 4 on the Riemann surface $S/<f^4>$.

The automorphism has fixed points if and only if f^q has also fixed points. If f_4^2 is fixed point free then there are two possible topological types for f_4 and then there exists two possible topological types for f with the same rotation indices for $\overline{f}^r$, $r = 1, ..., q$. Furthermore, the classification is reduced to the case of anticonformal automorphisms of order 4 with square fixed point free.

Let us show in an example the two topological types of order 4 anticonformal automorphisms with fixed point free square. Other examples of the same situation can be found in [2].

EXAMPLE 1. *Let Δ be an NEC group of signature (g; + ; [-]; {(2,4,4)}), with g>1 and let a_i, b_i, $i = 1, ..., g$, e, c_1, c_2, c_3, be a set of generators of Δ obeying the relations: $(\prod[a_i, b_i]).e = 1$, $c_i^2 = 1$, $(c_1c_2)^4 = (c_2c_3)^4 = (c_3ec_1e^{-1})^2 = 1$. Consider the group $< g_1, g_2, g_3 : g_1^2 = g_2^2 = g_3^2 = (g_1g_2)^4 = (g_2g_3)^4 = (g_3g_1)^4 = (g_3g_2g_1g_2)^4 = (g_1g_3g_2g_3)^4 = 1 >$ of order 64. We define the epimorphism $w : \Delta \to G$ by $w(a_i) = w(b_i) = w(e) = 1$, $w(c_i) = g_i$. If D is the complex disc then $D/Kerw = S$ is a surface of genus $64g + 1$. The surface S can be represented topologically by the rectangle of figure 1 where we attach to each circle the boundary of an orientable surface of genus g with one boundary component. The Riemann surface S admits two anticonformal automorphisms of order 4, $f_1$and f_2. The automorphism f_i , $i = 1, 2$, is given by the translation of vector t_i on the figure composed with the reflection on the line s_i. It is clear that f_i^2 is fixed point free, $i = 1, 2$. The quotient $S/<f_i>$ is the nonorientable surface K_i. Let γ_i be a closed curve with the property that when we cut along it the surface K_i is transformed in an orientable one. Then γ_1 lifts in S by $S \to S/<f_1>$ to 2 closed curves but γ_2 lifts to four curves. Since the homological class of the curves with the property of γ_i is well defined then the two automorphisms are not topologically conjugated.*

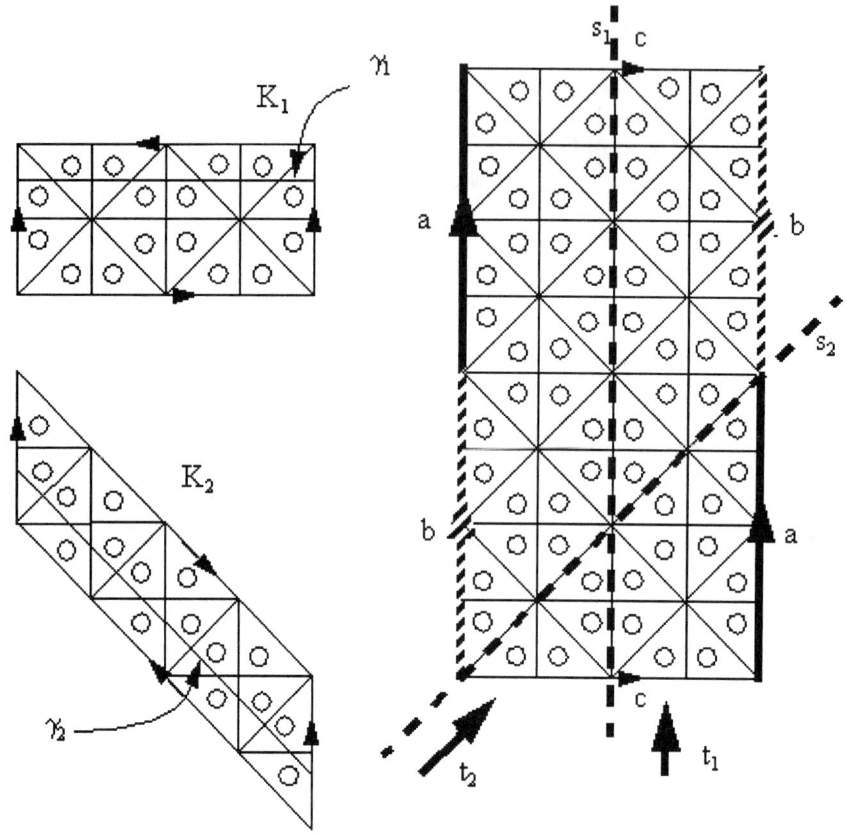

FIGURE 1

2. Anticonformal automorphisms of order four and homology.

We shall now study anticonformal automorphisms of order 4 with fixed point free square. We have seen in the last example that the rotation indices are not able to determine completely such automorphisms. We shall show how to characterize the topological types of these automorphisms using the action on the homology.

PROPOSITION 3. *Assume that S is a Riemann surface of genus $4k+1$ and f is an anticonformal automorphism of S of order 4 such that f^2 is fixed point free. Then f has two possible topological types and there is a basis of $H_1(S,Z)$ such that the automorphism induced by f in $H_1(S,Z)$ can be reduced to:*

$$\left(\bigoplus^{2k}\begin{pmatrix}0&0&0&1\\1&0&0&0\\0&1&0&0\\0&0&1&0\end{pmatrix}\right)\oplus\begin{pmatrix}-1&0\\0&1\end{pmatrix} \text{ or } \left(\bigoplus^{2k}\begin{pmatrix}0&0&0&1\\1&0&0&0\\0&1&0&0\\0&0&1&0\end{pmatrix}\right)\oplus\begin{pmatrix}0&1\\1&0\end{pmatrix},$$

depending on the topological type of f.

PROOF. By Theorem 0.2 of [1] we know that there are two topological types of automorphisms of order 4 with fixed point free squares. The two topological types are represented by the models in figure 2, where in each of the 4 circles of each

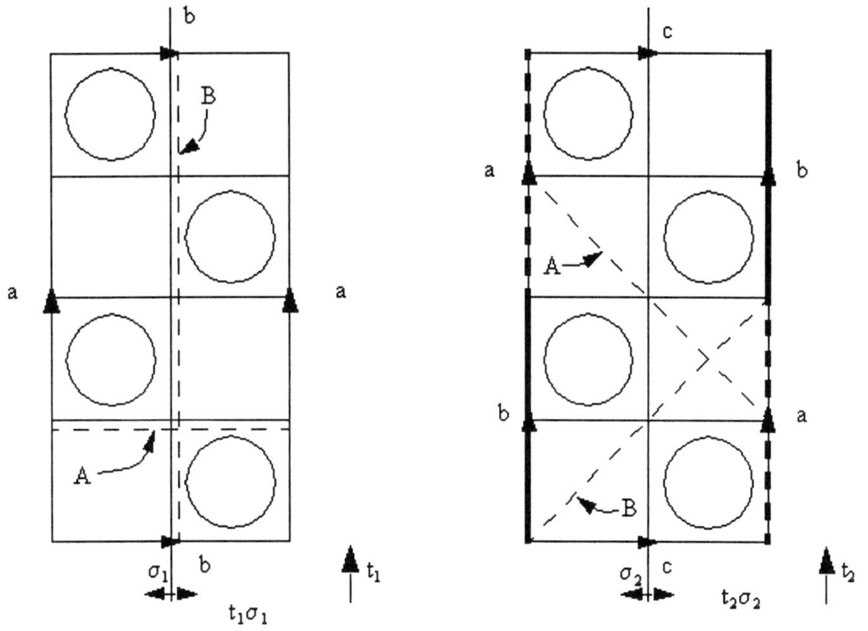

FIGURE 2

torus we attach the boundary of an orientable surface of genus k with a boundary component . Taking a canonical basis $(A_1, B_1, A_2, B_2, ..., A_k, B_k)$ of the surface of genus k that we attach on each circle, then the basis

$(..., A_i, f(A_i), f^2(A_i), f^3(A_i), B_i, f(B_i), f^2(B_i), f^3(B_i), ..., A, B)$

of $H_1(S, Z)$ gives the matrix of f as presented in the Proposition. ∎

COROLLARY 1. *Let S be a Riemann surface of genus $4k+1$ and f an anticonformal automorphism of order 4 such that f^2 is fixed point free. Let f_{*2} be the automorphism induced by f in $H_1(S, Z_2)$. The automorphism f can have two topological types, for one of them $\dim_{Z_2} Fix(f_{*2}) = 8k+2$ and for the other $\dim_{Z_2} Fix(f_{*2}) = 8k+1$.*

3. Anticonformal square roots of fixed point free conformal involutions.

In [3] it is shown that if two anticonformal automorphisms g_1 and g_2 of order 4 satisfy that g_1^2 and g_2^2 are fixed point free and conformally conjugate then g_1 and g_2 are topologically conjugate. Note that the condition that g_1^2 and g_2^2 are conformally conjugate can be replaced by assuming $g_1^2 = g_2^2$, substituting g_2 by a conjugate.

PROPOSITION 4. *([3]) Let S be a Riemann surface and f a fixed point free conformal involution. If g_1 and g_2 are anticonformal and $g_1^2 = g_2^2 = f$ then g_1 and g_2 have the same topological type.*

PROOF. The proof in [3] uses the group structure of Miller groups and of binary dihedral groups, we shall give now a more geometrical proof.

Since $g_1^2 = g_2^2 = f$ then there are two anticonformal fixed point free involutions $\widehat{g_1}$ and $\widehat{g_2}$ in $S/<f>$ induced by g_1 and g_2. Hence there is a dihedral group

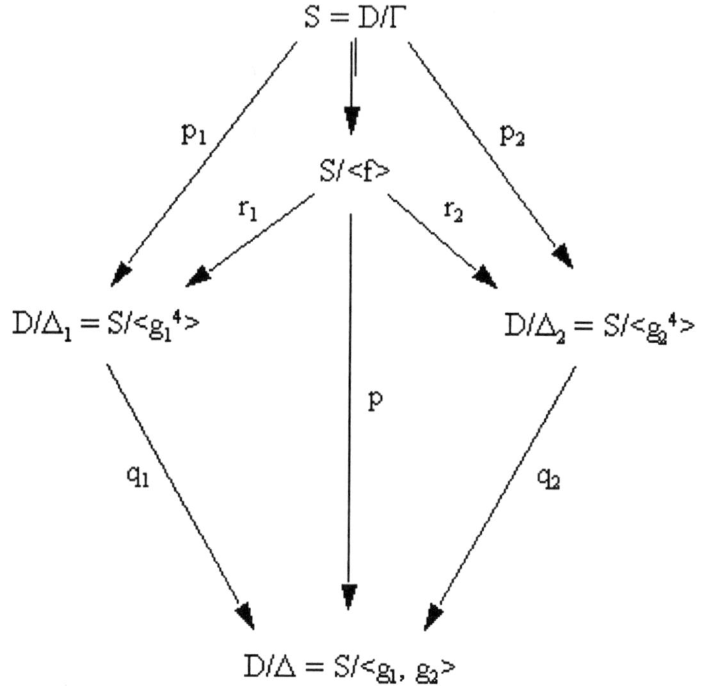

FIGURE 3

$D_\ell =< \widehat{g_1}, \widehat{g_2} >$ acting on $S/ < f >$ and $S/ < f > / < \widehat{g_1}, \widehat{g_2} >= S/ < g_1, g_2 >$. So there is an NEC group Δ with signature $(h; -; [n_1, ..., n_r]; \{-\})$ such that $D/\Delta = S/ < g_1, g_2 >$ and Δ has a surface subgroup Γ such that D/Γ uniformizes S. So we have the diagram of figure 3.

Let $\Delta_i \subset \Delta$, $i = 1, 2$, such that $S/ < g_i >= D/\Delta_i$ and $\eta_i : \Delta \to P(\Delta/\Delta_i)$ be the epimorphism given by the action of Δ on Δ/Δ_i by left multiplication, $i = 1, 2$. The epimorphism η_i is the monodromy epimorphism for the covering $q_i : S/ < g_i >\to S/ < g_1, g_2 >, i = 1, 2$.

Let $< d_1, ..., d_h, x_1, ..., x_r : \left(\prod^h d_i^2\right)\left(\prod^r x_i\right) = 1, x_i^{n_i} = 1 >$ be a canonical presentation of Δ. Let $\{1, ..., \ell\}$ denote the cosets of Δ/Δ_i. Let $\{\omega_j^i, j = 2, ..., \ell\}$, $i = 1, 2$, be Schreier transversals, i.e. elements of Δ such that $\eta_i(\omega_j^i)(1) = j$. A set of generators of Δ_i is:

$d_{ts}^i = \omega_s^i d_t [\omega^i_{\eta_i(d_t)(s)}]^{-1}$, $t = 1, ..., h$, $s = 1, ..., \ell$.

$x_{ts}^i = \omega_s^i x_t [\omega^i_{\eta_i(x_t)(s)}]^{-1}$, $t = 1, ..., r$, $s = 1, ..., \ell$.

By abelianization we obtain the following system of generators of $H_1(S/ < g_i >)$: D_{ts}^i, $t = 1, ..., h$, $s = 1, ..., \ell$, X_{ts}^i, $t = 1, ..., h$, $s = 1, ..., \ell$, $i = 1, 2$. Lifting the relations of Δ it is possible to show that $T_i = \sum^{\ell}\sum^{h} D_{ts}^i$ is the element of order two in $H_1(S/ < g_i >)$, $i = 1, 2$, see [3] pgs 95-96. Note that

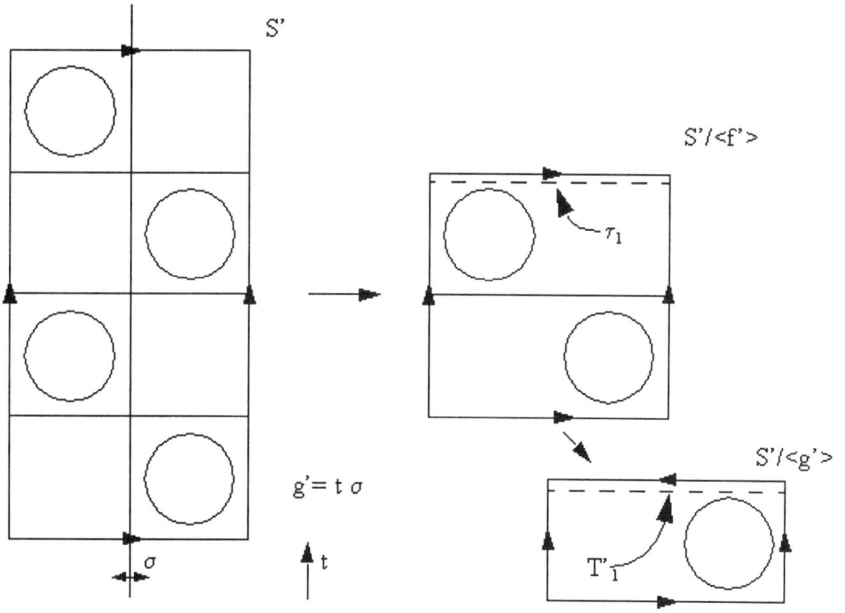

FIGURE 4

$T_i = \sum^{\ell}\sum^{h} D^i_{ts} = [q_i^{-1}(d_1,...,d_h)]$ (where $[q_i^{-1}(d_1,...,d_h)]$ means the element of $H_1(S/<g_i>)$ determined by $q_i^{-1}(d_1,...,d_h)$), since $q_i \circ r_i = p$, $i = 1,2$, then $r_1^{-1}(T_1) = r_2^{-1}(T_2) = p^{-1}(d_1,...,d_h)$.

In order to prove that g_1 and g_2 are topologically conjugated, it is necessary to show that if $\omega_i : H_1(S/<g_i>) \to Z_4$ is the monodromy epimorphism of p_i then $\omega_1(T_1) = \omega_2(T_2)$.

We shall prove that if $\omega_i(T_i) = 0$ then $\omega_j(T_j) = 0$, $\{i,j\} = \{1,2\}$.

Assume $\omega_1(T_1) = 0$. Then the action of g_1 is topologically conjugate by a homeomorphism h to the action of the glide reflection g_1' of the figure 4 on the surface S' obtained from the torus of the figure attaching a surface of genus $\frac{h-2}{2}$ on each circle on such torus.

Then T_1 can be represented by T_1' on the figure. Let τ_1 be one of the components of $r_1^{-1}(T_1)$. If $\gamma \in H_1(S/<g_i>)$ and γ preserves the orientation, then $\omega_i(\gamma)$ is 0 or 2 depending if $< h(r_i^{-1}(\gamma), \tau_1) >$ is 0 or 1, where $< .,. >$ is the intersection number modulo 2 and h is a homeomorphism from $S/<f>$ to $S'/<g_i'^2>$ giving the topological equivalence between g_1 and g_1'.

Since $r_1^{-1}(T_1) = r_2^{-1}(T_2)$ then
$< h(r_2^{-1}(T_2)), \tau_1) > = < h(r_1^{-1}(T_1)), \tau_1) > = < r_1'^{-1}(T_1'), \tau_1) > = 0$.

Hence $\omega_2(T_2) = 0$. Then $\omega_1(T_1) = \omega_2(T_2)$ and so g_1 and g_2 are topologically equivalent. ∎

References

[1] Costa, A.F., Classification of the orientation reversing homeomorphism of finite order of surfaces, Topology and its Applications 62 (1995) 145-162.

[2] Costa, A.F., Sobre automorfismos anticonformes y libres de superficies de Riemann, Revista de la Real Academia de Ciencias Exactas, Físicas y Naturales de Madrid 87 (1993) 489-496.

[3] Costa, A. F., On the topological type of anticonformal square roots of automorphisms of Riemann surfaces, Manuscripta Math. 89 (1996) 87-102.

[4] Nielsen, J., Die Struktur periodischer Transformationen von Flächen, Danske Vid Selsk. Mat-Fys. Medd. 1 (1937) 1-77.

[5] Yokoyama, K., Complete classification of periodic maps on compact surfaces, Tokio J. Math. 15 (1992) 247-279.

Departamento de Matemáticas Fundamentales, Facultad de Ciencias, UNED, 28040 Madrid, SPAIN

E-mail address: `acosta@mat.uned.es`

E-mail address: `asilva@mat.uned.es`

Contemporary Mathematics
Volume **240**, 1999

Isomorphisms between generalized Teichmüller spaces

C. J. Earle, F. P. Gardiner, and N. Lakic

ABSTRACT. Extending our previous results in [**5**] and [**14**] we determine the holomorphic isomorphisms between some generalized Teichmüller spaces.

1. Introduction

In our papers [**5**] and [**14**] we developed techniques that determine the holomorphic isomorphisms between many infinite dimensional Teichmüller spaces. In this paper we extend these techniques to study isomorphisms between some generalized Teichmüller spaces. Specifically, Theorems 2.1 and 2.2 in §2 describe the isomorphisms between Lieb's product Teichmüller spaces, and Theorem 3.1 in §3 describes the isomorphisms between the Teichmüller spaces of closed sets in the Riemann sphere. These spaces are among several new types of Teichmüller space that have been introduced recently. They have interesting applications to dynamics and to the study of holomorphic motions, and their theory is still under development (see [**6**], [**10**], [**11**], [**16**], and [**17**]). The final version of our preliminary report [**6**], which is still in preparation, will contain a discussion of isomorphisms of more of these spaces.

2. Isomorphisms between product spaces

2.1. Lieb's product Teichmüller space. For each j in an index set J, let S_j be a Riemann surface whose universal covering surface is isomorphic to the open unit disk. (We shall call such Riemann surfaces hyperbolic.) Let $T(S_j)$ be the Teichmüller space of S_j, 0_j be the basepoint of $T(S_j)$, and d_j be the Teichmüller metric on $T(S_j)$.

Let S be the disjoint union of the S_j. By definition the product Teichmüller space $T(S)$ is the set of points x in the cartesian product of the spaces $T(S_j)$ such that $\{d_j(0_j, x_j) : j \in J\}$ is a bounded set. The basepoint of $T(S)$ is the point x such that $x_j = 0_j$ for each j; we shall usually denote it by 0. The Teichmüller metric

$$d(x,y) = \sup\{d_j(x_j, y_j) : j \in J\}$$

on $T(S)$ is complete. In his 1990 Ph.D. dissertation [**16**], G. Lieb proves that $T(S)$ is a complex manifold modelled on a complex Banach space $B(S)$. He also shows

1991 *Mathematics Subject Classification.* Primary 32G15; Secondary 58F23.
The third author was supported in part by NSF Grant DMS 9706769.

that the Kobayashi metric on $T(S)$ equals the Teichmüller metric. The infinitesimal Kobayashi and Teichmüller metrics also coincide (see [**10**]). Lieb assumes that the index set J is countable, but that restriction is unnecessary (see [**10**]) and we shall not impose it here. This increased generality allows our results to be applied to the disjoint union of the leaves of a Riemann surface lamination if all leaves are hyperbolic.

Our first main result is

THEOREM 2.1. *Let S and $\widetilde{S}$ be the disjoint unions of hyperbolic Riemann surfaces S_j and $\widetilde{S}_{\tilde{j}}$ with the respective index sets J and $\widetilde{J}$. Let J' and $\widetilde{J}'$ be the sets of indices whose corresponding Riemann surface is not isomorphic to $\mathbb{C} \setminus \{0,1\}$. For any biholomorphic map $F\colon T(S) \to T(\widetilde{S})$ there are a bijection $\alpha\colon J' \to \widetilde{J}'$ and biholomorphic maps $F_j\colon T(S_j) \to T(\widetilde{S}_{\alpha(j)})$, $j \in J'$, such that*

$$F(x)_{\alpha(j)} = F_j(x_j) \quad \textit{for all } x \in T(S) \textit{ and } j \in J'. \tag{2.1}$$

We can draw an even stronger conclusion by putting more restrictions on the Riemann surfaces S_j and $\widetilde{S}_{\tilde{j}}$. Recall that a Riemann surface X has finite genus if there is a one-to-one holomorphic map f of X into a compact Riemann surface Y. If f and Y can be chosen so that the set $Y \setminus f(X)$ is finite we say that X has finite analytic type. More precisely, we say that X has finite type (g, n) if there is a one-to-one holomorphic map f of X into a compact Riemann surface Y of genus g such that $Y \setminus f(X)$ contains exactly n points.

We say that X is exceptional if it has finite type (g, n) and $2g + n$ is less than 5. For example, $\mathbb{C} \setminus \{0, 1\}$ is exceptional since it has finite type $(0, 3)$. Every non-exceptional Riemann surface is hyperbolic.

THEOREM 2.2. *Let S and $\widetilde{S}$ be disjoint unions of non-exceptional Riemann surfaces of finite genus. Every biholomorphic map of $T(S)$ onto $T(\widetilde{S})$ is the geometric isomorphism induced by some quasiconformal map of S onto $\widetilde{S}$.*

We shall derive Theorem 2.2 from Theorem 2.1 in the next subsection, after a discussion of the quasiconformal maps of S onto $\widetilde{S}$ and the geometric isomorphisms of $T(S)$ onto $T(\widetilde{S})$ that they induce.

REMARK 2.3. The finite dimensional version of Theorem 2.1, in which the sets J and $\widetilde{J}$ are finite and the Riemann surfaces S_j and $\widetilde{S}_{\tilde{j}}$ have finite analytic type, was proved by Gentilesco in [**12**]. Our proof in the general case combines ideas from [**12**] (see Lemma 2.9 below) with ideas and techniques from [**5**] and [**7**].

2.2. Proof of Theorem 2.2. Let J and $\widetilde{J}$ be the index sets for S and $\widetilde{S}$ respectively. By definition, a quasiconformal map from S to $\widetilde{S}$ is a homeomorphism f of S onto $\widetilde{S}$ such that the restriction f_j of f to S_j is a K-quasiconformal map of S_j onto its image for each j in J, with K independent of j.

Let $f\colon S \to \widetilde{S}$ be a quasiconformal map. Since f is a homeomorphism there is a bijection $\alpha\colon J \to \widetilde{J}$ such that $f_j(S_j)$ equals $\widetilde{S}_{\alpha(j)}$ for every j. The quasiconformal map f_j induces a biholomorphic map ρ_{f_j} of $T(S_j)$ onto $T(\widetilde{S}_{\alpha(j)})$ in a standard way:

$$\rho_{f_j}([g]) = \left[g \circ f_j^{-1}\right], \tag{2.2}$$

where $[g]$ is the Teichmüller equivalence class of the quasiconformal map g with domain S_j and $[g \circ f_j^{-1}]$ is the Teichmüller equivalence class of the quasiconformal map $g \circ f_j^{-1}$ with domain $\widetilde{S}_{\alpha(j)}$ (see [**5**]). The geometric isomorphism ρ_f of $T(S)$ onto $T(\widetilde{S})$ induced by f is defined by the formula

$$\rho_f(x)_{\alpha(j)} = \rho_{f_j}(x_j) \quad \text{for all } x \in T(S) \text{ and } j \in J. \tag{2.3}$$

It is easily verified that ρ_f is a biholomorphic map of $T(S)$ onto $T(\widetilde{S})$ (see [**10**]).

Now assume that S and $\widetilde{S}$ satisfy the hypotheses of Theorem 2.2, and consider an arbitrary isomorphism (biholomorphic map) F of $T(S)$ onto $T(\widetilde{S})$. By hypothesis none of the Riemann surfaces S_j or $\widetilde{S}_{\tilde{j}}$ has finite type $(0,3)$, so Theorem 2.1 gives a bijection $\alpha\colon J \to \widetilde{J}$ and biholomorphic maps F_j from $T(S_j)$ to $T(\widetilde{S}_{\alpha(j)})$ such that equation (2.1) holds for each j. Since S_j and $\widetilde{S}_{\alpha(j)}$ are non-exceptional Riemann surfaces of finite genus, we apply the main theorem of [**8**] or Theorem 4 of [**14**] according as S_j has finite analytic type or not, concluding in either case that F_j is the biholomorphic map ρ_{f_j} induced by a quasiconformal map f_j of S_j onto $S_{\alpha(j)}$. We choose f_j to be extremal in its Teichmüller equivalence class.

The basepoints 0_j and $0_{\alpha(j)}$ of $T(S_j)$ and $T(S_{\alpha(j)}$ are the Teichmüller classes of the identity maps id_j and $id_{\alpha(j)}$ on S_j and $\widetilde{S}_{\alpha(j)}$ respectively. By equation (2.2),

$$F_j(0_j) = \rho_{f_j}([id_j]) = [f_j^{-1}],$$

so f_j^{-1} and f_j are K_j-quasiconformal, with $K_j = \exp(2d_{\alpha(j)}(0_{\alpha(j)}, F_j(0_j)))$. Since $F(0)$ belongs to $T(\widetilde{S})$, equation (2.1) implies that $K = \sup\{K_j : j \in J\}$ is finite. Therefore the map $f\colon S \to \widetilde{S}$ whose restriction to S_j is f_j for all j is quasiconformal. Equations (2.1) and (2.3) show that ρ_f equals the given isomorphism F. □

2.3. Some complex Banach spaces. Theorem 2.1 is a consequence of an analogous statement about isometric isomorphisms between the tangent spaces to $T(S)$ and $T(\widetilde{S})$ at their basepoints. To describe these tangent spaces we need some Banach spaces that are discussed in more detail in Section 2 of [**7**]. (In [**7**] it was assumed that the set J is finite or countable. Again that restriction is unnecessary.)

By definition a Beltrami form on S is a tensor μ whose restriction to each S_j is a bounded measurable Beltrami form with L^∞ norm less than a constant M independent of j in J. For each j we set μ_j equal to μ times the characteristic function χ_j of S_j. The norm $\|\mu\|_\infty$ of μ is defined as the supremum of the L^∞ norms of the μ_j. We denote the Banach space of Beltrami forms on S by $Belt(S)$.

We say that μ in $Belt(S)$ is supported in S_j if $\mu = \mu_j$ almost everywhere in S_k for each index k in J.

A holomorphic quadratic differential on S is a tensor φ whose restrictions to the S_j are holomorphic quadratic differentials. Again we set $\varphi_j = \varphi\chi_j$ for each j in J. We say that φ is integrable if the norm

$$\|\varphi\| = \iint_S |\varphi| = \sum_j \iint_{S_j} |\varphi_j|$$

is finite. We denote the Banach space of integrable holomorphic quadratic differentials by $A(S)$. Observe that if $\varphi \in A(S)$ the finiteness of $\|\varphi\|$ implies that $\varphi_j = 0$ for all but countably many j. In addition, the set of φ in $A(S)$ such that $\varphi_j = 0$

for all but finitely many j is a dense subspace. We say that φ in $A(S)$ is supported in S_j if $\varphi = \varphi_j$.

The dual space $B(S)$ of $A(S)$ is fundamental in the theory of Teichmüller spaces because it is the tangent space to $T(S)$ at its basepoint. (See [**16**], [**17**], or [**10**].) For each v in $B(S)$ and each j in J we define v_j in $B(S)$ by the formula

$$v_j(\varphi) = v(\varphi_j) \qquad \text{for all } \varphi \in A(S).$$

We say that v is supported in S_j if $v = v_j$.

For every linear functional v in $B(S)$ there is a Beltrami form μ on S such that

$$v(\varphi) = \sum_j \iint_{S_j} \mu_j \varphi_j \qquad \text{for all } \varphi \text{ in } A(S). \tag{2.4}$$

If (2.4) holds we say that μ represents v. Clearly $\|v\| \leq \|\mu\|_\infty$.

The boundary seminorms b on $B(S)$ and b^* on $Belt(S)$ are defined by

$$b^*(\mu) = \inf\{\|\mu|S \setminus E\|_\infty : E \subset S,\ E \text{ compact}\}$$

and

$$b(v) = \inf\{b^*(\mu) : \mu \text{ represents } v\}.$$

Both seminorms are continuous since $b^*(\mu) \leq \|\mu\|_\infty$ and $b(v) \leq \|v\|$ for all μ and v. If $b^*(\mu)$ or $b(v)$ equals zero we say that μ or v vanishes at infinity.

The set of v that vanish at infinity is a closed subspace $B_0(S)$ of $B(S)$. For each v in $B_0(S)$ and each $\epsilon > 0$, the set of indices j such that $\|v_j\| \geq \epsilon$ is finite. Thus $v_j = 0$ for all but countably many j, and the set of v in $B_0(S)$ such that $v_j = 0$ for all but finitely many j is a dense subspace.

2.4. A duality theorem. In Lemma 6.3 of [**5**] we showed that for a hyperbolic Riemann surface S the pairing $(v, \varphi) \mapsto v(\varphi)$ establishes an isometric isomorphism between $A(S)$ and the dual space of $B_0(S)$. In [**7**] we indicated how to extend this result to a countable disjoint union of such surfaces. We shall now prove that the countability assumption is unnecessary.

LEMMA 2.4. *Let S be an arbitrary disjoint union of hyperbolic Riemann surfaces. The pairing $(v, \varphi) \mapsto v(\varphi)$ establishes an isometric isomorphism between $A(S)$ and the dual space of $B_0(S)$.*

PROOF. For each φ in $A(S)$ the function $\ell_\varphi(v) = v(\varphi)$ is obviously a bounded linear functional on $B_0(S)$ of norm at most $\|\varphi\|$. To see that the norm of ℓ_φ equals $\|\varphi\|$, let $\epsilon > 0$ be given. Choose a compact set E in S so that

$$\iint_E |\varphi| > \|\varphi\| - \epsilon \tag{2.5}$$

and $\|\varphi_j\| > 0$ for each of the finitely many indices j such that $E \cap S_j$ is not empty. The linear functional

$$v(\psi) = \iint_E \frac{|\varphi|}{\varphi} \psi, \qquad \psi \in A(S),$$

belongs to $B_0(S)$, has norm at most one, and satisfies $|\ell_\varphi(v)| > \|\varphi\| - \epsilon$ because of (2.5). Therefore the map $\varphi \mapsto \ell_\varphi$ is a linear isometry of $A(S)$ into the dual space of $B_0(S)$. We must show that this isometry is surjective.

Consider any bounded linear functional ℓ on $B_0(S)$. For each j Lemma 6.3 of [**5**] provides φ_j in $A(S)$ supported in S_j and satisfying

$$\ell(v_j) = v_j(\varphi_j)(= v(\varphi_j)) \qquad \text{for all } v \in B_0(S). \tag{2.6}$$

Now $\varphi_F = \sum_{j\in F} \varphi_j$ belongs to $A(S)$ for any finite subset F of S, and (2.6) gives

$$\ell_{\varphi_F}(v) = \sum_{j\in F} v(\varphi_j) = \ell\left(\sum_{j\in F} v_j\right) \qquad \text{for all } v \in B_0(S), \tag{2.7}$$

so $\|\varphi_F\| = \|\ell_{\varphi_F}\| \leq \|\ell\|$. Hence there is a unique φ in $A(S)$ such that $\varphi\chi_j$ equals φ_j for each j, and $\|\varphi\| \leq \|\ell\|$. Since elements of the form $\sum_{j\in F} v_j$, F finite, are dense in $B_0(S)$, equation (2.7) implies that $\ell_\varphi = \ell$. □

2.5. An identity theorem. In this section we shall prove a lemma that will be useful in our study of isometries of $B(S)$ and related spaces. We call v a Reich-Strebel functional if it is supported in some S_j and satisfies $b(v) < \|v\|$.

LEMMA 2.5. *If $R\colon B(S) \to B(S)$ is a $\mathbb{C}$-linear map of norm one and $R(v) = v$ for every Reich-Strebel functional v, then R is the identity operator on $B(S)$.*

PROOF. For the moment let us call φ in $A(S)$ *simple* if it is supported in some S_j. Since the finite sums of simple quadratic differentials are dense in $A(S)$ it suffices to show that $R(v)(\varphi) = v(\varphi)$ for all v in $B(S)$ and all simple φ.

Fix any v and any index j. We must prove that $R(v)(\varphi) = v(\varphi)$ for all φ in $A(S)$ supported in S_j. We may assume that the space of such φ has positive dimension. Let u be any Reich-Strebel functional supported in S_j. Consider the functions

$$f(t) = \|R(u + tv)\| = \|u + tR(v)\| \quad \text{and} \quad g(t) = \|u + tv\|, \quad t \in \mathbb{R}.$$

By Theorem 2 of [**7**], both $f'(0)$ and $g'(0)$ exist. Since $f(t) - g(t)$ has a maximum at 0, $f'(0) = g'(0)$. The formula for these derivatives in the cited theorem gives

$$\operatorname{Re}(R(v)(\varphi^u)) = \operatorname{Re}(v(\varphi^u)),$$

where φ^u is the unique simple quadratic differential such that $u(\varphi^u) = \|u\|$ and $\|\varphi^u\| = 1$.

To complete the proof it suffices to show that these quadratic differentials φ^u are dense in the set of simple φ of norm one supported in S_j. We shall deduce the density from a theorem of Lakic [**15**] with the help of degenerating Hamilton sequences.

Let μ be a Beltrami differential supported in S_j. By definition a degenerating Hamilton sequence for μ is a sequence $\{\varphi(n)\}_{n=1}^{\infty}$ in $A(S)$ supported in S_j such that $\|\varphi(n)\| \leq 1$ for all n and $\iint_{S_j} \mu\phi(n) \to \|\mu\|_\infty$ but $\iint_E |\varphi(n)| \to 0$ as $n \to \infty$ for every compact set E in S_j.

Now Theorem 7 of [**9**] tells us that the φ^u are precisely the simple φ of norm one supported in S_j such that the Beltrami form that equals $|\varphi|/\varphi$ on S_j and is supported in S_j has no degenerating Hamilton sequence. By Theorem 5 of [**15**] such quadratic differentials are dense among all simple φ of norm one supported in S_j. □

COROLLARY 2.6. *Let S and $\widetilde{S}$ be disjoint unions of hyperbolic Riemann surfaces, and let $L\colon B(S) \to B(\widetilde{S})$ be an invertible $\mathbb{C}$-linear isometry. Suppose L is sequentially continuous with respect to the weak* topologies of $B(S)$ and $B(\widetilde{S})$. If L maps $B_0(S)$ into $B_0(\widetilde{S})$ then L is the adjoint of a $\mathbb{C}$-linear map of $A(\widetilde{S})$ into $A(S)$.*

PROOF. Let L_0 be the $\mathbb{C}$-linear map from $B_0(S)$ to $B_0(\widetilde{S})$ obtained by restricting L to $B_0(S)$. Its adjoint L_0^* is a $\mathbb{C}$-linear map of $A(\widetilde{S})$ into $A(S)$. We claim that L equals the adjoint L_0^{**} of L_0^*. To prove this claim we must show that every v in $B(S)$ satisfies the condition

$$v(L_0^*(\widetilde{\varphi})) = (Lv)(\widetilde{\varphi}) \qquad \text{for all } \varphi \in A(\widetilde{S}). \tag{2.8}$$

Equation (2.8) holds for v in $B_0(S)$, by definition of L_0 and L_0^*. Because of the weak* sequential continuity of L, (2.8) also holds if v is the weak* limit of some sequence in $B_0(S)$. If v in $B(S)$ is supported in some S_j, it is easy to produce a sequence of elements of $B_0(S)$, all supported in S_j, that converges to v in the weak* topology. (See for example Lemma 6.2 in [**5**].) Therefore (2.8) holds for all such v. In particular $Lv = L_0^{**}(v)$ for every Reich-Strebel functional v in $B(S)$.

Now $R = L^{-1}L_0^{**}$ maps $B(S)$ into itself and satisfies the hypotheses of Lemma 2.5, so R is the identity map. □

2.6. Isometries of $B(S)$ onto $B(\widetilde{S})$. The promised infinitesimal version of Theorem 2.1 will be a corollary of the next two lemmas

LEMMA 2.7. *Let S and $\widetilde{S}$ be disjoint unions of hyperbolic Riemann surfaces, and let $L\colon B(S) \to B(\widetilde{S})$ be a surjective $\mathbb{C}$-linear isometry. Then*
(a) $L(B_0(S)) = B_0(\widetilde{S})$ and
(b) L is the adjoint of a $\mathbb{C}$-linear isometry of $A(\widetilde{S})$ onto $A(S)$.

PROOF. Statement (a) is given in the remark after Corollary 1 of Theorem 3 in [**7**]. The proof of that corollary also shows that L is sequentially continuous with respect to the weak* topologies of $B(S)$ and $B(\widetilde{S})$. Therefore, by the corollary to Lemma 2.5 above, L is the adjoint of a $\mathbb{C}$-linear map F from $A(\widetilde{S})$ to $A(S)$. Since $L = F^*$ is a surjective isometry, so is F. □

REMARK 2.8. In [**7**] we assumed that the index set J was finite or countable. That assumption makes $A(S)$ separable, so in our proof of Corollary 1 to Theorem 3 we were able to consider weak* limits of subsequences of bounded sequences in $B(\widetilde{S})$. To allow for the possibility that J is uncountable we must instead consider the weak* cluster points of bounded sequences. Apart from that the proof is unchanged.

The finite dimensional version of our next lemma was proved by J. A. Gentilesco in [**12**].

LEMMA 2.9. *Let S and $\widetilde{S}$ be the disjoint unions of hyperbolic Riemann surfaces S_j and $\widetilde{S}_{\tilde{j}}$ with the respective index sets J and $\widetilde{J}$. Let J' and $\widetilde{J}'$ be the sets of indices whose corresponding Riemann surface is not isomorphic to $\mathbb{C} \setminus \{0, 1\}$. For any surjective $\mathbb{C}$-linear isometry $F\colon A(S) \to A(\widetilde{S})$ there is a bijection $\alpha\colon J' \to \widetilde{J}'$ such that*

$$(F\varphi)_{\alpha(j)} = F(\varphi_j) \quad \text{for all } \varphi \in A(S) \text{ and } j \in J'. \tag{2.9}$$

PROOF. Since J' is empty if and only if $A(S)$ has dimension zero, we may assume that J' is not empty. It is easy to verify that a unit vector φ in $A(S)$ is an extreme point of the closed unit ball if and only if it is supported in some S_j, j in J'. Since $F(\varphi)$ is an extreme point of the closed unit ball of $A(\widetilde{S})$, it is supported in some $\widetilde{S}_{\widetilde{j}}$, $\widetilde{j}$ in $\widetilde{J}'$. It is easy to see that $\widetilde{j}$ depends only on j, not on φ. This gives us a map α from J' to $\widetilde{J}'$ satisfying (2.9). It is easy to see that α is bijective by considering the isometry F^{-1}. □

COROLLARY 2.10. *Let S, $\widetilde{S}$, J', and $\widetilde{J}'$ be as above. If $L\colon B(S) \to B(\widetilde{S})$ is a surjective $\mathbb{C}$-linear isometry there is a bijection $\alpha\colon J' \to \widetilde{J}'$ such that*

$$(Lv)_{\alpha(j)} = L(v_j) \quad \textit{for all } v \in B(S) \textit{ and } j \in J'. \tag{2.10}$$

PROOF. By Lemmas 2.7 and 2.9 there are a bijection $\alpha\colon J' \to \widetilde{J}'$ and a surjective $\mathbb{C}$-linear isometry $F\colon A(S) \to A(\widetilde{S})$ such that $F^* = L^{-1}$ and F satisfies (2.9). Now for any v in $B(S)$, φ in $A(S)$, and j in J' we have

$$L(v_j)(F\varphi) = v_j(\varphi) = v(\varphi_j) \quad \text{and}$$

$$(Lv)_{\alpha(j)}(F\varphi) = Lv((F\varphi)_{\alpha(j)}) = Lv(F(\varphi_j)) = v(\varphi_j).$$

That proves (2.10) since F is surjective. □

2.7. Proof of Theorem 2.1. For each j in J let $B(S_j)$ be the closed subspace of $B(S)$ consisting of the v that are supported in S_j. The Bers embeddings (see [**10**]) allow us to identify $T(S)$ and $T(S_j)$ with regions $\widetilde{T}(S)$ and $\widetilde{T}(S_j)$ in $B(S)$ and $B(S_j)$ respectively so that

(B1) $\widetilde{T}(S_j) = \widetilde{T}(S) \cap B(S_j)$ for all j,
(B2) if $\widetilde{x} \in \widetilde{T}(S)$ then $\widetilde{x}_j \in \widetilde{T}(S_j)$ for all j,
(B3) the natural projection $x \mapsto x_j$ from $T(S)$ to $T(S_j)$ is identified with the map $\widetilde{x} \mapsto \widetilde{x}_j$ from $\widetilde{T}(S)$ to $\widetilde{T}(S_j)$ for all j,
(B4) the basepoints of $T(S)$ and the $T(S_j)$ are identified with 0 in $B(S)$,
(B5) the infinitesimal Kobayashi metric on the tangent space to $\widetilde{T}(S)$ at 0 is given by the natural norm on $B(S)$ as the dual space of $A(S)$.

We turn now to the proof of the theorem. All geometric isomorphisms F satisfy equation (2.1), and any biholomorphic map between product Teichmüller spaces can be composed with a geometric isomorphism so that the composed map carries basepoint to basepoint (see [**10**]). In proving Theorem 2.1 we may therefore assume that the given biholomorphic map $F\colon T(S) \to T(\widetilde{S})$ preserves basepoints. Let J and $\widetilde{J}$ be the respective index sets and let J' and $\widetilde{J}'$ be the subsets defined in the statement of Theorem 2.1.

The Bers embeddings allow us to identify $T(S)$ and $T(\widetilde{S})$ with regions $\widetilde{T}(S)$ and $\widetilde{T}(\widetilde{S})$ in $B(S)$ and $B(\widetilde{S})$ so that $\widetilde{T}(S)$ has properties (B1) through (B5) and $\widetilde{T}(\widetilde{S})$ has corresponding properties with respect to its index set $\widetilde{J}$. Note that $\widetilde{T}(S_j)$ and $\widetilde{T}(\widetilde{S}_{\widetilde{j}})$ equal $\{0\}$ for j in $J \setminus J'$ and $\widetilde{j}$ in $\widetilde{J} \setminus \widetilde{J}'$.

Now F gives rise to a biholomorphic map $\widetilde{F}\colon \widetilde{T}(S) \to \widetilde{T}(\widetilde{S})$, and each Fréchet derivative map $\widetilde{F}'(\widetilde{x})\colon B(S) \to B(\widetilde{S})$, $\widetilde{x} \in \widetilde{T}(S)$, is a surjective $\mathbb{C}$-linear isometry with respect to the infinitesimal Kobayashi metrics at $\widetilde{x}$ and $\widetilde{F}(\widetilde{x})$. Corollary 2.10 implies that for each $\widetilde{x}$ there is a bijection $\alpha_{\widetilde{x}}$ from J' to $\widetilde{J}'$ such that

$$(2.11)\qquad (\widetilde{F}'(\widetilde{x})v)_{\alpha_{\widetilde{x}}(j)} = \widetilde{F}'(\widetilde{x})(v_j) \quad \text{for all } v \in B(S) \text{ and } j \in J'.$$

Indeed (2.11) follows immediately from Corollary 2 and (B5) if $\widetilde{x}$ and $\widetilde{F}(\widetilde{x})$ equal zero. The general case is reduced to that special case by pre- and post-composing $\widetilde{F}$ with suitable geometric isomorphisms.

Equation (2.11) and the continuous dependence of $\widetilde{F}'(\widetilde{x})$ on $\widetilde{x}$ imply that the set of $\widetilde{x}$ with $\alpha_{\widetilde{x}}(j) = \widetilde{j}$ is both open and closed in $\widetilde{T}(S)$ for each fixed j in J' and $\widetilde{j}$ in $\widetilde{J}'$. Therefore $\alpha_{\widetilde{x}}$ does not depend on $\widetilde{x}$ in $\widetilde{T}(S)$, and we can replace (2.11) by

$$(2.12)\qquad (\widetilde{F}'(\widetilde{x})v)_{\alpha(j)} = \widetilde{F}'(\widetilde{x})(v_j) \quad \text{for all } \widetilde{x} \in \widetilde{T}(S), v \in B(S), \text{ and } j \in J'$$

with $\alpha\colon J' \to \widetilde{J}'$ a fixed bijection. This equation in turn implies

$$(2.13)\qquad \widetilde{F}(\widetilde{x})_{\alpha(j)} = \widetilde{F}(\widetilde{x}_j) \qquad \text{for all } \widetilde{x} \in \widetilde{T}(S) \text{ and } j \in J'.$$

To derive (2.13) from (2.12) we consider first the case when $\widetilde{x} = \widetilde{x}_j$ and belongs to $\widetilde{T}(S_j) \subset \widetilde{T}(S)$. If $\gamma\colon [0,1] \to \widetilde{T}(S_j)$ is a smooth path joining 0 to $\widetilde{x}$ in $\widetilde{T}(S_j)$, then $\gamma'(t) = \gamma'(t)_j$ for all t in $[0,1]$, so

$$\begin{aligned}\widetilde{F}(\widetilde{x}) &= \int_0^1 \widetilde{F}'(\gamma(t))\gamma'(t)\,dt = \int_0^1 \widetilde{F}'(\gamma(t))(\gamma'(t)_j)\,dt \\ &= \int_0^1 (\widetilde{F}'(\gamma(t))\gamma'(t))_{\alpha(j)}\,dt = \widetilde{F}(\widetilde{x})_{\alpha(j)}\end{aligned}$$

as required.

Now for each j the level sets of the map $x \mapsto x_j$ from $T(S)$ to $T(S_j)$ are connected closed submanifolds of $T(S)$ (see [**10**]). Therefore, by (B1), (B2), and (B3), for any $\widetilde{x}$ in $\widetilde{T}(S)$ and any j in J' there is a smooth path $\gamma\colon [0,1] \to \widetilde{T}(S)$ such that $\gamma(0) = \widetilde{x}_j$, $\gamma(1) = \widetilde{x}$, and $\gamma(t)_j = \widetilde{x}_j$ for all t in $[0,1]$. It follows that $\gamma'(t)_j = 0$ for all t in $[0,1]$ and

$$\begin{aligned}\widetilde{F}(\widetilde{x})_{\alpha(j)} - \widetilde{F}(\widetilde{x}_j) &= (\widetilde{F}(\widetilde{x}) - \widetilde{F}(\widetilde{x}_j))_{\alpha(j)} \\ &= \int_0^1 (\widetilde{F}'(\gamma(t))\gamma'(t))_{\alpha(j)}\,dt = \int_0^1 \widetilde{F}'(\gamma(t))(\gamma'(t)_j)\,dt = 0,\end{aligned}$$

which is equation (2.13).

By (B1), (B2), and (B3), equation (2.13) implies the existence of holomorphic maps F_j from $T(S_j)$ to $T(\widetilde{S}_{\alpha(j)})$ satisfying equation (2.1). These maps are biholomorphic because the same reasoning applied to the map F^{-1} from $T(\widetilde{S})$ to $T(S)$ shows that $F^{-1}(y)_j = F_j^{-1}(y_{\alpha(j)})$ for all y in $T(\widetilde{S})$ and j in J'. □

3. Isomorphisms of Teichmüller spaces of closed sets.

3.1. The spaces $T(E)$. Let E be a closed subset of the Riemann sphere $\widehat{\mathbb{C}}$ that contains the points 0, 1, and ∞, and let S be its complement. Each connected component S_j of S is a hyperbolic Riemann surface, and S is the disjoint union of these surfaces, so we can consider the Teichmüller space $T(S)$ and the Banach spaces $A(S)$, $B(S)$, and $B_0(S)$ defined in §2.

In his dissertation [**16**], Lieb defines the Teichmüller space $T(E)$ of the closed set E, shows that it is intimately related to the set of holomorphic motions of E (see also [**17**]), and proves that it is a complex manifold, naturally isomorphic to the product of $T(S)$ and the open unit ball $M(E)$ in $L^\infty(E)$ (see also [**10**]). We take 0 as the basepoint of $M(E)$. The basepoint of $T(E)$ is the point that corresponds to $(0,0)$ in $T(S) \times M(E)$.

Our next main result is

THEOREM 3.1. *Let E and $\widetilde{E}$ be closed subsets of $\widehat{\mathbb{C}}$, each containing 0, 1, and ∞, and let S and $\widetilde{S}$ be their respective complements. Suppose F is a biholomorphic map of $T(E) = T(S) \times M(E)$ onto $T(\widetilde{E}) = T(\widetilde{S}) \times M(\widetilde{E})$. There are biholomorphic maps F_1 of $T(S)$ onto $T(\widetilde{S})$ and F_2 of $M(E)$ onto $M(\widetilde{E})$ such that*

$$F(x,\mu) = (F_1(x), F_2(\mu)) \quad \textit{for all } x \in T(S) \textit{ and } \mu \in M(E). \tag{3.1}$$

COROLLARY 3.2. *The complex manifolds $T(E)$ and $T(\widetilde{E})$ are isomorphic if and only if the following two conditions are both satisfied.*
(I-1) Either E and $\widetilde{E}$ both have area zero or they both have positive area.
(I-2) There is a quasiconformal map of S onto $\widetilde{S}$.

REMARK 3.3. Here, as in §2.2, a quasiconformal map of S onto $\widetilde{S}$ is a homeomorphism whose restriction to each S_j is K-quasiconformal, with K independent of j.

3.2. Proof of Corollary 3.2. By Theorem 3.1, $T(E)$ and $T(\widetilde{E})$ are isomorphic if and only if $M(E)$ is isomorphic to $M(\widetilde{E})$ and $T(S)$ is isomorphic to $T(\widetilde{S})$. Since $M(E)$ is a single point when E has area zero and is infinite dimensional when E has positive area, (I-1) is a necessary condition for $M(E)$ and $M(\widetilde{E})$ to be isomorphic. It is also sufficient because $L^\infty(E)$ is isometrically isomorphic to $L^\infty([0,1])$ when E has positive area. (See Theorem 2.3 and the final paragraph of §1 in [**13**].)

As we explained in §2.2, if f is a quasiconformal map from S to $\widetilde{S}$, then the geometric isomorphism $\rho_f \colon T(S) \to T(\widetilde{S})$ is biholomorphic. Therefore (I-2) is a sufficient condition for $T(S)$ and $T(\widetilde{S})$ to be isomorphic.

To see that (I-2) is also necessary, consider a biholomorphic map F of $T(S)$ onto $T(\widetilde{S})$. If $T(S)$ (and therefore $T(\widetilde{S})$) has finite dimension, then E and $\widetilde{E}$ are finite sets of equal cardinality, S and $\widetilde{S}$ are connected, and (I-2) obviously holds. If $T(S)$ and $T(\widetilde{S})$ are infinite dimensional, then E and $\widetilde{E}$ are infinite sets and each component of S or $\widetilde{S}$ is a non-exceptional Riemann surface of finite genus. By Theorem 2.2, $F = \rho_f$ for some quasiconformal map f of S onto $\widetilde{S}$. □

REMARK 3.4. Lieb gives an alternative geometric description of $T(E)$ as the set of E-equivalence classes of quasiconformal maps $f\colon \widehat{\mathbb{C}} \to \widehat{\mathbb{C}}$ that fix the points 0, 1, and ∞. Two such maps f and g are E-equivalent if $f^{-1} \circ g$ is isotopic to the identity on $\widehat{\mathbb{C}}$ by an isotopy that leaves E pointwise fixed (see [**16**] or [**10**]).

This description leads to a natural class of "geometric isomorphisms" from $T(E)$ to $T(\widetilde{E})$. Let f be any quasiconformal map of $\widehat{\mathbb{C}}$ onto itself such that $f(E)$ equals $\widetilde{E}$. The geometric isomorphism $\rho_f\colon T(E) \to T(\widetilde{E})$ induced by f maps the E-equivalence class of g to the $\widetilde{E}$-equivalence class of the map $A_g \circ g \circ f^{-1}$, where A_g is the unique Möbius transformation such that $A_g \circ g \circ f^{-1}$ fixes the points 0, 1, and ∞.

Each such geometric isomorphism ρ_f is biholomorphic (see [**10**]), but Corollary 3.2 implies that not all isomorphisms between spaces $T(E)$ and $T(\widetilde{E})$ are geometric. For example, let E and $\widetilde{E}$ be closed sets in $\widehat{\mathbb{C}}$ that contain 0, 1, and ∞ and have zero area. Suppose both E and $\widetilde{E}$ and their complements S and $\widetilde{S}$ are connected. Then S and $\widetilde{S}$ are both conformally equivalent to the open unit disk, so $T(E)$ and $T(\widetilde{E})$ are isomorphic to each other. (In fact they are both isomorphic to the universal Teichmüller space.) If E and $\widetilde{E}$ are not homeomorphic, there is certainly no geometric isomorphism of $T(E)$ onto $T(\widetilde{E})$.

3.3. Proof of Theorem 3.1. Let $T(E) = T(S) \times M(E)$, where S is the complement of E in $\widehat{\mathbb{C}}$. The tangent spaces to $T(S)$ and $M(S)$ at their basepoints are $B(S)$ and $L^\infty(E)$ respectively, so the tangent space to $T(E)$ at its basepoint is naturally isomorphic to the Banach space $B(S) \oplus L^\infty(E)$. We give that space its sup norm $\|(v, \mu)\| = \max\{\|v\|, \|\mu\|_\infty\}$. This norm coincides with the infinitesimal Kobayashi metric on $T(E)$ at its basepoint $(0, 0)$ (see [**10**]).

Let $\widetilde{S}$ be the complement of the closed set $\widetilde{E}$ in $\widehat{\mathbb{C}}$, and let 0, 1, and ∞ belong to $\widetilde{E}$. Give the Banach space $B(\widetilde{S}) \oplus L^\infty(\widetilde{E})$ its sup norm. We shall prove in §4 that every surjective $\mathbb{C}$-linear isometry L from $B(S) \oplus L^\infty(E)$ to $B(\widetilde{S}) \oplus L^\infty(\widetilde{E})$ has the form

$$L(v, \mu) = (L_1(v), L_2(\mu)),$$

where $L_1 \colon B(S) \to B(\widetilde{S})$ and $L_2 \colon L^\infty(E) \to L^\infty(\widetilde{E})$ are surjective $\mathbb{C}$-linear isometries. This fact allows us to prove Theorem 3.1 by essentially the same method that we used to prove Theorem 2.1. We outline the steps, leaving easy details to the reader.

Using Bers embeddings, we replace $T(S)$ by $\widetilde{T}(S)$ and $T(\widetilde{S})$ by $\widetilde{T}(\widetilde{S})$. Thus we have a biholomorphic map $\widetilde{F} \colon \widetilde{T}(S) \times M(E) \to \widetilde{T}(\widetilde{S}) \times M(\widetilde{E})$. Each Fréchet derivative map $\widetilde{F}'(\widetilde{x}, \mu)$ is a surjective $\mathbb{C}$-linear isometry with respect to the appropriate infinitesimal Kobayashi metrics. Theorem 4.1 in §4 implies that for each $(\widetilde{x}, \mu)$ there are $\mathbb{C}$-linear maps $L_1(\widetilde{x}, \mu)$ from $B(S)$ to $B(\widetilde{S})$ and $L_2(\widetilde{x}, \mu)$ from $L^\infty(E)$ to $L^\infty(\widetilde{E})$ such that

$$\widetilde{F}'(\widetilde{x}, \mu)(v, \nu) = (L_1(\widetilde{x}, \mu)(v), L_2(\widetilde{x}, \mu)(\nu)) \tag{3.2}$$

for all v in $B(S)$ and ν in $L^\infty(E)$. Indeed (3.2) follows immediately from Theorem 4.1 when $(\widetilde{x}, \mu)$ and $\widetilde{F}(\widetilde{x}, \mu)$ equal $(0, 0)$, and the general case follows by pre- and post-composition with geometric isomorphisms (see [**10**]).

Now define holomorphic maps $\widetilde{F}_1$ and $\widetilde{F}_2$ by the formula

$$\widetilde{F}(\widetilde{x}, \mu) = (\widetilde{F}_1(\widetilde{x}, \mu), \widetilde{F}_2(\widetilde{x}, \mu)).$$

Since there is a smooth path joining any two points in either $\widetilde{T}(S)$ or $M(E)$, (3.2) implies easily that $\widetilde{F}_1$ depends only on $\widetilde{x}$ and $\widetilde{F}_2$ depends only on μ. That implies the formula (3.1). The maps F_1 and F_2 in that formula are biholomorphic because $F^{-1}(y, \nu) = (F_1^{-1}(y), F_2^{-1}(\nu))$ for all (y, ν) in $T(\widetilde{S}) \times M(\widetilde{E})$. □

4. The isometries of $B(S) \oplus L^\infty(E)$

4.1. The isometry theorem. Our proof of Theorem 3.1 used the following theorem, which we shall prove in this section.

THEOREM 4.1. *Let E and $\widetilde{E}$ be closed subsets of $\widehat{\mathbb{C}}$, each containing* 0, 1, *and* ∞, *and let S and $\widetilde{S}$ be their respective complements. Every surjective $\mathbb{C}$-linear isometry L from $B(S) \oplus L^\infty(E)$ to $B(\widetilde{S}) \oplus L^\infty(\widetilde{E})$ has the form*

$$L(v, \mu) = (L_1(v), L_2(\mu)),$$

where $L_1 \colon B(S) \to B(\widetilde{S})$ and $L_2 \colon L^\infty(E) \to L^\infty(\widetilde{E})$ are surjective $\mathbb{C}$-linear isometries.

4.2. The case $E = \widehat{\mathbb{C}}$. Let $L \colon L^\infty(\mathbb{C}) \to B(\widetilde{S}) \oplus L^\infty(\widetilde{E})$ be an invertible $\mathbb{C}$-linear isometry. We must prove that $\widetilde{E} = \widehat{\mathbb{C}}$. For this purpose we examine the Fréchet differentiability of the norm function at points on the unit sphere.

It is well known that the norm function of $L^\infty(\mathbb{C})$ is nowhere Fréchet differentiable on the unit sphere. Indeed, Fréchet differentiability at the point μ on the unit sphere would imply that for any $\epsilon > 0$ there is a number $\delta > 0$ such that if f and g are functions on the unit sphere of $L^1(\mathbb{C})$ satisfying

$$(4.1) \qquad |1 - \iint_{\mathbb{C}} f(z)\mu(z)\,dx\,dy| < \delta \text{ and } |1 - \iint_{\mathbb{C}} g(z)\mu(z)\,dx\,dy| < \delta,$$

then $\|f - g\|_1 < \epsilon$. (See the discussion of Šmulyan's test on pages 2–5 of [**2**].) But for any $\delta > 0$ it is easy to find functions f and g on the unit sphere of $L^1(\mathbb{C})$ that have disjoint supports (so that $\|f - g\|_1 = 2$) and satisfy the condition (4.1).

On the other hand, if $B(\widetilde{S})$ has positive dimension it contains a Reich-Strebel functional v of norm one. (These functionals were defined at the beginning of §2.5.) By Theorem 2 of [**7**], the norm function of $B(\widetilde{S})$ is Fréchet differentiable at v, so the norm function of $B(\widetilde{S}) \oplus L^\infty(\widetilde{E})$ is Fréchet differentiable at $(v, 0)$.

Now, if $L \colon L^\infty(\mathbb{C}) \to B(\widetilde{S}) \oplus L^\infty(\widetilde{E})$ is a surjective $\mathbb{C}$-linear isometry, the norm function of $B(\widetilde{S}) \oplus L^\infty(\widetilde{E})$ must be nowhere Fréchet differentiable on the unit sphere, so $B(\widetilde{S})$ must have dimension zero. The only possibilities are $\widetilde{E} = \widehat{\mathbb{C}}$ (so that $\widetilde{S}$ is empty) or $\widetilde{E} = \{0, 1, \infty\}$ (so $\widetilde{S} = \mathbb{C} \setminus \{0, 1\}$). The second possibility is excluded by the fact that $L^\infty(\mathbb{C})$ is infinite dimensional, so $\widetilde{E} = \widehat{\mathbb{C}}$. □

4.3. The case where E has area zero. Let $L \colon B(S) \to B(\widetilde{S}) \oplus L^\infty(\widetilde{E})$ be a surjective $\mathbb{C}$-linear isometry. We must show that $\widetilde{E}$ has area zero. This is clear if $B(S)$ (and therefore $B(\widetilde{S}) \oplus L^\infty(\widetilde{E})$) is finite dimensional, so we shall assume $B(S)$ is infinite dimensional. In this case each subspace $B(S_j)$ (see §2.7) is infinite dimensional.

Choose v in $B(S)$ so that v_j is a Reich-Strebel functional of norm one for each j. Each v_j is an extreme point of the closed unit ball of $B(S_j)$ (see Lemmas 7.1 and 7.3 of [**5**]). Fix j, set $v'_j = v - v_j$, and observe that $v_j + cv'_j$ is an extreme point of the closed unit ball of $B(S)$ for every complex number c on the unit circle.

Set $L(v_j) = (\widetilde{u}, \mu)$ and $L(v'_j) = (\widetilde{v}, \nu)$. Suppose $\widetilde{E}$ has positive area. Then $\mu + c\nu$ is an extreme point of the closed unit ball of $L^\infty(\widetilde{E})$ for every c on the unit circle. It follows that the product $\mu\nu$ equals zero almost everywhere, so either μ is zero almost everywhere or else there is a set $Q \subset \widetilde{E}$ such that Q has positive area and μ has absolute value one almost everywhere on Q. In the latter case the norm function on $B(\widetilde{S}) \oplus L^\infty(\widetilde{E})$ is not Fréchet differentiable at $(\widetilde{u}, \mu) = L(v_j)$. That is impossible because the norm on $B(S)$ is Fréchet differentiable at the Reich-Strebel functional v_j. We conclude that μ is zero almost everywhere and that $L(v_j) = (\widetilde{u}, 0)$. We derived this formula under the assumption that $\widetilde{E}$ has positive area; obviously it also holds when the area of $\widetilde{E}$ is zero.

Now define $L_1\colon B(S) \to B(\widetilde{S})$ and $L_2\colon B(S) \to L^\infty(\widetilde{E})$ by the formula

$$L(v) = (L_1(v), L_2(v))$$

and consider the map $R(v) = L^{-1}(L_1(v), 0)$ of $B(S)$ to itself. We have proved that $R(v_j) = v_j$ for every Reich-Strebel functional v_j. Since

$$\|R(v)\| = \|L_1(v)\| \leq \|v\|$$

for all v in $B(S)$, R has norm one and Lemma 2.5 implies that R is the identity map. Therefore $L(v) = (L_1(v), 0)$ for all v in $B(S)$. Since L is surjective $L^\infty(\widetilde{E}) = \{0\}$ and $\widetilde{E}$ has zero area. $\square$

4.4. The general case. The arguments in §§4.2 and 4.3 can be applied equally well to L^{-1}, so it remains only to consider the case when both E and $\widetilde{E}$ are proper subsets of $\widehat{\mathbb{C}}$ with positive area. In this case both $B(S)$ and $B(\widetilde{S})$ have infinite dimension.

As in §4.3, choose v in $B(S)$ so that each v_j is a Reich-Strebel functional of norm one, and set $v'_j = v - v_j$ for each j. Let λ be the characteristic function of E. Then $(v_j, 0) + c(v'_j, \lambda)$ is an extreme point of the unit ball of $B(S) \oplus L^\infty(E)$ for each c in the unit circle.

Set $L(v_j, 0) = (\widetilde{u}, \mu)$ and $L(v'_j, \lambda) = (\widetilde{v}, \nu)$. Then $\mu + c\nu$ is an extreme point of the closed unit ball of $L^\infty(\widetilde{E})$ for every c on the unit circle. We conclude as in §4.3 that either μ is zero almost everywhere or else there is a set $Q \subset \widetilde{E}$ such that Q has positive area and μ has absolute value one almost everywhere on Q. Again the latter case is excluded because the norm of $B(S) \oplus L^\infty(E)$ is Fréchet differentiable at $(v_j, 0)$. Therefore $L(v_j, 0) = (\widetilde{u}, 0)$.

For the moment define $L_1\colon B(S) \to B(\widetilde{S})$ and $L_2\colon B(S) \to L^\infty(\widetilde{E})$ by the formula

$$L(v, 0) = (L_1(v), L_2(v)).$$

Let P_1 be the projection $(v, \mu) \mapsto v$ of $B(S) \oplus L^\infty(E)$ onto $B(S)$ and let R be the map $v \mapsto P_1(L^{-1}(L_1(v), 0))$. Once again $R(v_j) = v_j$ for every Reich-Strebel functional v_j, so Lemma 2.5 says that R is the identity map.

If we define $\Lambda_1\colon B(\widetilde{S}) \to B(S)$ and $\Lambda_2\colon B(\widetilde{S}) \to L^\infty(E)$ by the formula

$$L^{-1}(\widetilde{v}, 0) = (\Lambda_1(\widetilde{v}), \Lambda_2(\widetilde{v})),$$

we have $\Lambda_1(L_1(v)) = R(v) = v$ for all v in $B(S)$. Interchanging the roles of L and L^{-1} in the reasoning above, we obtain $L_1(\Lambda_1(\widetilde{v})) = \widetilde{v}$ for all $\widetilde{v}$ in $B(\widetilde{S})$. Therefore L_1 is invertible and Λ_1 is its inverse. Since L_1 and Λ_1 have norm at most one, both are surjective isometries.

Now choose any μ in $L^\infty(E)$ with $\|\mu\|_\infty < 1$ and let $L(0, \mu) = (\widetilde{u}, \widetilde{\nu})$. Choose any extreme point v in the closed unit ball of $B(S)$. Then

$$1 = \|(v, \pm\mu)\| = \|L(v, \pm\mu)\| = \|(L_1(v) \pm \widetilde{u}, L_2(v) \pm \widetilde{\nu})\| \geq \|L_1(v) \pm \widetilde{u}\|.$$

Since L_1 is a surjective isometry, $L_1(v)$ is an extreme point in the closed unit ball of $B(\widetilde{S})$, so $\widetilde{u} = 0$.

We have proved that $L(0, \mu) = (0, \widehat{L}(\mu))$ for some $\mathbb{C}$-linear isometry $\widehat{L}$ mapping $L^\infty(E)$ into $L^\infty(\widetilde{E})$. Since this identity holds for any μ in the open unit ball it obviously holds everywhere in $L^\infty(E)$. The same argument applied to L^{-1} shows that $L^{-1}(0,\widetilde{\mu}) = (0,\widehat{\Lambda}(\widetilde{\mu}))$ for some isometry $\widehat{\Lambda}$ of $L^\infty(\widetilde{E})$ into $L^\infty(E)$. Obviously $\widehat{L}$ and $\widehat{\Lambda}$ are inverse maps, so both are surjective.

It remains to prove that $L(v,0) = (L_1(v),0)$ (equivalently, $L_2(v) = 0$) for all v in $B(S)$. We follow the method of the preceding two paragraphs. Assume that v has norm less than one and that μ is an extreme point in the closed unit ball of $L^\infty(E)$. Then $1 = \|L(\pm v,\mu)\| \geq \|\pm L_2(v) + \widehat{L}(\mu)\|$. Since $\widehat{L}(\mu)$ is an extreme point of the closed unit ball of $L^\infty(\widetilde{E})$, $L_2(v) = 0$ as required. □

5. A problem about holomorphic motions

We continue to assume that E is a closed set in $\widehat{\mathbb{C}}$ and that 0, 1, and ∞ belong to E. In his dissertation [**16**], Lieb shows that the holomorphic motions of E over the open unit disk Δ are in natural bijective correspondence with the holomorphic maps $f\colon \Delta \to T(E)$ such that $f(0)$ is the basepoint of $T(E)$. This correspondence can be described as follows.

Let $\phi\colon \Delta \times E \to \Delta \times \widehat{\mathbb{C}}$ be a holomorphic motion of E over Δ (see [**3**] or [**4**]). By Slodkowski's extension theorem (see [**3**]) there is a holomorphic motion $\widetilde{\phi}\colon \Delta \times \widehat{\mathbb{C}} \to \Delta \times \widehat{\mathbb{C}}$ that extends ϕ. Let π_2 be the projection of $\Delta \times \widehat{\mathbb{C}}$ onto $\widehat{\mathbb{C}}$. By the lambda lemma (see [**3**]) for each t in Δ the map $z \mapsto \widetilde{\phi}_t(z) = \pi_2(\widetilde{\phi}(t,z))$ is a quasiconformal map of $\widehat{\mathbb{C}}$ onto itself. Let $\widetilde{f}\colon \Delta \to M(\mathbb{C})$ be the map that sends t in Δ to the Beltrami coefficient of $\widetilde{\phi}_t$. By Theorem 2 of [**1**], $\widetilde{f}$ is a holomorphic map. According to Lieb [**16**] (see also [**10**] and [**17**]) there is a natural holomorphic split submersion Φ_E from $M(\mathbb{C})$ to $T(E)$ such that the holomorphic map $f = \Phi_E \circ \widetilde{f}$ from Δ to $T(E)$ depends only on the motion ϕ (not on its extension), and conversely f determines ϕ. Since $\Phi_E(0)$ is the basepoint of $T(E)$ and $\widetilde{\phi}_0$ is the identity map on $\widehat{\mathbb{C}}$, f maps 0 to the basepoint of $T(E)$.

Now suppose $F\colon T(E) \to T(E)$ is a biholomorphic map that fixes the basepoint. If $f\colon \Delta \to T(E)$ is the holomorphic map determined by a holomorphic motion ϕ, then $F \circ f$ is also a holomorphic map from Δ to $T(E)$ mapping 0 to the basepoint, so it determines a new holomorphic motion $F_*(\phi)$. In this way F induces a bijective map F_* on the set of holomorphic motions of E over Δ. The map F_* can be difficult to compute, as the following example shows.

EXAMPLE 5.1. Let E be the extended real axis. Its complement S is the (disjoint) union of the upper and lower half planes, and $T(E)$ is isomorphic to $T(S)$, which is the product of the Teichmüller spaces of the upper and lower half planes. The bijective conformal map of S onto itself given by the identity map in the lower half plane and by $z \mapsto z+1$ in the upper half plane induces a basepoint preserving biholomorphic map F from $T(E)$ to itself. (Define the biholomorphic map $\widetilde{F}$ of $M(\mathbb{C})$ onto itself by $\widetilde{F}(\mu)(z) = \mu(z-1)$ if $y > 0$ and $\widetilde{F}(\mu)(z) = \mu(z)$ otherwise. Then $F \circ \Phi_E = \Phi_E \circ \widetilde{F}$.)

We do not know an explicit formula for $F_*(\phi)$ even when ϕ is the relatively simple motion corresponding to the map $f(t) = \Phi_E(t\mu)$ from Δ to $T(E)$, where $\mu(z) = z/\bar{z}$ for $z \neq 0$. The motion ϕ itself is given explicitly by $\phi(t,x) = (t,x)$ when $t \in \Delta$ and $x = 0$, 1, or ∞, and

$$\phi(t,x) = (t, |x|^{\frac{2t}{1-t}}x) = \left(t, x\exp\left(\frac{2t}{1-t}\log|x|\right)\right), \; t \in \Delta,\, x \in \mathbb{R} \setminus \{0,1\}.$$

REMARK 5.2. Let E be a closed subset of $\widehat{\mathbb{C}}$ that contains 0, 1, and ∞, and let V be any simply connected complex Banach manifold with a basepoint. In [**17**], Mitra shows that the holomorphic motions of E over V are in bijective correspondence with the basepoint preserving holomorphic maps from V to $T(E)$. This is a substantial generalization of Lieb's theorem.

References

[1] L. Bers and H. Royden, *Holomorphic families of injections*, Acta Math. **157** (1986), 259–286.

[2] R. Deville, G. Godefroy, and V. Zizler, *Smoothness and renormings in Banach spaces*, Pitman Monographs and Surveys in Pure and Applied Mathematics 64, Longman Scientific & Technical, England, 1993.

[3] A. Douady, *Prolongement de mouvements holomorphes [d'aprés Slodkowski et autres]*, Séminaire N. Bourbaki, Vol. 1993/1994, Astérisque No. 227 (1995) Exp. No. 775, pp. 3, 7–20.

[4] C. J. Earle, *Some maximal holomorphic motions*, in *Lipa's legacy*, Contemp. Math. **211** (1997), 183–192.

[5] C. J. Earle and F. P. Gardiner, *Geometric isomorphisms between infinite dimensional Teichmüller spaces*, Trans. Amer. Math. Soc. **348** (1996), 1163–1190.

[6] C. J. Earle, F. P. Gardiner, and N. Lakic, *Teichmüller spaces with asymptotic conformal equivalence*, I.H.E.S. preprint, June, 1995.

[7] ———, *Vector fields for holomorphic motions of closed sets*, in *Lipa's legacy*, Contemp. Math. **211** (1997), 193–225.

[8] C. J. Earle and I. Kra, *On isometries between Teichmüller spaces*, Duke Math. J. **41** (1974), 583–591.

[9] C. J. Earle and Li Zhong, *Isometrically embedded polydisks in infinite dimensional Teichmüller spaces*, J. Geom. Anal., to appear.

[10] C. J. Earle and S. Mitra, *Analytic dependence of conformal invariants on parameters*, in preparation.

[11] F. P. Gardiner and D. P. Sullivan, *Symmetric structures on a closed curve*, Amer. J. Math. **114** (1992), 683–736.

[12] J. A. Gentilesco, *Automorphisms of the deformation space of a Kleinian group*, Trans. Amer. Math. Soc. **248** (1979), 207–220.

[13] H. E. Lacey and S. J. Bernau, *Characterizations and classifications of some classical Banach spaces*, Adv. Math. **12** (1974), 367–401.

[14] N. Lakic, *An isometry theorem for quadratic differentials on Riemann surfaces of finite genus*, Trans. Amer. Math. Soc. **349** (1997), 2951–2967.

[15] ———,*Strebel points*, in *Lipa's legacy*, Contemp. Math. **211** (1997), 417–431.

[16] G. S. Lieb, *Holomorphic motions and Teichmüller space*, Ph.D. Dissertation, Cornell University, 1990.

[17] S. Mitra, *Teichmüller spaces and holomorphic motions*, preprint.

(Earle) DEPARTMENT OF MATHEMATICS, CORNELL UNIVERSITY, ITHACA, NEW YORK 14853
E-mail address: `cliff@math.cornell.edu`

(Gardiner) DEPARTMENT OF MATHEMATICS, BROOKLYN COLLEGE, CUNY, BROOKLYN, NEW YORK 11210
E-mail address: `fgardine@email.gc.cuny.edu`

(Lakic) DEPARTMENT OF MATHEMATICS, CORNELL UNIVERSITY, ITHACA, NEW YORK 14853
Current address: Department of Mathematics and Computer Science, Herbert H Lehman College, CUNY, Bronx, New York 10468
E-mail address: `nlakic@alpha.lehman.cuny.edu`

Contemporary Mathematics
Volume **240**, 1999

Ramanujan partition identities

Hershel M. Farkas and Irwin Kra

1. Introduction and summary of some results

1.1. The *classical Ramanujan partition function P* defined by

$$\prod_{n=1}^{\infty} \frac{1}{1-x^n} = \sum_{n=0}^{\infty} P(n)x^n \tag{1}$$

has been the subject of much study since Ramanujan made many conjectures and proved some remarkable theorems concerning congruence properties of the numbers $P(n)$ appearing above. Ramanujan proved that[1]

$$P(5m+4) \equiv 0 \bmod 5, \quad P(7m+5) \equiv 0 \bmod 7 \quad \text{and } P(11m+6) \equiv 0 \bmod 11.$$

He also more or less conjectured (see [**21**] and [**12**, Ch. 7]) and partially proved the more general statements; one for each fixed $n \in \mathbb{Z}^+$.
(a) If $24m \equiv 1 \bmod 5^n$, then $P(m) \equiv 0 \bmod 5^n$.
(b) If $24m \equiv 1 \bmod 7^n$, then $P(m) \equiv 0 \bmod 7^{\left[\frac{n+2}{2}\right]}$.
(c) If $24m \equiv 1 \bmod 11^n$, then $P(m) \equiv 0 \bmod 11^n$.
We shall refer to these three results as the *(Ramanujan level n) congruences for the primes 5, 7 and 11*, respectively. Formula (1) defines, of course, the value at x in the unit disc of a nonvanishing holomorphic function $\mathbf{P}$.

1.2. The great interest in the function P is due in part to the combinatorial interpretation of $P(n)$ as the number of unrestricted partitions of $n \in \mathbb{Z}^+$. In principle, the following class of functions, the *partition functions* P_N, $N \in \mathbb{Z}$, defined by

$$\mathbf{P}^N(x) = \prod_{n=1}^{\infty} \frac{1}{(1-x^n)^N} = \sum_{n=0}^{\infty} P_N(n)x^n,$$

1991 MSC: 30F10, 11F12, 11F30, 14H05, 30F30, 30F35, 14H45.
Research by IK supported in part by NSF Grant DMS 9500557. Research by HMF supported in part by the Gabriella and Paul Rosenbaum Foundation and the Edmund Landau Center for Research in Mathematical Analysis sponsored by the Minerva Foundation Germany. Research of HMF and IK supported in part by a US-Israel BSF Grant 95-348.

[1]The December 1997 issue of *The American Mathematical Monthly* contains four articles in honor of Ramanujan on the 110th anniversary of his birth. One of these [**2**] in the *Notes* section gives what may be the shortest combinatorial proof of the first of these congruences.

which are clearly related to the classical partition function, are just as interesting and have been investigated in their own right (see, for example, [**1**]). Further, in order to prove the above congruences for P we need information about the congruence properties of series of the form $P_N(km+\alpha)$ for various fixed integers k, N and α. For positive N, the function P_N can be thought of as counting the number of partitions of the integers using N sets of colored numbers.

If N is negative, then we are in a perhaps more interesting situation. For $n \in \mathbb{Z}^+$, $P_N(n)$ represents the difference between the number of even and odd partitions of n where we are using positive integers of $-N$ distinct colors and no repetitions are permitted. In this sense a partition is called *even* provided it consists of an even number of parts and *odd* otherwise.

The case $N=-1$ was in fact considered by Euler; the case $N=-3$ by Jacobi. These are the only cases with a closed expression for the partition function. The classical partition function corresponds to $N=1$ and we always write P for P_1.

1.3. The Ramanujan congruences are part of a vast literature on combinatorial number theory. Exposition of various topics appear, for example, in [**12**]. Our work in this area proceeds from function theoretic considerations involving Riemann surfaces defined by the action on the upper half plane $\mathbb{H}^2$ of subgroups of the modular group, $\mathrm{PSL}(2,\mathbb{Z})$. The well known congruences for the primes $k=5$ and 7 are established in [**9**] and [**13**] as consequences of the theta constant identities

$$\sum_{l=0}^{4}\left(\frac{\theta'\begin{bmatrix}1\\1\end{bmatrix}(0,5(\tau+l))}{\theta'\begin{bmatrix}1\\1\end{bmatrix}(0,\frac{\tau+l}{5})}\right)^{\frac{1}{3}} = 25\left(\frac{\theta'\begin{bmatrix}1\\1\end{bmatrix}(0,5\tau)}{\theta'\begin{bmatrix}1\\1\end{bmatrix}(0,\tau)}\right)^{2},$$

and

$$\sum_{l=0}^{6}\left(\frac{\theta'\begin{bmatrix}1\\1\end{bmatrix}(0,7(\tau+l))}{\theta'\begin{bmatrix}1\\1\end{bmatrix}(0,\frac{\tau+l}{7})}\right)^{\frac{1}{3}} =$$

$$49\left(\frac{\theta'\begin{bmatrix}1\\1\end{bmatrix}(0,7\tau)}{\theta'\begin{bmatrix}1\\1\end{bmatrix}(0,\tau)}\right)^{\frac{4}{3}}\left(7\left(\frac{\theta'\begin{bmatrix}1\\1\end{bmatrix}(0,7\tau)}{\theta'\begin{bmatrix}1\\1\end{bmatrix}(0,\tau)}\right)^{\frac{4}{3}}+1\right).$$

In terms of the local coordinate $x=\exp 2\pi\imath\tau$ at P_∞ on the surface $\overline{\mathbb{H}^2/\Gamma_o(k)}$, the above theta identities become

$$\sum_{m=1}^{\infty} P(5m-1)x^m = 5\,x\,\frac{\prod_{n=1}^{\infty}(1-x^{5n})^5}{\prod_{n=1}^{\infty}(1-x^n)^6} \tag{2}$$

for $k=5$ and

$$\sum_{m=1}^{\infty} P(7m-2)x^m = 7\,x\,\frac{\prod_{n=1}^{\infty}(1-x^{7n})^3}{\prod_{n=1}^{\infty}(1-x^n)^4}\left[7\,x\,\frac{\prod_{n=1}^{\infty}(1-x^{7n})^4}{\prod_{n=1}^{\infty}(1-x^n)^4}+1\right] \tag{3}$$

for $k=7$, respectively. The last two equations appear as early as 1919 in the papers of Ramanujan, [**19**]. The level one congruences for the primes 5 and 7 are immediate consequences of these identities. The proof of the level one congruence for $k=11$ uses a different method which on the one hand is, in a sense to be explained, weaker than the method used above but on the other hand gives us the desired result. For this k, our proof in [**7**] proceeds along the following lines. We show that the $\Gamma_o(11)$-invariant function

$$\sum_{l=0}^{10}\left(\frac{\theta'\begin{bmatrix}1\\1\end{bmatrix}(0,\frac{\tau+l}{11})}{\theta'\begin{bmatrix}1\\1\end{bmatrix}(0,11(\tau+l))}\right)^{\frac{10}{3}}$$

which in terms of the local coordinate x at P_∞ is given by

$$11\prod_{n=1}^{\infty}(1-x^{11n})^{-10}\sum_{m=-4}^{\infty}P_{-10}(11m+50)x^m$$

is constant (the apparent pole of order 4 is not there). From this we conclude that

$$11^4P_{-10}\left(\frac{m-4}{11}\right)=P_{-10}(11m+6),\ m\in\mathbb{Z}^+\cup\{0\}.$$

One easily shows that if $N\equiv 10 \mod 11$ and $n\equiv 6 \mod 11$, then $P_{-N}(n)\equiv 0 \mod 11$. Taking $N=-1$ gives the congruence for $P(11m+6)$.

The purpose of this paper is to give a more or less unified treatment, based on this last described method, which gives us the three level one congruences and in fact allows us to obtain the level two congruences for the primes 5 and 7. This paper reports essentially on only known results; the reader may consult [**12**] for historical references and an alternate derivation of these results. Furthermore the reader can find in [**1**], [**10**], [**11**], and in [**17**], [**18**] techniques which are very close to the ideas we employ here. The reason for this note is that in our opinion, our methods yield slightly more general results than other known methods and our approach explains the congruences in a very simple unified manner from a function theoretic point of view. The congruences are essentially consequences of the fact that certain procedures produce constant functions on compact Riemann surfaces. A similar unified proof, based on function and number theoretic arguments in [**14**] and [**15**], of the three level one Ramanujan congruences appeared in [**16**][2]. The methods and results of that paper should be contrasted with those of the current manuscript.

In the above papers where the ideas are similar to ours, the function theory tends to be hidden and mysterious. We believe that we expose it here in a manner that non-experts in number theory can appreciate. In this way we feel that we are opening this area to non experts. Furthermore, the principal actor in our investigations is the theta constant with rational characteristics as opposed to the η-function which appears in almost all the other presentations. It seems to us that since the theta function is a more fundamental object which appears in many other places, its occurrence in this theory should be highlighted rather than suppressed. However, our work on partition identities involves only the z-derivative of the theta function with integral odd characteristic evaluated at $z=0$ (that is, the classical η-function). In previous publications [**5**], [**3**], [**4**] we have also seen that certain identities when expressed in terms of theta constants tend to be simpler than when expressed in other forms. Finally, the theta function used here is the "one variable" function; however the theta function has a several variables generalization which

[2]We learned of this paper only after much of the work reported here was completed.

plays a central role in the theory of compact Riemann surfaces of genus greater than one . It is hoped that ultimately some of the ideas used here will be generalized to the several variable case.

2. Construction of meromorphic functions

In [**5**], we discussed a procedure for the construction of meromorphic functions and differentials on the Riemann surfaces $\mathbb{H}^2/\Gamma(k)$ and the calculation of their divisors. Here $k \in \mathbb{Z}^+$ is arbitrary. The technique, using theta functions, considered in that paper is useful for other subgroups of $\Gamma = \mathrm{PSL}(2,\mathbb{Z})$. In particular, it works nicely for the subgroups[3] $G(k)$, $\Gamma_o(k)$ and $\Gamma^o(k)$ as well as for the intersection of the last two listed groups which we denote by $\Gamma(k,k)$. Let us now fix for the rest of this paper, a positive prime k.

The idea for constructing interesting functions is to start with a fixed Riemann surface and a meromorphic function on the surface, average the function over a group (not always a group of automorphisms of the surface) and thus obtain a meromorphic function on the quotient of the original surface by the group. This is not a new idea, however the reason it works so well in our case is that we have fairly good control over the divisors of the functions in question. In fact, in many cases we are constructing in this way constant functions and we can evaluate the constant.

We begin with the meromorphic $\Gamma(k)$-invariant function

$$\frac{\theta'\begin{bmatrix}1\\1\end{bmatrix}(0,k\tau)}{\theta'\begin{bmatrix}1\\1\end{bmatrix}(0,\frac{\tau}{k})},$$

whose divisor on $\overline{\mathbb{H}^2/\Gamma(k)}$ is

$$\left(\frac{P_{\frac{1}{k}}P_{\frac{2}{k}}...P_{\frac{k-1}{2k}}}{A\left(P_{\frac{1}{k}}P_{\frac{2}{k}}...P_{\frac{k-1}{2k}}\right)}\right)^{\frac{k^2-1}{8}}.$$

In the above, the point P_x is the puncture on the Riemann surface $\mathbb{H}^2/\Gamma(k)$ represented by the point x on the (extended) real axis and A is the motion $\tau \mapsto \frac{-1}{\tau}$. Thus our function is holomorphic and nonzero on the surface $\mathbb{H}^2/\Gamma(k)$ and has a meromorphic extension to its natural compactification $\overline{\mathbb{H}^2/\Gamma(k)}$. The above function is not, in general, $\Gamma(k,k)$-invariant. However,

$$\left(\frac{\theta'\begin{bmatrix}1\\1\end{bmatrix}(0,k\tau)}{\theta'\begin{bmatrix}1\\1\end{bmatrix}(0,\frac{\tau}{k})}\right)^8$$

[3]These families of groups are defined in [**8**], where we also introduced the notation used here.

is and its divisor on $\overline{\mathbb{H}^2/\Gamma(k,k)}$ is $\left(\frac{P_\infty}{P_0}\right)^{k^2-1}$. Our final observation is that

$$\left(\frac{\theta'\begin{bmatrix}1\\1\end{bmatrix}(0,k\tau)}{\theta'\begin{bmatrix}1\\1\end{bmatrix}(0,\tau)}\right)^8$$

is $\Gamma_o(k)$-invariant and its divisor on $\overline{\mathbb{H}^2/\Gamma_o(k)}$ is $\left(\frac{P_\infty}{P_0}\right)^{k-1}$. We are now faced with the following question: what is the least positive integer N for which

$$\left(\frac{\theta'\begin{bmatrix}1\\1\end{bmatrix}(0,k\tau)}{\theta'\begin{bmatrix}1\\1\end{bmatrix}(0,\frac{\tau}{k})}\right)^{\frac{8N}{k^2-1}} \tag{4}$$

is a meromorphic function on $\mathbb{H}^2/\Gamma(k,k)$? Another way of phrasing this question is: what is the least $N \in \mathbb{Z}^+$ for which the divisor $\left(\frac{P_\infty}{P_0}\right)^N$ is principal on $\overline{\mathbb{H}^2/\Gamma(k,k)}$?

Our strategy is now as follows: having found a minimal N described above, that we call $\beta(\Gamma(k,k))$, then all N for which (4) defines a $\Gamma(k,k)$-invariant function are multiples of this minimal one. We have shown in [**7**] that $\beta(\Gamma(k,k))$ is a factor of the maximum of 1 and $\frac{k^2-1}{24}$. We conjecture that $\beta(\Gamma(k,k)) = \beta(k)$, this maximum. We have established the conjecture for $k \leq 17$. Hence we define for each $N \in \mathbb{Z}^+$ the function

$$F_{k,-N}(\tau) = \frac{1}{k}\sum_{l=0}^{k-1}\left(\frac{\theta'\begin{bmatrix}1\\1\end{bmatrix}(0,k(\tau+l))}{\theta'\begin{bmatrix}1\\1\end{bmatrix}(0,\frac{\tau+l}{k})}\right)^{-\frac{\alpha(k)N}{3}},$$

where

$$\alpha(k) = \begin{cases} 8 \text{ for } k=2 \\ 3 \text{ for } k=3 \\ 1 \text{ for } k \geq 5 \end{cases}.$$

We then easily see that $F_{k,-N}$ is a meromorphic function on $\overline{\mathbb{H}^2/\Gamma_o(k)}$ with a pole of order at most $[\frac{\beta(k)N}{k}]$ at P_∞ that is regular elsewhere. Notice that the above construction imposes severe restrictions on the function $F_{k,-N}$. Namely, if the genus of $\mathbb{H}^2/\Gamma_o(k)$ is positive, then any N with $\beta(k)N < 2k$ will give rise to a constant function automatically while if the genus of $\mathbb{H}^2/\Gamma_o(k)$ is zero any N with $\beta(k)N < k$ produces a constant.

We need to recall some standard facts (whose proofs can be found, for example in [**9**]). In terms of the local coordinate $x = \exp(2\pi\imath\tau)$ at P_∞ on $\overline{\mathbb{H}^2/\Gamma_o(k)}$, we have

$$\theta'\begin{bmatrix}1\\1\end{bmatrix}(0,\tau) = -2\pi x^{\frac{1}{8}}\prod_{n=1}^{\infty}(1-x^n)^3,$$

and for all $l \in \mathbb{Z}$,

$$\theta'\begin{bmatrix}1\\1\end{bmatrix}(0,\tau+l) = \left(\exp\left\{\frac{\pi\imath l}{4}\right\}\right)\theta'\begin{bmatrix}1\\1\end{bmatrix}(0,\tau).$$

It thus follows that

$$\left(\frac{\theta'\begin{bmatrix}1\\1\end{bmatrix}(0,\frac{\tau+l}{k})}{\theta'\begin{bmatrix}1\\1\end{bmatrix}(0,k(\tau+l))}\right)^{\frac{8N}{k^2-1}} = \frac{(\epsilon_l x^{\frac{1}{k}})^{-N}\prod_{n=1}^{\infty}(1-(\epsilon_l x^{\frac{1}{k}})^n)^{\frac{24N}{k^2-1}}}{\prod_{n=1}^{\infty}(1-x^{nk})^{\frac{24N}{k^2-1}}},$$

where $\epsilon_l = \exp\left(\frac{2\pi\imath l}{k}\right)$. Hence in terms of the above local coordinate

$$\sum_{l=0}^{k-1}\left(\frac{\theta'\begin{bmatrix}1\\1\end{bmatrix}(0,\frac{\tau+l}{k})}{\theta'\begin{bmatrix}1\\1\end{bmatrix}(0,k(\tau+l))}\right)^{\frac{8N}{k^2-1}} =$$

$$\sum_{n=0}^{\infty}P_{\frac{24N}{k^2-1}}(n)x^{kn}\left(\sum_{l=0}^{k-1}(\epsilon_l x^{\frac{1}{k}})^{-N}\prod_{n=1}^{\infty}(1-(\epsilon_l x^{\frac{1}{k}})^n)^{\frac{24N}{k^2-1}}\right)$$

$$= k\left(\sum_{n=0}^{\infty}P_{\frac{24N}{k^2-1}}(n)x^{kn}\right)\left(\sum_{m=-[\frac{N}{k}]}^{\infty}P_{\frac{24N}{1-k^2}}(mk+N)x^m\right). \tag{5}$$

All of our results concerning the functions $F_{k,-N}$ and when they are constant (in particular, all our proofs of the level one Ramanujan congruences) now flow from the above calculations.

THEOREM 1. *Let let k be a positive prime and let $N \in \mathbb{Z}^+$. Then $F_{k,-N}$ is a meromorphic function on $\overline{\mathbb{H}^2/\Gamma_o(k)}$ with a pole at P_∞ of order at most $\left[\frac{\beta(k)N}{k}\right]$.*

PROOF. Our definition of $\beta(k)$ guarantees that $\left(\frac{\theta'\begin{bmatrix}1\\1\end{bmatrix}(0,\frac{\tau}{k})}{\theta'\begin{bmatrix}1\\1\end{bmatrix}(0,k\tau)}\right)^{\frac{8N\beta(k)}{k^2-1}}$ is a meromorphic function on $\overline{\mathbb{H}^2/\Gamma(k,k)}$ with divisor $\left(\frac{P_0}{P_\infty}\right)^{\beta(k)N}$. We average this function over the group (viewed as a subgroup of $\mathrm{PSL}(2,\mathbb{Z}_k)$) generated by the $\tau \mapsto \tau+1$ which gives us a function on $\overline{\mathbb{H}^2/\Gamma_o(k)}$. The statement about the order of this function at P_∞ is clear as soon as one realizes that if x is a local coordinate at P_∞ on $\overline{\mathbb{H}^2/\Gamma(k,k)}$ then x^k is a local coordinate at P_∞ on $\overline{\mathbb{H}^2/\Gamma_o(k)}$. Obviously formula (5) tells us what $\mathrm{ord}_\infty F_{k,-N}$ is. □

In terms of the correct local coordinate x at P_∞ on $\overline{\mathbb{H}^2/\Gamma_o(k)}$, the function $F_{k,-N}(\tau)$ is given by (5). We have already remarked that the function $F_{k,-N}$ is constant for many choices of k and N. Since the only possible pole of this function is at P_∞, it will be constant if and only if it is regular at P_∞. The condition that the function be regular at P_∞ is that the negative Laurent coefficients in (5) vanish. This is our next

THEOREM 2. *Let k and N be as in the previous theorem. The following three conditions are equivalent:*
(a) *$F_{k,-N}$ is a constant function.*

(b) $P_{-\alpha(k)N}(mk+\beta(k)N)=0$ *for* $m=-\left[\frac{\beta(k)N}{k}\right]$, ..., -2, -1.
(c) *the function* $F_{k,-N}$ *is invariant under the map* $A_k:\tau\mapsto -\frac{1}{k\tau}$.
Under these conditions, the constant value of $F_{k,-N}(\tau)$ *is* $kP_{-\alpha(k)N}(\beta(k)N)$.

PROOF. Only that (c) implies (a) requires some comment. If $F_{k,-N}$ is invariant under the automorphism A_k of $\overline{\mathbb{H}^2/\Gamma_o(k)}$, then (since this function is holomorphic at P_0) it is holomorphic at P_∞; hence constant. □

REMARK 1. The list of $\left[\frac{\beta(k)N}{k}\right]$ arguments for the function $P_{-\alpha(k)N}$

$$-\left[\frac{\beta(k)N}{k}\right]k+\beta(k)N,\ -\left(\left[\frac{\beta(k)N}{k}\right]+1\right)k+\beta(k)N...,\ -k+\beta(k)N,$$

can also be written as

$$R\left(\frac{\beta(k)N}{k}\right),\ R\left(\frac{\beta(k)N}{k}\right)+k,\ ...,\ \beta(k)N-k,$$

where $R\left(\frac{n}{k}\right)$ is the remainder of dividing k into n. This list is to be considered as the empty set for integers N with $1\le\beta(k)N<k$.

REMARK 2. It is a corollary of the last theorem that

$$P_{-\alpha(k)N}\left(R\left(\frac{\beta(k)N}{k}\right)\right)=0\Longleftrightarrow \operatorname{ord}_\infty F_{k,-N}>-\left[\frac{(\beta(k)}{k}\right].$$

This observation provides a reformulation of Lehmer's question on the vanishing of coefficients of the Fourier series expansion of the discriminant function from the elliptic theory (see our discussion in [**9**]).

3. The level one congruences

THEOREM 3. *For k and N satisfying one (hence all three) of the conditions of the Theorem 2,*

$$\frac{\sum_{m=0}^\infty P_{-\alpha(k)N}(km+\beta(k)N)x^m}{\sum_{m=0}^\infty P_{-\alpha(k)N}(m)x^{km}}=P_{-\alpha(k)N}(\beta(k)N).$$

COROLLARY 1. *Let* $m\in\mathbb{Z}^+\cup\{0\}$. *For k and N as in the theorem*[4],

$$P_{-\alpha(k)N}(km+\beta(k)N)=P_{-\alpha(k)N}(\beta(k)N)\ P_{-\alpha(k)N}\left(\frac{m}{k}\right).$$

We can now prove the level one congruences for the primes 5, 7 and 11. We use (see, for example, [**12**, Ch. 4]) the known results that

$$\beta(\Gamma(5,5))=1,\quad \beta(\Gamma(7,7))=2,\quad \beta(\Gamma(11,11))=5.$$

We give function theoretic proofs, without relying on number theoretic arguments, of these three equalities in [**7**]. Of course it is trivial that $\beta(\Gamma(5,5))=1$ because $\overline{\mathbb{H}^2/\Gamma(5,5)}$ is conformally equivalent to the Riemann sphere. We enumerate the constant functions we need for our results. (A more complete list of constant functions is found in the appendix §5.4.)

[4]We use the convention that $P_N(n)=0$ if $n\notin\mathbb{Z}^+\cup\{0\}$.

(1) $F_{5,-4}(\tau)$ is constant and thus it follows that

$$\sum_{m=0}^{\infty} P_{-4}(5m+4)x^m = P_{-4}(4)\sum_{m=0}^{\infty} P_{-4}(m)x^{5m} = -5\sum_{m=0}^{\infty} P_{-4}(m)x^{5m};$$

so that, in particular, $P_{-4}(5m+4) \equiv 0 \mod 5$.
(2) $F_{7,-3}(\tau)$ is constant and thus

$$\sum_{m=0}^{\infty} P_{-3}(7m+6)x^m = P_{-3}(6)\sum_{m=0}^{\infty} P_{-3}(m)x^{7m} = -7\sum_{m=0}^{\infty} P_{-3}(m)x^{7m};$$

so $P_{-3}(7m+6) \equiv 0 \mod 7$.
(3) $F_{7,-6}(\tau)$ is constant since $P_{-6}(5) = 0$. We thus have

$$\sum_{m=0}^{\infty} P_{-6}(7m+12)x^m = P_{-6}(12)\sum_{m=0}^{\infty} P_{-6}(m)x^{7m} = 7^2\sum_{m=0}^{\infty} P_{-6}(m)x^{7m}$$

and $P_{-6}(7m+5) \equiv 0 \mod 7^2$.
(4) $F_{11,-10}(\tau)$ is constant since $P_{-10}(6) = P_{-10}(17) = P_{-10}(28) = P_{-10}(39) = 0$. Therefore

$$\sum_{m=0}^{\infty} P_{-10}(11m+50)x^m = P_{-10}(50)\sum_{m=0}^{\infty} P_{-10}(m)x^{11m} = 11^4\sum_{m=0}^{\infty} P_{-10}(m)x^{11m}$$

and $P_{-10}(11m+6) \equiv 0 \mod 11^4$.

The above congruences are immediate consequences of the theorem at the beginning of this section. The remarkable fact is that they allow us to conclude

THEOREM 4. *For all* $m \in \mathbb{Z}^+ \cup \{0\}$, *we have*

$$P(5m+4) \equiv 0 \mod 5,\ P(7m+5) \equiv 0 \mod 7 \text{ and } P(11m+6) \equiv 0 \mod 11.$$

PROOF. Clearly for each $N \in \mathbb{Z}$, $\mathbf{P} = \mathbf{P}^{-N+1}\mathbf{P}^N$; so that

$$\mathbf{P}(x) = \prod_{n=1}^{\infty} \frac{(1-x^n)^{N-1}}{(1-x^n)^N}.$$

The Frobenius automorphism tells us that for each positive prime k,

$$(1-x^n)^k \equiv 1 - x^{kn} \mod k.$$

Taking the Cauchy product of the appropriate power series leads to the identity

$$P(n) \equiv \sum_{j=0}^{n} P\left(\frac{j}{k}\right) P_{1-k}(n-j) \mod k,$$

for all primes k. Since $P(\frac{j}{k}) = 0$ unless j is a multiple of k, we can replace the previous identity by

$$P(n) \equiv \sum_{l=0}^{[\frac{n}{k}]} P(l)P_{1-k}(n-kl) \mod k.$$

If we now take respectively for the primes $k = 5$, 7 and 11, and for $n = 5m+4$, $7m+5$, and $11m+6$, we obtain the congruences of the theorem from the constant functions described in **(1)**, **(3)** and **(4)** above. It is important to observe that even though the congruence obtained for $P_{-6}(7m+5)$ was mod 7^2 and the congruence for $P_{-10}(11m+6)$ was mod 11^4, our result is only mod the appropriate k, since we are relying on the Frobenius automorphism. □

REMARK 3. Many of the ideas of this paper can be traced back to Ramanujan. For example, the congruences

$$P_{-4}(5m+4) \equiv 0 \mod 5 \text{ and } P_{-6}(7m+5) \equiv 0 \mod 7,$$

used above were proved in [**19**] using the Euler and Jacobi identities.

It should be apparent to the reader that we have not enumerated all the possible constant functions $F_{k,-N}$ and have therefore neglected many other possible congruences. Our aim to show how to get the level one congruences in a unified and simple way motivated our presentation. In the event there are readers who would like to see the extent of the myriad of possibilities afforded by these ideas we include in an appendix at the end of this paper some more possibilities. There are still many unanswered questions which we shall be discussing in [**7**].

4. The level two congruences for the primes 5 and 7

Equation (2) is much stronger than the level one congruence for the prime 5. It is equivalent to

$$\sum_{m=0}^{\infty} P(5m+4)x^{m+1} = 5\,x \sum_{m=0}^{\infty} P_{-5}(m)x^{5m} \sum_{m=0}^{\infty} P_6(m)x^m$$

and implies the recurrence formula

$$P(5m+4) = 5\sum_{j=0}^{[\frac{m}{5}]} P_{-5}(j)P_6(m-5j). \tag{6}$$

Starting with the obvious relation $\mathbf{P}^6 = \mathbf{P}\mathbf{P}^5$, we conclude from the Frobenius automorphism that $P_6(5m+4) \equiv 0 \mod 5$. (This is one of the $k=5$, $N=1$ congruences of Theorem 7.) Hence replacing m by $5m+4$ in (6) tells us that $5^2|P(25m+24)$. This is the level two congruence for the prime 5. Similarly, (3) can be rewritten as

$$\sum_{m=0}^{\infty} P(7m+5)x^m =$$

$$7\sum_{m=0}^{\infty} P_{-3}(m)x^{7m} \sum_{m=0}^{\infty} P_4(m)x^m + 7^2\,x \sum_{m=0}^{\infty} P_{-7}(m)x^{7m} \sum_{m=0}^{\infty} P_8(m)x^m;$$

that implies the recurrence formula

$$P(7m+5) = 7\sum_{j=0}^{[\frac{m}{7}]} P_{-3}(j)P_4(m-7j) + 7^2 \sum_{j=0}^{[\frac{m-1}{7}]} P_{-7}(j)P_8(m-1-7j). \tag{7}$$

In the previous section we have seen that $P_{-3}(7m+6) \equiv 0 \mod 7$. This implies (using $\mathbf{P}^4 = \mathbf{P}^{-3}\mathbf{P}^7$) that $P_4(7m+6) \equiv 0 \mod 7$. (This is one of the $k=7$, $N=0$, congruences of Theorem 7.) Replacing m by $7m+6$ in (7), we obtain the level two congruence for the prime 7. We have established

THEOREM 5. *For all* $m \in \mathbb{Z}^+ \cup \{0\}$,

$$P(5^2m+24) \equiv 0 \mod 5^2 \text{ and } P(7^2m+47) \equiv 0 \mod 7^2.$$

5. Two appendices

We discuss in the first appendix (first subsection) a family of subgroups of the modular group that is useful for our work, and we list in the second (last two subsections) more consequences of Theorem 3. Both appendices will be important in future reports.

5.1. The groups $\Gamma(k,k)$ are special cases of a more general family of subgroups of $\mathrm{PSL}(2,\mathbb{Z})$. Let p and q be two positive integers. We define

$$\Gamma(p,q) = \left\{ \begin{bmatrix} a & b \\ c & d \end{bmatrix} \in \Gamma;\ p|b \text{ and } q|c \right\} = \Gamma^o(p) \cap \Gamma_o(q).$$

We let $A_q = \begin{bmatrix} 0 & -1 \\ q & 0 \end{bmatrix}$. Our first observation is that for every motion $\gamma \in \Gamma(p,q)$, $A_q \circ \gamma \circ A_q^{-1} \in \Gamma_o(pq)$ as is easily seen from the fact that A_q conjugates $\begin{bmatrix} a & b \\ c & d \end{bmatrix}$ to $\begin{bmatrix} d & -\frac{c}{q} \\ -qb & a \end{bmatrix}$. Hence $\Gamma(p,q) \cong \Gamma_o(pq)$. It follows that

$$[\Gamma_o(q) : \Gamma(p,q)] = \frac{\mu(\Gamma_o(pq))}{\mu(\Gamma_o(q))} = p \prod_{\mathcal{P}|p, \mathcal{P}\nmid q} \left(1 + \frac{1}{\mathcal{P}}\right).$$

Fix a prime k, and let m and $n \in \mathbb{Z}^+ \cup \{0\}$. Then

$$[\Gamma_o(k^n) : \Gamma(k^m, k^n)] = \begin{cases} k^m \text{ if } n > 0 \text{ or } m = 0 \\ k^{m-1}(1+k) \text{ if } n = 0 \text{ and } m > 0 \end{cases}.$$

In particular, if $n > 0$, then the k^m motions

$$B^l;\ l = 0,\ ...,\ k^m - 1,$$

provide representatives for a complete set of left or right $\Gamma(k^m, k^n)$-cosets for $\Gamma_o(k^n)$, and if

$$P_\infty,\ P_0,\ ...,\ P_r,$$

is a list of punctures on $\mathbb{H}^2/\Gamma_o(k^n)$, then

$$P_\infty,\ P_{B^l(0)},\ ...,\ P_{B^l(r)},\ l = 0,\ 1,\ ...,\ k^m - 1,$$

is the corresponding list on $\mathbb{H}^2/\Gamma(k^m, k^n)$. In particular,

$$P_\infty,\ P_0,\ ...,\ P_{k^m-1},$$

are the punctures on $\mathbb{H}^2/\Gamma(k^m, k)$. In the last two lists, there are duplications. We know that the number of punctures on $\mathbb{H}^2/\Gamma_o(k^n)$ is given by

$$\nu_\infty(\Gamma_o(k^n)) = \begin{cases} 2k^{\frac{n-1}{2}} \text{ if } n \text{ is odd} \\ k^{\frac{n}{2}-1}(k+1) \text{ if } n \text{ is even} \end{cases}.$$

Thus $\mathbb{H}^2/\Gamma(9,3)$ has 6 punctures; they are

$$P_\infty,\ P_0,\ P_3,\ P_6,\ P_1 = P_4 = P_7,\ P_2 = P_5 = P_8.$$

The motion $C = \begin{bmatrix} 13 & -9 \\ 3 & -2 \end{bmatrix} \in \Gamma(9,3)$ maps 1 to 4, for example.

If $n = 0$, then P_∞ is the only puncture on $\mathbb{H}^2/\Gamma$; the punctures on $\mathbb{H}^2/\Gamma^o(k^m)$ are described in [**20**, Ch. I]. If $m = 1$, we can take

$$B^l,\ l = 0,\ ...,\ k-1, \text{ and } \begin{bmatrix} k & -1 \\ 1 & 0 \end{bmatrix}$$

as right coset representatives for $\Gamma/\Gamma^o(k)$.

A second case of interest is $(p, q) = 1$. In this case,

$$[\Gamma_o(q) : \Gamma(p,q)] = p \prod_{\mathcal{P}|p} \left(1 + \frac{1}{\mathcal{P}}\right) = \mu(\Gamma_o(p)).$$

The ideas encountered here will be explored further in [**7**] (see also [**6**]); in particular, in our study of the higher level Ramanujan congruences.

5.2. We start with a well known fact. Its proof is reproduced for the reader's convenience.

PROPOSITION 1. [**16**], [**12**, Th. 8 of Ch. 7] *Let k be a positive prime and let N, s and $t \in \mathbb{Z}$. Assume that $N \equiv s \mod k$. If for all $n \in \mathbb{Z}^+$ with $n \equiv t \mod k$, $P_{-s}(n) \equiv 0 \mod k$, then for all such n, $P_{-N}(n) \equiv 0 \mod k$.*

PROOF.

$$\sum_{n=0}^{\infty} P_{-N}(n)x^n = \prod_{n=1}^{\infty}(1-x^n)^N = \prod_{n=1}^{\infty}(1-x^n)^{Qk+s} = \prod_{n=1}^{\infty}(1-x^n)^{Qk}(1-x^n)^s$$

$$\equiv \prod_{n=1}^{\infty}(1-x^{kn})^Q\,(1-x^n)^s = \sum_{n=0}^{\infty} P_{-Q}(n)x^{kn} \sum_{n=0}^{\infty} P_{-s}(n)x^n \mod k$$

(the next to last equality is a consequence of the Frobenius automorphism). Hence from the last equality

$$\sum_{n=0}^{\infty} P_{-N}(n)x^n \equiv \sum_{n=0}^{\infty} P_{-Q}\left(\frac{n}{k}\right) x^n \sum_{n=0}^{\infty} P_{-s}(n)x^n \mod k;$$

comparing coefficients in this last equation yields

$$P_{-N}(n) \equiv \sum_{j=0}^{n} P_{-Q}\left(\frac{j}{k}\right) P_{-s}(n-j) = \sum_{j=0}^{\left[\frac{n}{k}\right]} P_{-Q}(j)\; P_{-s}(n-kj) \mod k.$$

The last equality is the consequence of the fact that only those j's that are multiples of k contribute to the sum. $\square$

5.3. We are interested, because some of the cases are needed in this work and others will be used in the future, in determining for which primes k and which positive integers N is $F_{k,-N}$ constant. When $F_{k,-N}$ is a constant function we need to know (to get good applications) when does $k|P_{-\alpha(k)N}(N)$.

For $k = 2,\ 3,\ 5$, $\beta(\Gamma(k,k)) = 1$, hence for $N \in \mathbb{Z}^+$, Theorem 2 yields a function with a single pole at P_∞ of order $\leq \left[\frac{N}{k}\right]$. Similarly, for $k \geq 5$, $\beta(k) = \frac{k^2-1}{24}$ and hence for every $N \in \mathbb{Z}^+$, we have a function with a pole of order at most $\left[\frac{(k^2-1)N}{24k}\right]$ (the results, in these cases, apply to the partition functions $P_{-\alpha(k)N}$). There are two reasons why a particular function $F_{k,-N}$ reduces to be a constant. The first, trivial, reason is because a function of degree $\leq \left[\frac{\beta(k)N}{k}\right]$ does not exist on the surface $\overline{\mathbb{H}^2/\Gamma_o(k)}$; the second, computational, reason is that the combinatorial number theory of the partition of function $P_{-\alpha(k)N}$ forces enough Laurent series coefficients to vanish.

We have not made an exhaustive study of when we obtain the constant functions. We summarize some of our conclusions below; more details and more special cases will be found in our monograph [**7**]. We have included here many of the most interesting cases.

(a) As mentioned earlier, for $k = 2$, 3 and 5, $\beta(\Gamma(k,k)) = 1$. Hence in these cases for $N = 1, 2, \ldots, k-1$, $F_{k,-N}$ is constant since it is regular at P_∞.

(a1) For $k = 2$, only $F_{2,-1}$ is constant. This follows from the fact that $P_{-8N}(1) = -8N$ (used for odd N) and $P_{-8N}(0) = 1$ (used for even N).

(a2) For $k = 3$, only $F_{3,-1}$ and $F_{3,-2}$ are constant. This follows from the fact that $P_{-3N}(1) = -3N$ (used for $N \equiv 1 \mod 3$), $P_{-3N}(2) = \frac{9}{2}N(N-1)$ (used for $N \equiv 2 \mod 3$) and $P_{-3N}(0) = 1$ (used for $N \equiv 0 \mod 3$).

(a3) For $k = 5$, the functions $F_{5,-N}$ for $N = 1, 2, 3, 4$, are constant for trivial reasons. In addition, only $F_{5,-8}$ and $F_{5,-14}$ are constant. This follows from two facts. The first of these are the general formulae $P_{-N}(1) = -N$ (used for $N \equiv 1 \mod 5$), $P_{-N}(2) = \frac{N(N-3)}{2}$ (used for $N \equiv 2 \mod 5$), $P_{-N}(3) = -\frac{N(N-1)(N-8)}{3!}$ (used for $N \equiv 3 \mod 5$), $P_{-N}(4) = \frac{N(N-1)(N-3)(N-14)}{4!}$ (used for $N \equiv 4 \mod 5$) and $P_{-N}(0) = 1$ (used for $N \equiv 0 \mod 5$). The second fact we need is the calculation $0 = P_{-14}(4) = P_{-14}(9)$. We will also use below that $P_{-8}(8) = -5^3$ and $P_{-14}(14) = 5^6$.

(b) For $k = 7$, we have $\beta(\Gamma(7,7)) = 2$ and $p(\overline{\mathbb{H}^2/\Gamma_o(7)})$[5] $= 0$. The functions $F_{7,-N}$, $N = 1, 2, 3$, are constant for trivial reasons. The functions $F_{7,-6}$ and $F_{7,-10}$ are constant as a result of the calculations $0 = P_{-6}(5) = P_{-10}(6) = P_{-10}(13)$. We also will need to know that $P_{-6}(12) = 7^2$ and $P_{-10}(20) = 7^4$. There are no other constant functions $F_{7,-N}$ as a consequence of the tabulation of $P_N(n)$ in [**15**].

(c) For $k = 11$, $\beta(\Gamma(11,11)) = 5$ and $p(\overline{\mathbb{H}^2/\Gamma_o(11)}) = 1$. Hence for $N = 1, 2, 3, 4$, the functions $F_{11,-N}$ are constant for trivial reasons. The calculations

$$0 = P_{-6}(8) = P_{-6}(19),\ 0 = P_{-8}(7) = P_{-8}(18) = P_{-8}(29),$$
$$0 = P_{-10}(6) = P_{-10}(17) = P_{-10}(28) = P_{-10}(39),$$
$$0 = P_{-26}(9) = P_{-26}(20) = \ldots = P_{-26}(120),$$

show that $F_{11,-N}$ are constant functions for $N = 6, 8, 10$ and 26. We also have

$$P_{-6}(30) = 11^2,\ P_{-8}(40) = -11^3,\ P_{-10}(50) = 11^4 \text{ and } P_{-26}(130) = 11^{12}.$$

(d) For $k = 13$ ($\beta(\Gamma(13,13)) = 7$ and $p(\overline{\mathbb{H}^2/\Gamma_o(13)}) = 0$) there are two constant functions $F_{13,-1}$ (for trivial reasons) and $F_{13,-3}$ (because $P_{-3}(8) = 0$ but $P_{-3}(21) = 13$).

(e) For $k = 17$ ($\beta(\Gamma(17,17)) = 12$ and $p(\overline{\mathbb{H}^2/\Gamma_o(17)}) = 1$) there are three constant functions $F_{17,-1}$, $F_{17,-2}$ (for trivial reasons) and $F_{17,-3}$ (because $P_{-3}(2) = 0 = P_{-3}(19)$ but $P_{-3}(36) = 17$).

(f) For $k = 19$ ($\beta(19) = 15$, $\beta(\Gamma(19,19))$ is a divisor of 15 and $p(\overline{\mathbb{H}^2/\Gamma_o(19)}) = 1$) there are three constant functions $F_{19,-1}$, $F_{19,-2}$ (for trivial reasons) and $F_{19,-3}$ (because $P_{-3}(7) = 0 = P_{-3}(26)$ but $P_{-3}(45) = -19$). So is $F_{19,-10}$ (with $P_{10}(150) = 19^4$), but $F_{19,-18}$ is not constant.

(g) For $k = 23$ ($\beta(23) = 22$ and $p(\overline{\mathbb{H}^2/\Gamma_o(23)}) = 2$) there are three constant functions $F_{23,-1}$, $F_{23,-2}$ (for trivial reasons) and $F_{23,-3}$ (because $P_{-3}(20) = 0 = P_{-3}(43)$ but $P_{-3}(66) = -23$).

(h) For $k = 29$ ($\beta(29) = 35$ and $p(\overline{\mathbb{H}^2/\Gamma_o(19)}) = 1$) there are three constant

[5]We denote the genus of the closed surface X by $p(X)$.

functions $F_{29,-1}$ (for trivial reasons), $F_{29,-2}$ (because $P_{-2}(12) = 0 = P_{-2}(41)$ but $P_{-2}(70) = -1$) and $F_{29,-3}$ (because $P_{-3}(18) = 0 = P_{-3}(47) = P_{-3}(76)$ but $P_{-3}(105) = 29$).

(i) It is easy to determine at this point all the primes k for which $F_{k,-N}$ is constant for $N = 1$, 2 and 3; for example, $F_{67,-1}$ is constant.

5.4. From Theorem 3, its corollary and the calculations of the previous paragraph, we conclude the following congruences for arbitrary $m \in \mathbb{Z}^+ \cup \{0\}$.

(a) For $k = 2$, 3, 5, and $N = 1$, 2, ..., $k - 1$, we have

$$P_{\frac{24N}{1-k^2}}(km + N) = P_{\frac{24N}{1-k^2}}(N)\ P_{\frac{24N}{1-k^2}}\left(\frac{m}{k}\right).$$

Hence (after evaluating appropriate coefficients $P_{\frac{24N}{1-k^2}}(N)$) :

(a1) $(k = 2)$

$$P_{-8}(2m+1) = -2^3 P_{-8}\left(\frac{m}{2}\right).$$

(a2) $(k = 3)$

$$P_{-3}(3m+1) = -3P_{-3}\left(\frac{m}{3}\right),\ P_{-6}(3m+2) = 3^2 P_{-6}\left(\frac{m}{3}\right).$$

(a3) $(k = 5)$ Here we also have nontrivial congruences for $N = 8$ and 14.

$$P_{-1}(5m+1) = -P_{-1}\left(\frac{m}{5}\right),\ P_{-2}(5m+2) = -P_{-2}\left(\frac{m}{5}\right),$$

$$P_{-3}(5m+3) = 5P_{-3}\left(\frac{m}{5}\right),\ P_{-4}(5m+4) = -5P_{-4}\left(\frac{m}{5}\right),$$

$$P_{-8}(5m+3) = -5^3 P_{-8}\left(\frac{m-1}{5}\right),\ P_{-14}(5m+4) = 5^6 P_{-4}\left(\frac{m-2}{5}\right).$$

(b) (The case $k = 7$.) For $N = 1$, 2, 3, 6 and 10, and all $m \in \mathbb{Z}^+$, we have

$$P_{-N}(7m+2N) = P_{-N}(2N)P_{-N}\left(\frac{m}{7}\right);$$

that is,

$$P_{-1}(7m+2) = -P_{-1}\left(\frac{m}{7}\right),\ P_{-2}(7m+4) = P_{-2}\left(\frac{m}{7}\right),\ P_{-3}(7m+6) = -7P_{-3}\left(\frac{m}{7}\right),$$

$$P_{-6}(7m+5) = 7^2 P_{-6}\left(\frac{m-1}{7}\right),\ P_{-10}(7m+6) = 7^4 P_{-10}\left(\frac{m-2}{7}\right).$$

(c) (The case $k = 11$.) For $N = 1$, 2, 3, 4, 6, 8, 10 and 26, and all $m \in \mathbb{Z}^+ \cup \{0\}$, we have

$$P_{-N}(11m+5N) = P_{-N}(5N)\ P_{-N}\left(\frac{m}{11}\right);$$

that is,

$$P_{-1}(11m+5) = P_{-1}\left(\frac{m}{11}\right),\ P_{-2}(11m+10) = P_{-2}\left(\frac{m}{11}\right),$$

$$P_{-3}(11m+4) = -11P_{-3}\left(\frac{m-1}{11}\right),\ P_{-4}(11m+9) = -11P_{-4}\left(\frac{m-1}{11}\right),$$

$$P_{-6}(11m+8) = 11^2 P_{-6}\left(\frac{m-2}{11}\right),\ P_{-8}(11m+7) = -11^3 P_{-8}\left(\frac{m-3}{11}\right),$$

$$P_{-10}(11m+6) = 11^4 P_{-10}\left(\frac{m-4}{11}\right),\ P_{-26}(11m+9) = 11^{12} P_{-26}\left(\frac{m-11}{11}\right).$$

(d) (The case $k = 13$.) For $N = 1$ and 3, and all $m \in \mathbb{Z}^+ \cup \{0\}$, we have

$$P_{-N}(13m + 7N) = P_{-N}(7N)\ P_{-N}\left(\frac{m}{13}\right);$$

that is,

$$P_{-1}(13m + 7) = P_{-1}\left(\frac{m}{13}\right),\ P_{-3}(13m + 8) = 13P_{-3}\left(\frac{m-1}{13}\right).$$

(e) (The case $k = 17$.) For $N = 1,\ 2$ and 3, and all $m \in \mathbb{Z}^+ \cup \{0\}$, we have

$$P_{-N}(17m + 12N) = P_{-N}(12N)\ P_{-N}\left(\frac{m}{17}\right);$$

that is,

$$P_{-1}(17m + 12) = -P_{-1}\left(\frac{m}{17}\right),\ P_{-2}(17m + 7) = -P_{-2}\left(\frac{m-1}{17}\right),$$

$$P_{-3}(17m + 2) = 17P_{-3}\left(\frac{m-2}{17}\right).$$

(f) (The case $k = 19$.) For $N = 1,\ 2,\ 3$ and 10, and all $m \in \mathbb{Z}^+ \cup \{0\}$, we have

$$P_{-N}(19m + 15N) = P_{-N}(15N)P_{-N}\left(\frac{m}{19}\right);$$

that is,

$$P_{-1}(19m + 15) = -P_{-1}\left(\frac{m}{19}\right),\ P_{-2}(19m + 11) = P_{-2}\left(\frac{m-1}{19}\right),$$

$$P_{-3}(19m + 7) = 19P_{-3}\left(\frac{m-2}{19}\right),$$

$$P_{-10}(19m + 17) = 19^4 P_{-10}\left(\frac{m-7}{19}\right).$$

(g) (The case $k = 23$.) For $N = 1,\ 2$ and 3, and all $m \in \mathbb{Z}^+ \cup \{0\}$, we have

$$P_{-N}(23m + 22N) = P_{-N}(22N)P_{-N}\left(\frac{m}{23}\right);$$

that is,

$$P_{-1}(23m + 22) = P_{-1}\left(\frac{m}{23}\right),\ P_{-2}(23m + 21) = P_{-2}\left(\frac{m-1}{23}\right),$$

$$P_{-3}(23m + 20) = -23P_{-3}\left(\frac{m-2}{23}\right).$$

(h) (The case $k = 29$.) For $N = 1,\ 2$ and 3, and all $m \in \mathbb{Z}^+ \cup \{0\}$, we have

$$P_{-N}(29m + 35N) = P_{-N}(35N)P_{-N}\left(\frac{m}{29}\right);$$

that is,

$$P_{-1}(29m + 35) = -P_{-1}\left(\frac{m}{29}\right),\ P_{-2}(29m + 12) = P_{-2}\left(\frac{m-2}{29}\right),$$

$$P_{-3}(29m + 18) = 29P_{-3}\left(\frac{m-3}{29}\right).$$

(i) There are many more similar identities for P_{-N} with $N = 1$, 2 and 3 involving primes $k \geq 31$; for example[6],

$$P_{-1}(67m + 187) = -P_{-1}\left(\frac{m}{67}\right).$$

REMARK 4. The above equalities carry much information. For example, the statement $P_{-4}(5m+4) = -5P_{-4}\left(\frac{m}{5}\right)$ tells us that

$$P_{-4}(5m+4) \equiv 0 \mod 5 \text{ for all } m \in \mathbb{Z}^+ \cup \{0\}$$

and also that

$$P_{-4}(5m+4) = 0 \text{ for all } m \in \mathbb{Z}^+ \cup \{0\} \not\equiv 0 \mod 5;$$

in particular,

$$0 = P_{-4}(25m+9) = P_{-4}(25m+14) =$$
$$P_{-4}(25m+19) = P_{-4}(25m+24) \text{ for all } m \in \mathbb{Z}^+ \cup \{0\}.$$

We rewrite our basic equality as

$$P_{-4}(5^2m+4) = -5P_{-4}(m)$$

and define for each $n \in \mathbb{Z}^+ \cup \{0\}$,

$$l_n = 4(5^{2n} + \ldots + 5^2 + 1) = \frac{5^{2n+2}-1}{6}.$$

Then we conclude that

$$P_{-4}(5^{2n}m + l_{n-1} = (-5)^n P_{-4}(m)$$
$$\text{and } P_{-4}(5^{2n+1}m + l_n) = (-5)^{n+1}P_{-4}\left(\frac{m}{5}\right) \text{ for all } m \in \mathbb{Z}^+ \cup \{0\}.$$

The proof is by induction on n. We have established the first claim for $n = 1$ and the second for $n = 0$. The induction steps read

$$P_{-4}(5^{2n+2}m + l_n) = P_{-4}(5^2(5^{2n}m + l_{n-1}) + 4) = -5P_{-4}(5^{2n}m + l_{n-1}) =$$
$$(-5)^{n+1}P_{-4}(m)$$

and

$$P_{-4}(5^{2n+3}m + l_{n+1}) = P_{-4}(5^2(5^{2n+1}m + l_n) + 4) = -5P_{-4}(5^{2n+1}m + l_n) =$$
$$(-5)^{n+2}P_{-4}\left(\frac{m}{5}\right).$$

The next theorem summarizes some of the equalities we have obtained.

THEOREM 6. *For all $n \in \mathbb{Z}^+$ and all $m \in \mathbb{Z}^+ \cup \{0\}$,*
($k = 2$)

$$P_{-8}\left(2^{2n}m + \frac{2^{2n}-1}{3}\right) = (-2)^{3n}P_{-8}(m)$$
$$\text{and } P_{-8}\left(2^{2n+1}m + \frac{2^{2n+2}-1}{3}\right) = (-2)^{3(n+1)}P_{-8}\left(\frac{m}{2}\right),$$

($k = 3$)

$$P_{-3}\left(3^{2n}m + \frac{3^{2n}-1}{8}\right) = (-3)^n P_{-3}(m)$$
$$\text{and } P_{-3}\left(3^{2n+1}m + \frac{3^{2n+2}-1}{8}\right) = (-3)^{n+1}P_{-3}\left(\frac{m}{3}\right),$$

[6]Such an identity certainly exists whenever $N < \frac{48k}{k^2-1}$.

$$P_{-6}\left(3^{2n}m+\frac{3^{2n}-1}{4}\right)=3^{2n}P_{-6}(m)$$

$$\textit{and } P_{-6}\left(3^{2n+1}m+\frac{3^{2n+2}-1}{4}\right)=3^{2(n+1)}P_{-6}\left(\frac{m}{3}\right),$$

(k = 5)

$$P_{-1}\left(5^{2n}m+\frac{5^{2n}-1}{24}\right)=(-1)^{n}P_{-1}(m)$$

$$\textit{and } P_{-1}\left(5^{2n+1}m+\frac{5^{2n+2}-1}{24}\right)=(-1)^{n+1}P_{-1}\left(\frac{m}{5}\right),$$

$$P_{-2}\left(5^{2n}m+\frac{5^{2n}-1}{12}\right)=(-1)^{n}P_{-2}(m)$$

$$\textit{and } P_{-2}\left(5^{2n+1}m+\frac{5^{2n+2}-1}{12}\right)=(-1)^{n+1}P_{-2}\left(\frac{m}{5}\right),$$

$$P_{-3}\left(5^{2n}m+\frac{5^{2n}-1}{8}\right)=5^{n}P_{-3}(m)$$

$$\textit{and } P_{-3}\left(5^{2n+1}m+\frac{5^{2n+2}-1}{8}\right)=5^{n+1}P_{-3}\left(\frac{m}{5}\right),$$

$$P_{-4}\left(5^{2n}m+\frac{5^{2n}-1}{6}\right)=(-5)^{n}P_{-4}(m)$$

$$\textit{and } P_{-4}\left(5^{2n+1}m+\frac{5^{2n+2}-1}{6}\right)=(-5)^{n+1}P_{-4}\left(\frac{m}{5}\right),$$

$$P_{-8}\left(5^{2n}m+\frac{5^{2n}-1}{3}\right)=(-5)^{3n}P_{-8}(m)$$

$$\textit{and } P_{-8}\left(5^{2n+1}m+\frac{5^{2n+2}-1}{3}\right)=(-5)^{3(n+1)}P_{-8}\left(\frac{m}{5}\right),$$

$$P_{-14}\left(5^{2n}m+7\frac{5^{2n}-1}{12}\right)=5^{6n}P_{-14}(m)$$

$$\textit{and } P_{-14}\left(5^{2n+1}m+7\frac{5^{2n+2}-1}{12}\right)=5^{6(n+1)}P_{-14}\left(\frac{m}{5}\right)$$

(k = 7)

$$P_{-1}\left(7^{2n}m+\frac{7^{2n}-1}{24}\right)=(-1)^{n}P_{-1}(m)$$

$$\textit{and } P_{-1}\left(7^{2n+1}m+\frac{7^{2n+2}-1}{24}\right)=(-1)^{n+1}P_{-1}\left(\frac{m}{7}\right),$$

$$P_{-2}\left(7^{2n}m+\frac{7^{2n}-1}{12}\right)=P_{-2}(m)\ \textit{and } P_{-2}\left(7^{2n+1}m+\frac{7^{2n+2}-1}{12}\right)=P_{-2}\left(\frac{m}{7}\right),$$

$$P_{-3}\left(7^{2n}m+\frac{7^{2n}-1}{8}\right)=(-7)^{n}P_{-3}(m)$$

$$\textit{and } P_{-3}\left(7^{2n+1}m+\frac{7^{2n+2}-1}{8}\right)=(-7)^{n+1}P_{-3}\left(\frac{m}{7}\right),$$

$$P_{-6}\left(7^{2n}m+\frac{7^{2n}-1}{4}\right)=7^{2n}P_{-6}(m)$$

$$\text{and } P_{-6}\left(7^{2n+1}m + \frac{7^{2n+2}-1}{4}\right) = 7^{2(n+1)}P_{-6}\left(\frac{m}{7}\right),$$

$$P_{-10}\left(7^{2n}m + 5\frac{7^{2n}-1}{4}\right) = 7^{4n}P_{-10}(m)$$

$$\text{and } P_{-10}\left(7^{2n+1}m + 5\frac{7^{2n+2}-1}{4}\right) = 7^{4(n+1)}P_{-10}\left(\frac{m}{7}\right),$$

$(k = 11)$

$$P_{-1}\left(11^{2n}m + \frac{11^{2n}-1}{24}\right) = P_{-1}(m)$$

$$\text{and } P_{-1}\left(11^{2n+1}m + \frac{11^{2n+2}-1}{24}\right) = P_{-1}\left(\frac{m}{11}\right),$$

$$P_{-2}\left(11^{2n}m + \frac{11^{2n}-1}{12}\right) = P_{-2}(m)$$

$$\text{and } P_{-2}\left(11^{2n+1}m + \frac{11^{2n+2}-1}{12}\right) = P_{-2}\left(\frac{m}{11}\right),$$

$$P_{-3}\left(11^{2n}m + \frac{11^{2n}-1}{8}\right) = (-11)^{n}P_{-3}(m)$$

$$\text{and } P_{-3}\left(11^{2n+1}m + \frac{11^{2n+2}-1}{8}\right) = (-11)^{n+1}P_{-3}\left(\frac{m}{11}\right),$$

$$P_{-4}\left(11^{2n}m + \frac{11^{2n}-1}{6}\right) = (-11)^{n}P_{-4}(m)$$

$$\text{and } P_{-4}\left(11^{2n+1}m + \frac{11^{2n+2}-1}{6}\right) = (-11)^{n+1}P_{-4}\left(\frac{m}{11}\right),$$

$$P_{-6}\left(11^{2n}m + \frac{11^{2n}-1}{4}\right) = 11^{2n}P_{-6}(m)$$

$$\text{and } P_{-6}\left(11^{2n+1}m + \frac{11^{2n+2}-1}{4}\right) = 11^{2(n+1)}P_{-6}\left(\frac{m}{11}\right),$$

$$P_{-8}\left(11^{2n}m + \frac{11^{2n}-1}{3}\right) = (-11)^{3n}P_{-8}(m)$$

$$\text{and } P_{-8}\left(11^{2n+1}m + \frac{11^{2n+2}-1}{3}\right) = (-11)^{3(n+1)}P_{-8}\left(\frac{m}{11}\right),$$

$$P_{-10}\left(11^{2n}m + 5\frac{11^{2n}-1}{12}\right) = 11^{4n}P_{-10}(m)$$

$$\text{and } P_{-10}\left(11^{2n+1}m + 5\frac{11^{2n+2}-1}{12}\right) = 11^{4(n+1)}P_{-10}\left(\frac{m}{11}\right),$$

$$P_{-26}\left(11^{2n}m + 13\frac{11^{2n}-1}{12}\right) = 11^{12n}P_{-26}(m)$$

$$\text{and } P_{-26}\left(11^{2n+1}m + 13\frac{11^{2n+2}-1}{12}\right) = 11^{12(n+1)}P_{-26}\left(\frac{m}{11}\right),$$

$(k = 13)$

$$P_{-1}\left(13^{2n}m + \frac{13^{2n}-1}{24}\right) = P_{-1}(m)$$

$$\text{and } P_{-1}\left(13^{2n+1}m+\frac{13^{2n+2}-1}{24}\right)=P_{-1}\left(\frac{m}{13}\right),$$

$$P_{-3}\left(13^{2n}m+\frac{13^{2n}-1}{8}\right)=13^{n}P_{-3}(m)$$

$$\text{and } P_{-3}\left(13^{2n+1}m+\frac{13^{2n+2}-1}{8}\right)=13^{n+1}P_{-3}\left(\frac{m}{13}\right),$$

(k = 17)

$$P_{-1}\left(17^{2n}m+\frac{17^{2n}-1}{24}\right)=(-1)^{n}P_{-1}(m)$$

$$\text{and } P_{-1}\left(17^{2n+1}m+\frac{17^{2n+2}-1}{24}\right)=(-1)^{n+1}P_{-1}\left(\frac{m}{17}\right),$$

$$P_{-2}\left(17^{2n}m+\frac{17^{2n}-1}{12}\right)=(-1)^{n}P_{-2}(m)$$

$$\text{and } P_{-2}\left(17^{2n+1}m+\frac{17^{2n+2}-1}{12}\right)=(-1)^{n+1}P_{-2}\left(\frac{m}{17}\right),$$

$$P_{-3}\left(17^{2n}m+\frac{17^{2n}-1}{8}\right)=17^{n}P_{-3}(m)$$

$$\text{and } P_{-3}\left(17^{2n+1}m+\frac{17^{2n+2}-1}{8}\right)=17^{n+1}P_{-3}\left(\frac{m}{17}\right),$$

(k = 19)

$$P_{-1}\left(19^{2n}m+\frac{19^{2n}-1}{24}\right)=(-1)^{n}P_{-1}(m)$$

$$\text{and } P_{-1}\left(19^{2n+1}m+\frac{19^{2n+2}-1}{24}\right)=(-1)^{n+1}P_{-1}\left(\frac{m}{19}\right),$$

$$P_{-2}\left(19^{2n}m+\frac{19^{2n}-1}{12}\right)=P_{-2}(m)$$

$$\text{and } P_{-2}\left(19^{2n+1}m+\frac{19^{2n+2}-1}{12}\right)=P_{-2}\left(\frac{m}{19}\right),$$

$$P_{-3}\left(19^{2n}m+\frac{19^{2n}-1}{8}\right)=19^{n}P_{-3}(m)$$

$$\text{and } P_{-3}\left(19^{2n+1}m+\frac{19^{2n+2}-1}{8}\right)=19^{n+1}P_{-3}\left(\frac{m}{19}\right),$$

$$P_{-10}\left(19^{2n}m+5\frac{19^{2n}-1}{12}\right)=19^{4n}P_{-10}(m)$$

$$\text{and } P_{-10}\left(19^{2n+1}m+5\frac{19^{2n+2}-1}{12}\right)=19^{4(n+1)}P_{-10}\left(\frac{m}{19}\right).$$

REMARK 5. In contrast (see, for example, [**6**]), the results for positive N analogous to the equalities in the last theorem are congruences; nontrivial only for some of the cases.

REMARK 6. We compare our results with those of Newman [**15**]. The main result of [**15**] may be restated as follows:
Restrict N to 2, 4, 6, 8, 10, 14, and 26. Let $k > 3$ be a prime such that

$$N(k+1) \equiv 0 \mod 24.$$

Then

$$P_{-N}\left(km + \frac{N(k^2-1)}{24}\right) = (-k)^{\frac{1}{2}N-1} P_{-N}\left(\frac{m}{k}\right),$$

for all $m \in \mathbb{Z} \cup \{0\}$.

Our work applies also to the primes 2 and 3. In addition, neither of the two lists of congruences is a proper subset of the other. For example, the congruences for $P_{-14}(5m+4)$ and $P_{-10}(7m+6)$ are on our list but not on Newman's; whereas the congruences for $P_{-4}(17m+14)$ and $P_{-8}(17m+11)$ are on Newman's list but not ours. Since we have a necessary and sufficient condition for a function $F_{k,-N}$ to be constant, any congruence on Newman's list can easily be added to ours by computing a finite set of partition coefficients.

The above calculations together with Proposition 1 yield immediately the following

THEOREM 7. *For all $N \in \mathbb{Z}$ and all $m \in \mathbb{Z}^+ \cup \{0\}$,*

$$P_{2N}(2m+1) \equiv 0 \mod 2,$$
$$P_{3N}(3m+1) \equiv 0 \equiv P_{3N}(3m+2) \mod 3,$$
$$P_{5N+2}(5m+3) \equiv 0 \equiv P_{5N+1}(5m+4) \mod 5,$$
$$P_{7N+4}(7m+6) \equiv 0 \equiv P_{7N+1}(7m+5) \mod 7,$$
$$P_{11N+8}(11m+4) \equiv 0 \equiv P_{11N+7}(11m+9) \equiv P_{11N+5}(11m+8) \mod 11,$$
$$P_{11N+3}(11m+7) \equiv 0 \equiv P_{11N+1}(11m+6) \equiv P_{11N+7}(11m+9) \mod 11,$$
$$P_{13N+10}(13m+8) \equiv 0 \mod 13,$$
$$P_{17N+14}(17m+2) \equiv 0 \mod 17$$

and

$$P_{19N+16}(19m+7) \equiv 0 \equiv P_{19N+9}(19m+17) \mod 19.$$

The mysterious patterns described above are clearly in need of further investigation and explanation. As remarked earlier, the last theorem contains, of course, the three Ramanujan level one congruences for the primes 5, 7 and 11.

References

[1] A.O.L. Atkin, *Ramanujan congruences for $p_{-k}(n)$*, Can. J. Math. **20** (1968), 67–78.

[2] J.L. Drost, *A shorter proof of the Ramanujan congruence modulo 5*, Amer. Math. Monthly **104** (1997), 963–964.

[3] H.M. Farkas and Y. Kopeliovich, *New theta constant identities*, Israel J. Math. **82** (1993), 133–140.

[4] ———, *New theta constant identities II*, Proc. Amer. Math. Soc. **123** (1995), 1009–1020.

[5] H.M. Farkas, Y. Kopeliovich, and I. Kra, *Uniformization of modular curves*, Comm. Anal. Geom. **4** (1996), 207–259 and 681.

[6] H.M. Farkas and I. Kra, *A function theoretic approach to the Ramanujan partition identities with applications to combinatorial number theory*, Proceedings of I Iberoamerican Congress on Geometry, January 1998, Chile, 75–106 and in this volume.

[7] ———, *Theta constants, Riemann surfaces and the modular group*, monograph in preparation.

[8] ———, *Automorphic forms for subgroups of the modular group. II: Groups containing congruence subgroups*, J. d'Analyse Math. **70** (1996), 91–156.

[9] ———, *Theta constant identities with applications to combinatorial number theory*, Contemporary Math. **211** (1997), 227–252.

[10] B. Gordon, *Ramanujan congruences for $p_{-k}(n) \equiv 11^r$*, Glasgow Math. J. **24** (1983), 107–123.

[11] K. Hughes, *Ramanujan congruences for $p_{-k}(n)$ modulo powers of 17*, Can. J. Math. **43** (1991), 506–525.

[12] M.I. Knopp, *Modular Functions in Analytic Number Theory*, Chelsea Publishing Company, 1993.

[13] I. Kra, *On a Ramanujan partition identity*, Proceedings of the Ashkelon Conference on Complex Function Theory (1996), Israel Conference Math. Proceedings, volume **11**, 1997, pp. 167–172.

[14] M. Newman, *Remarks on some modular identities*, Trans. Amer. Math. Soc. **73** (1952), 313–320.

[15] ———, *An identity for the coefficients of certain modular forms*, J. Lond. Math. Soc. **30** (1955), 488–493.

[16] ———, *Some theorems about $p_r(n)$*, Can. J. Math. **9** (1957), 68–70.

[17] ———, *Construction and application of a class of modular functions II*, Proc. Lond. Math. Soc. **9** (1959), 373–387.

[18] ———, *Modular functions revisited*, Analytic Number Theory, Springer Lecture Notes 899, 1980, pp. 396–421.

[19] S. Ramanujan, *Some properties of $p(n)$, the number of partitions of n*, Proc. Camb. Phil. Soc. **19** (1919), 207–210.

[20] G. Shimura, *Introduction to the Arithmetic Theory of Automorphic Functions*, Publications of the Mathematical Society of Japan, Kanô memorial lectures 1, vol. 11, Iwannami Shoten and Princeton University Press, 1971.

[21] G.N. Watson, *Ramanujans Vermutung über Zerfällungsanzahlen*, J. Reine Angew. Math. **179** (1938), 97–118.

The Hebrew University of Jerusalem, Jerusalem 91904, Israel
and
State University of New York at Stony Brook, Stony Brook, NY 11794, USA

Contemporary Mathematics
Volume **240**, 1999

A function theoretic approach to the Ramanujan partition identities with applications to combinatorial number theory

Hershel M. Farkas and Irwin Kra

1. Introduction

This article, an expanded version of talks delivered at the First Iberoamerican Congress on Geometry, has two parts. The first part (§§2–10) outlines the development of the theory of what we call *Ramanujan partition identities* from a function theoretic point of view. The second part (§11) deals with applications to combinatorial number theory. The mathematics involved is an interesting combination of core areas of the subject displaying a deep interplay between complex analysis (function theory), number theory, algebra and combinatorics. It has nontrivial connections to representation theory and mathematical physics. This development is based on beautiful AND simple concepts. It yields elegant formulae spanning the gambit from elementary to very complicated. It confronts the researcher with very basic questions; for example, when is a multivalued function on a Riemann surface single valued. Much is known about the subject. There is a rich literature and work in the field involves communing with giants of the past. However, it also reinforces ones sense of modesty since almost every discovery a new researcher in the area makes can be found some place in the works of the previous generations. The development of the subject may also be based on combinatorics (see, for example, [**1**]) or hypergeometric functions ([**13**], [**16**]).

We study the *Ramanujan partition functions*

$$P_N : \mathbb{Z}^+ \cup \{0\} \to \mathbb{Z},\ N \in \mathbb{Z},$$

defined by convergent products and sums on the unit disc Δ

$$\mathbf{P}^N(x) = \prod_{n=1}^{\infty} \frac{1}{(1-x^n)^N} = \sum_{n=0}^{\infty} P_N(n)x^n,\ x \in \Delta.$$

Write P for P_1. Ramanujan [**22**] proved that for all $m \in \mathbb{Z}^+ \cup \{0\}$,

$$P(5m+4) \equiv 0 \bmod 5,\ P(7m+5) \equiv 0 \bmod 7 \text{ and } P(11m+6) \equiv 0 \bmod 11,$$

1991 MSC: 30F10, 11F12, 11F30, 14H05, 30F30, 30F35, 14H45.

Research by IK supported in part by NSF Grant DMS 9500557. Research by HMF supported in part by the Gabriella and Paul Rosenbaum Foundation and the Edmund Landau Center for Research in Mathematical Analysis sponsored by the Minerva Foundation Germany. Research of HMF and IK supported in part by a US-Israel BSF Grant 95-348.

and more or less conjectured (see [**23**] and [**18**, Ch. 7]) and partially proved that (here $n \in \mathbb{Z}^+$):
(a) If $24m \equiv 1 \bmod 5^n$, then $P(m) \equiv 0 \bmod 5^n$.
(b) If $24m \equiv 1 \bmod 7^n$, then $P(m) \equiv 0 \bmod 7^{1+[\frac{n}{2}]}$.
(c) If $24m \equiv 1 \bmod 11^n$, then $P(m) \equiv 0 \bmod 11^n$.
We refer to these as the *Ramanujan level n congruences for the primes* 5, 7 and 11, respectively.

Several partition functions P_N have been investigated more than the general case; in part, because of various applications and/or interpretations. Among these are the cases $N = 1$ (studied by Ramanujan), $N = -1$ (Euler), $N = -3$ (Jacobi) and $N = -24$ (Lehmer). In particular, only for $N = -1$ and -3 do we have (see §2.5 below) a closed formula for the *partition coefficients* $P_N(n)$, $n \in \mathbb{Z}^+ \cup \{0\}$. Other significant contributions to the theory were made by G.H. Hardy, G.N. Watson, L.J. Mordel, J. Lehner, A.O.L. Atkin and M. Newman, among others.

There are also congruences for some other primes; for example, 2, 3 and 13.
(d) If $8m \equiv 1 \bmod 3^n$, then $P_3(m) \equiv 0 \bmod 3^{[\frac{3n}{2}]+1}$.
(e) If $3m \equiv 1 \bmod 2^n$, then $P_8(m) \equiv 0 \bmod 2^{3[[\frac{n+1}{2}]]}$.
The results for 13 are different in flavor from the previous ones. A sample formula is
(f)

$$P(13m + 6) \equiv 11P_{-11}(m) \bmod 13.$$

Several immediate questions should be addressed. What primes appear in the theory? Why? What is the significance of the integer 24 (and its factors 3 and 8)? There are many conjectures (see, for example, [**5**]) and some of these have been established (for example, [**19**, Ch. 6] contains Atkin's proofs of the conjectures of [**5**]). However, it is our opinion that much more remains to be done. In particular, this beautiful chapter in mathematics seems to have been diverted from a main stream in complex analysis. We hope that our contributions will in part revive the interests of those working on Riemann surfaces in applications to number theoretic questions.

We start with a combinatorial interpretation of the partition functions. It is, of course, obvious that for all $N \in \mathbb{Z}$, $P_N(0) = 1$ and $P_N(1) = N$. It is easily seen that in general for $n \in \mathbb{Z}^+$, $P_N(n)$ is a polynomial in N of degree n, zero constant term (thus a root at $N = 0$) and all the rest of the coefficients are positive rationals (hence no roots in $\mathbb{R}^+$) . These polynomials may, however, have roots in $-\mathbb{Z}^+$, the negative integers. It is a nontrivial matter of some interest to compute these polynomials and their real roots[1]. Assume that $n \in \mathbb{Z}^+$. For $N \in \mathbb{Z}^+$, $P_N(n)$ counts the total number of partitions of n using positive integers of N colors. For $-N \in \mathbb{Z}^+$, $P_N(n)$ counts the difference between the total number of even and odd partitions of n WITHOUT REPETITIONS using positive integers of $-N$ colors. A *partition* of $n \in \mathbb{Z}^+$ is a sum of positive integers totaling n. Thus the partitions of 4 are: 4, 3+1, 2+2, 2+1+1, 1+1+1+1. The *parity* of the partition is the parity of the number of summands. One easily checks from the definitions or the last enumeration that

$$P(4) = 5,\ P_{-1}(4) = 0,\ P_{-3}(4) = 0,\ P_2(4) = 20,\ P_{-2}(4) = 1.$$

[1]For $0 \le n \le 10$, Newman [**20**] computed the polynomials $P_{\cdot}(n)$ and their roots in $\mathbb{Z}$. Further results will be found in [**9**].

A sample formula, hinting at the underlying function theory, that already appears in [**22**] is

$$\sum_{m=1}^{\infty} P(5m-1)x^m = 5\ x\ \frac{\prod_{n=1}^{\infty}(1-x^{5n})^5}{\prod_{n=1}^{\infty}(1-x^n)^6}$$

which, for reasons that should become apparent in due course, is more conveniently rewritten as

$$\prod_{n=1}^{\infty}(1-x^{5n})\sum_{m=1}^{\infty} P(5m-1)x^m = 5\ x\ \left(\prod_{n=1}^{\infty}\frac{1-x^{5n}}{1-x^n}\right)^6.$$

To unify the various identities, we have been discussing we introduce

DEFINITION 1. Let

$$e : \mathbb{Z}^+ \cup \{0\} \to \mathbb{Z}^+ \cup \{0\}$$

be a function and let F be a meromorphic function on a Riemann surface X. Let $x \in X$ and let z be a local coordinate on X vanishing at x. We say the F *satisfies a level* n *congruence for the prime* k *(with respect to the local coordinate* z*) of type* e if the Laurent series expansion of F in terms of z is

$$z^{I_o}\sum_{i=0}^{\infty} a_i z^i, \text{ with } I_o \text{ and } a_i \in \mathbb{Z},\ a_i \equiv 0 \mod k^{e(n)}.$$

It turns out that all the congruences discussed so far arise in this way.

The aim of our research in this area is the study of function theory on Riemann surfaces represented by subgroups of the modular group $\mathrm{SL}(2,\mathbb{Z})$ acting on the upper half plane

$$\mathbb{H}^2 = \{z \in \mathbb{C};\ \Im z > 0\}.$$

The function theoretic approach to some problems in combinatorial number theory is a pleasant byproduct.

We are reporting on a body of work that originates with fundamental ideas and results of Srinivasa Ramanujan. The proceedings of a centenary conference in his honor [**2**] show the tremendous influence his work has had on 20th century mathematics; especially relevant for us are the articles [**7**], [**15**] and [**6**].

2. The needed tools

We summarize the ingredients in our development of this theory.

2.1. Compact Riemann surfaces of genus $p = 0$ and 1. We need to describe some properties of the field $\mathcal{K}(X)$ of meromorphic functions on a closed Riemann surface X of genus p. Any two elements of $\mathcal{K}(X)$ are, of course, algebraically dependent. It is a nontrivial task, in general, to determine the equations satisfied by the two functions. It is fortunate that for most applications, the cases $p = 0$ and 1 suffice. For $p = 0$, $\mathcal{K}(X) \cong \mathbb{C}(z)$, the field of complex valued rational functions. In particular, if we fix $x \in X$, and let $F_N \in \mathcal{K}(X)$ have pole at x of order $N > 0$ and no other poles, then there exist constants $c_i \in \mathbb{C}$ such that

$$F_N = \sum_{i=0}^{N} c_i F_1^i. \tag{1}$$

For $p = 1$, $\mathcal{K}(X) \cong \mathbb{C}(x)[y]/\mathcal{M}$, where $\mathcal{M}$ is the principal maximal ideal generated by

$$y^2 - 4(x-e_1)(x-e_2)(x-e_3),$$

with e_1, e_2, e_3 three distinct complex numbers. With x and F_N as above[2], now $(N \geq 2)$

$$F_N = \left(\sum_{i=0}^{\left[\frac{N-3}{2}\right]} c_i F_2^i\right) F_3 + \sum_{j=0}^{\left[\frac{N}{2}\right]} d_j F_2^j, \quad c_i \in \mathbb{C},\ d_i \in \mathbb{C}. \tag{2}$$

We may (but are not required to) use $x = \wp$ and $y = \wp'$, the Weierstrass $\wp$-function and its derivative (in this case we may take $e_1 = \wp\left(\frac{1}{2}\right)$, $e_2 = \wp\left(\frac{\tau}{2}\right)$ and $e_3 = \wp\left(\frac{1+\tau}{2}\right)$, where $X = \mathbb{C}/G$ and G is the group of translations by $n + m\tau$ with n, $m \in \mathbb{Z}$ and $\tau \in \mathbb{H}^2$.

The undetermined constants (c_i and d_i) in equations (1) and (2) are usually computed using simple linear algebra.

2.2. Theta functions. The (one variable) *theta function* with *characteristic* $\begin{bmatrix} \epsilon \\ \epsilon' \end{bmatrix} \in \mathbb{R}^2$, *argument* $z \in \mathbb{C}$ and *parameter* $\tau \in \mathbb{H}^2$ is defined by the infinite series

$$\theta\begin{bmatrix} \epsilon \\ \epsilon' \end{bmatrix}(z,\tau) = \sum_{n=-\infty}^{\infty} \exp 2\pi\imath\left(\frac{1}{2}\left(n+\frac{\epsilon}{2}\right)^2\tau + \left(n+\frac{\epsilon}{2}\right)\left(z+\frac{\epsilon'}{2}\right)\right)$$

that, for fixed characteristic, converges absolutely and uniformly on compact subsets of $\mathbb{C} \times \mathbb{H}^2$. Two groups act on θ-functions. For $(m,n) \in \mathbb{Z}^2$,

$$\theta\begin{bmatrix} \epsilon \\ \epsilon' \end{bmatrix}(z+m+n\tau,\tau) = \exp 2\pi\imath\left(\frac{m\epsilon - n\epsilon'}{2} - nz - n^2\frac{\tau}{2}\right)\theta\begin{bmatrix} \epsilon \\ \epsilon' \end{bmatrix}(z,\tau).$$

More interesting, for our applications, is the *transformation formula* obtained by studying the action of $\gamma = \begin{bmatrix} a & b \\ c & d \end{bmatrix} \in \mathrm{PSL}(2,\mathbb{Z})$ on our functions:

$$\frac{\exp\pi\imath\left\{\frac{-cz^2}{c\tau+d}\right\}\theta\begin{bmatrix} \epsilon \\ \epsilon' \end{bmatrix}\left(\frac{z}{c\tau+d},\frac{a\tau+b}{c\tau+d}\right)}{\theta\begin{bmatrix} a\epsilon + c\epsilon' - ac \\ b\epsilon + d\epsilon' + bd \end{bmatrix}(z,\tau)} = \kappa\left(\begin{bmatrix} \epsilon \\ \epsilon' \end{bmatrix},\gamma\right)(c\tau+d)^{\frac{1}{2}},$$

where $\kappa\left(\begin{bmatrix} \epsilon \\ \epsilon' \end{bmatrix},\gamma\right)$ is a constant of absolute value one that depends only on the motion γ and the characteristic $\begin{bmatrix} \epsilon \\ \epsilon' \end{bmatrix}$. It is a root of unity if the characteristic is rational. Its dependence on the motion and characteristic is rather complicated.

We are interested mostly in the case $z = 0$, and we hence abbreviate the θ-*constant* $\theta\begin{bmatrix} \epsilon \\ \epsilon' \end{bmatrix}(0,\cdot)$ by $\theta\begin{bmatrix} \epsilon \\ \epsilon' \end{bmatrix}$; the θ-constant is obviously a holomorphic function on $\mathbb{H}^2$. The transformation formula clearly simplifies in this case, and implies that for certain characteristics, a sufficiently high power of the θ-constant is a cusp form for an appropriate subgroup of the modular group.

[2]Throughout this paper, $[x]$ stands for the integral part of $x \in \mathbb{R}$ (the greatest integer $\leq x$), and $[[x]]$ for the smallest integer $\geq x$. Thus $[[x]] = -[-x]$.

2.3. Fuchsian groups. We use certain families of subgroups of the modular group $\Gamma = \text{PSL}(2,\mathbb{Z})$. The theory of θ-constants with rational characteristics is most suitable for the study of the function theory for the *level N principal congruence subgroup*, $\Gamma(N)$, of the modular group (here $N \in \mathbb{Z}^+$) defined by

$$\Gamma(N) = \left\{ \begin{bmatrix} a & b \\ c & d \end{bmatrix} \in \Gamma;\ a \text{ and } d \equiv 1 \mod N,\ b \text{ and } c \equiv 0 \mod N \right\}.$$

This point of view was pursued in [**10**] and [**11**]. In the present context, we rely mostly on the odd integral characteristic $\begin{bmatrix} 1 \\ 1 \end{bmatrix}$ and a different family of groups. Let p and $q \in \mathbb{Z}^+$. We define

$$\Gamma(p,q) = \left\{ \begin{bmatrix} a & b \\ c & d \end{bmatrix} \in \Gamma;\ p|b \text{ and } q|c \right\}.$$

The formulae

$$\Gamma(1,q) = \Gamma_o(q) \text{ and } \Gamma(p,1) = \Gamma^o(p)$$

relate this terminology to a more standard one. It follows readily that

$$\Gamma(p,q) = \Gamma^o(p) \bigcap \Gamma_o(q) \cong \Gamma_o(pq).$$

Since we are dealing with finite index subgroups G of the modular group Γ, that contains only 2 and 3 torsion, it makes sense to define

$\mu = \mu(G) = [\Gamma : G]$, the index of G in Γ,
$\nu_i = \nu_i(G)$, the number of G-inequivalent elliptic fixed points (in $\mathbb{H}^2$) of order $i = 2,\ 3$, and
$\nu_\infty = \nu_\infty(G)$, the number of G-inequivalent parabolic fixed points (in $\mathbb{Q} \cup \{\infty\}$).

While ν_2 and ν_3 are only nonnegative, ν_∞ must be positive (because of the finite index assumption). We call the quadruple $\{\mu;\ \nu_2,\ \nu_3,\ \nu_\infty\}$, the *branch schema* for G. This quadruple is a topological invariant for the orbifold $\mathbb{H}^2/G$; its genus $p = p(G)$ is computed, using Riemann-Hurwitz, as

$$p = 1 + \frac{\mu}{12} - \frac{\nu_2}{4} - \frac{\nu_3}{3} - \frac{\nu_\infty}{2}.$$

The canonical projection

$$P = P(G) : \mathbb{H}^2 \to \mathbb{H}^2/G$$

extends to the set of fixed points, $\mathbb{Q}_G \subset \mathbb{Q} \cup \{\infty\}$, of parabolic elements of G; we will use the same symbol P to denote the extension, and we will often write P_x for $P(x)$ with $x \in \mathbb{H}^2 \cup \mathbb{Q}_G$. The compactification of $\mathbb{H}^2/G$ obtained by filling in the punctures will be denoted by $\overline{\mathbb{H}^2/G}$. We thus also view

$$P : \mathbb{H}^2 \cup \mathbb{Q}_G \to \overline{\mathbb{H}^2/G}.$$

For most of our groups, $\mathbb{Q}_G = \mathbb{Q} \cup \{\infty\}$. The following two tables describe the topology of the most common surfaces encountered on our journey and hint (look for surfaces of genus 0) why congruences for some primes k are easier to describe.

Branch schema and genera for some groups $\Gamma_o(k)$.

k	μ	ν_2	ν_3	ν_∞	p
1	1	1	1	1	0
2	3	1	0	2	0
3	4	0	1	2	0
4	6	0	0	3	0
5	6	2	0	2	0
6	12	0	0	4	0
7	8	0	2	2	0
8	12	0	0	4	0
9	12	0	0	4	0
10	18	2	0	4	0
11	12	0	0	2	1
12	24	0	0	6	0
13	14	2	2	2	0
17	18	2	0	2	1
19	20	0	2	2	1
23	24	0	0	2	2

Branch schema and genera of some groups $\Gamma(k,k)$.

k	μ	ν_2	ν_3	ν_∞	p
1	1	1	1	1	0
2	6	0	0	3	0
3	12	0	0	4	0
4	24	0	0	6	0
5	30	2	0	6	0
6	72	0	0	10	2
7	56	0	2	8	1
8	96	0	0	12	3
9	108	0	0	12	4
10	180	0	0	18	6
11	132	0	0	12	6
12	288	0	0	28	10
13	182	2	2	14	8

To introduce the groups of primary concern to us, we fix a prime k. Let $m \in \mathbb{Z}^+ \cup \{0\}$. Then

$$[\Gamma_o(k) : \Gamma(k^m, k)] = k^m$$

and the motions

$$B^l = \begin{bmatrix} 1 & l \\ 0 & 1 \end{bmatrix}; \ l = 0,\ 1,\ \ldots,\ k^m - 1,$$

form representatives for the $\Gamma(k^m, k)$-cosets in $\Gamma_o(k)$. Thus the averaging operator

$$\frac{1}{k^m} \sum_{l=0}^{k^m-1} f \circ B^l$$

takes a $\Gamma(k^m, k)$-invariant function f to a $\Gamma_o(k)$-invariant function.

2.4. Automorphic functions and forms. The function

$$\tau \mapsto \frac{1}{(2\pi)^8}(\theta')^8\left[\begin{array}{c}1\\1\end{array}\right](0,\tau)$$

is a basis for the one dimensional space of cusp forms[3] $\mathbb{A}_6(\mathbb{H}^2,\Gamma)$ for the modular group. Its Fourier series expansion in terms of the variable $x=\exp(2\pi\imath\tau)$ is

$$x+\sum_{n=2}^{\infty}a_nx^n,$$

and it has no zeros in $\mathbb{H}^2$. It hence make sense to extract an appropriate root of this function; a 24-th root in this case. We have obtained the classical η-function. Two formulae relate the various functions under study. The first identity

$$\eta(\tau)=\exp\left(-\frac{\pi\imath}{6}\right)\theta\left[\begin{array}{c}\frac{1}{3}\\1\end{array}\right](0,3\tau)$$

is an easy consequence of the fact that $\mathbb{H}^2/\Gamma$ has genus zero (see, for example, [**12**]); the second

$$\theta'\left[\begin{array}{c}1\\1\end{array}\right](0,\tau)=2\pi\imath\theta^3\left[\begin{array}{c}\frac{1}{3}\\1\end{array}\right](0,3\tau)$$

follows directly from our definitions. The group Γ acts naturally on the η-function. We have

$$\eta(\tau+1)=\exp\left(\frac{\pi\imath}{12}\right)\eta(\tau)\text{ and }\eta\left(-\frac{1}{\tau}\right)=\exp\left(-\frac{\pi\imath}{4}\right)\eta(\tau)\sqrt{-\tau}.$$

The above invariance formulae (part of the multiplier system for the η-function) follow almost immediately from the transformation formula for the θ-function and the expressions for κ obtained in [**10**]. We need, however, complete information for the multiplier system of this function. It is given in the following proposition whose proof can be found in [**18**, Ch. 3], where the definitions of the Jacobi and Legendre symbols $\left(\frac{\cdot}{\cdot}\right)^*$ and $\left(\frac{\cdot}{\cdot}\right)_*$ (whose values are ±1) can also be found. The proof given there has a strong number theoretic flavor.

PROPOSITION 1. *The η-function satisfies the functional equations*

$$\eta(\gamma(\tau))=v(\gamma)(c\tau+d)^{\frac{1}{2}}\eta(\tau),\ \tau\in\mathbb{H}^2,\ \gamma=\left[\begin{array}{cc}a&b\\c&d\end{array}\right]\in\mathrm{SL}(2,\mathbb{Z}),$$

where

$$v(\gamma)=\left\{\begin{array}{r}\left(\frac{d}{c}\right)^*\exp\left\{\frac{\pi\imath}{12}[(a+d)c-bd(c^2-1)-3c]\right\},\ c\text{odd},\\ \left(\frac{c}{d}\right)_*\exp\left\{\frac{\pi\imath}{12}[(a+d)c-bd(c^2-1)+3(d-1-c]\right\},\ c\text{even}.\end{array}\right.$$

2.5. Products and sums. The only closed expressions for the partition functions P_N known to us are the *Jacobi identity* (the case $N=-3$)

$$\prod_{n=1}^{\infty}(1-x^n)^3=\sum_{n=0}^{\infty}(-1)^n(2n+1)x^{\frac{n(n+1)}{2}}\tag{3}$$

and the *Euler identity* ($N=-1$)

$$\prod_{n=1}^{\infty}(1-x^n)=\sum_{n=-\infty}^{\infty}(-1)^nx^{\frac{n(3n+1)}{2}}.\tag{4}$$

[3]For elements φ in this space $\varphi(z)dz^6$ is invariantly defined.

From any (in particular, from a combinatorial) proof (see [**17**, §19.11] for an argument based on [**14**]) of this second identity, one can obtain the first one. Euler's identity is also a consequence of the proof of the *Jacobi triple product*

$$\prod_{n=1}^{\infty}(1-x^{2n})(1+x^{2n-1}z)\left(1+\frac{x^{2n-1}}{z}\right)=\sum_{n=-\infty}^{\infty}x^{n^2}z^n. \tag{5}$$

We hence conclude that in terms of $x=\exp(2\pi\imath\tau)$,

$$\eta(\tau)=x^{\frac{1}{24}}\sum_{n=-\infty}^{\infty}(-1)^n x^{\frac{n(3n+1)}{2}}=x^{\frac{1}{24}}\prod_{n=1}^{\infty}(1-x^n)$$

for which one can give a purely function theoretic proof (see [**9**]).

2.6. The final ingredient in our recipe is the construction of the same meromorphic function on a compact Riemann surface in two or more ways. We then express the resulting equality in terms of "good" local coordinates.

2.7. We summarize the key points for the construction of partition identities. Start with a fixed Riemann surface Y and a meromorphic function on it. Choose (if it exists) a group G acting analytically on the surface. Average the function over the group. Obtain a meromorphic function on the quotient surface Y/G. Obtain an alternate description of the meromorphic function constructed. Rewrite the resulting equation in terms of appropriate local coordinates. In many cases, interesting formulae are consequences of this recipe.

3. Constructions of meromorphic functions

It is convenient to start with

DEFINITION 2. For the positive prime k, let

$$\alpha(k)=\begin{cases}8 \text{ for } k=2\\ 3 \text{ for } k=3\\ 1 \text{ for } k\geq 5\end{cases} \quad\text{and}\quad \beta(k)=\begin{cases}1 \text{ for } k\leq 5\\ \frac{k^2-1}{24} \text{ for } k\geq 5\end{cases}.$$

Note that for all primes k,

$$\frac{k^2-1}{24}=\frac{\beta(k)}{\alpha(k)}.$$

For a finite index subgroup G of Γ, we identify the field of meromorphic functions[4] $\mathcal{K}(\overline{\mathbb{H}^2/G})$ on $\overline{\mathbb{H}^2/G}$ with the field $\mathcal{K}(G)$ of G-invariant meromorphic functions on $\mathbb{H}^2$ with continuous extensions (with values in $\mathbb{C}\cup\{\infty\}$) to $\overline{\mathbb{H}^2\cup\mathbb{Q}_G}$.

If for some prime k, $\overline{\mathbb{H}^2/\Gamma_o(k)}$ has genus zero (this happens only for $k=2$, 3, 5, 7 and 13), then $f_k(\tau)=\left(\frac{\eta(k\tau)}{\eta(\tau)}\right)^{\frac{24}{k-1}}\in\mathcal{K}(\overline{\mathbb{H}^2/\Gamma_o(k)})$ with divisor $\frac{P_\infty}{P_0}$. It is also useful to observe that for $k=2$, 3 and 5, the surface $\overline{\mathbb{H}^2/\Gamma(k,k)}$ has genus zero, and $h_k(\tau)=\left(\frac{\eta(k\tau)}{\eta(\frac{\tau}{k})}\right)^{\frac{24}{k^2-1}}\in\mathcal{K}(\Gamma(k,k))$ with divisor $\frac{P_\infty}{P_0}$. The interplay between these functions is an important ingredient in our study.

[4]To avoid notational proliferation, we write $f(\tau)$ to indicate that f is a function defined for $\tau\in\mathbb{H}^2\cup\mathbb{Q}\cup\{\infty\}$. We are thus identifying, as in earlier classical notation, a function and its value at a point.

The next calculation (theorem below) is more involved. Before we state the result, it is instructive to analyze the role of $\alpha(k)$ and $\beta(k)$ in our theory. We are mostly concerned with the family of groups $\Gamma(k^m, k)$. The most important cases being $m = 0$ and 1. The surfaces represented by such groups have two distinguished punctures: P_∞ and P_0. We define two linear subspaces of the field $\mathcal{K}(\Gamma(k^m, k))$:

$$\mathcal{K}(\Gamma(k^m, k))_\infty = \{f \in \mathcal{K}(\Gamma(k^m, k));\ f \text{ is regular except at } P_\infty\}$$

and

$$\mathcal{K}(\Gamma(k^m, k))_0 = \{f \in \mathcal{K}(\Gamma(k^m, k));\ f(\infty) = 0 \text{ and } f \text{ is regular except at } P_0\}.$$

DEFINITION 3. We define $\beta(\Gamma(k^m, k))$ as the smallest positive integer β with $P_\infty^\beta P_0^{-\beta}$ a principal divisor on the compact Riemann surface $\overline{\mathbb{H}^2/\Gamma(k^m, k)}$. We let $\beta_{P_\infty}(\Gamma(k^m, k))$ (respectively, $\beta_{P_0}(\Gamma(k^m, k))$) be the smallest nongap at P_∞ (P_0).

At this point, it is easy to see (as a consequence of Proposition 1) that $\alpha(k)$ is the smallest positive integer α such that

$$\left(\frac{\eta(k\tau)}{\eta\left(\frac{\tau}{k^m}\right)}\right)^\alpha,\ m \text{ odd and } \left(\frac{\eta(\tau)}{\eta\left(\frac{\tau}{k^m}\right)}\right)^\alpha,\ m \text{ even},$$

is a $\Gamma(k^m, k)$-invariant function. For the β-invariants, note that

$$\beta(\Gamma(k^m, k)) \le \beta_{P_\infty}(\Gamma(k^m, k)),\ \beta(\Gamma(k^m, k)) \le \beta_{P_0}(\Gamma(k^m, k)) \text{ and } \beta(\Gamma(k, k))|\beta(k).$$

We have established that for small primes k and conjecture that for all primes k,

$$\beta(\Gamma(k, k)) = \beta(k).$$

The three numbers $\beta(\Gamma(k^m, k))$, $\beta_{P_\infty}(\Gamma(k^m, k))$ and $\beta_{P_0}(\Gamma(k^m, k))$ are important function theoretic invariants whose properties will be investigated in future studies.

THEOREM 1. *Let k be a prime and $N \in \mathbb{Z}$.*

(a) *For odd $n \in \mathbb{Z}^+$,* [5]

$$F_{k,n,N}(\tau) = \frac{1}{k^n} \sum_{l=0}^{k^n-1} \left(\frac{\eta(k(\tau + l))}{\eta(\frac{\tau+l}{k^n})}\right)^{\alpha(k)N}$$

defines a $\Gamma_o(k)$-invariant function that is holomorphic, except possibly at the punctures, with

$$\mathrm{ord}_{P_\infty} F_{k,n,N} \ge \left[\left[\frac{\alpha(k)N}{24}\left(k - \frac{1}{k^n}\right)\right]\right];$$

equality holds provided $N > 0$. Further

$$\mathrm{ord}_{P_0} F_{k,n,N} = \frac{\alpha(k)N}{24}(1 - k^{n+1}) \text{ for } N > 0.$$

(b) *For even $n \in \mathbb{Z}^+ \cup \{0\}$,*

$$F_{k,n,N}(\tau) = \frac{1}{k^n} \sum_{l=0}^{k^n-1} \left(\frac{\eta(\tau + l)}{\eta(\frac{\tau+l}{k^n})}\right)^{\alpha(k)N}$$

[5]Abbreviate $F_{k,1,N} = F_{k,N}$.

defines a $\Gamma_o(k)$-invariant function that is holomorphic, except possibly at the punctures, with

$$\mathrm{ord}_{P_\infty} F_{k,n,N} = \left[\left[\frac{\alpha(k)N}{24}\left(1-\frac{1}{k^n}\right)\right]\right],$$

and

$$\mathrm{ord}_{P_0} F_{k,n,N} = \frac{\alpha(k)Nk}{24}(1-k^n) \text{ for } N>0.$$

The basic fact needed to establish the theorem is that

$$\left(\frac{\eta(k\tau)}{\eta\left(\frac{\tau}{k^n}\right)}\right)^{\alpha(k)}, \ n \text{ odd and } \left(\frac{\eta(\tau)}{\eta\left(\frac{\tau}{k^n}\right)}\right)^{\alpha(k)}, \ n \text{ even,}$$

is a $\Gamma(k^n,k)$-invariant function with poles only on points $\Gamma_o(k)/\Gamma(k^n,k)$-equivalent to P_0. When its N-th power is averaged over $\Gamma_o(k)/\Gamma(k^n,k)$ it produces a $\Gamma_o(k)$-invariant function $F_{k,n,N}$ that has only singularities at P_0, if $N>0$.

The significance of the last theorem for partition identities becomes evident once we specialize to low values for the prime k and we obtain alternate expressions for the functions $F_{k,n,N}$.

COROLLARY 1. *For $k = 2, 3, 5, 7$ and 13, and $N \in \mathbb{Z}^+$,*

$$F_{k,n,N} = \sum_{i=[[\frac{\alpha(k)N}{24}(k-\frac{1}{k^n})]]}^{\frac{\alpha(k)N}{24}(k^{n+1}-1)} c_i f_k^i, \ n \text{ odd, and } F_{k,n,N} = \sum_{i=[[\frac{\alpha(k)N}{24}(1-\frac{1}{k^n})]]}^{\frac{\alpha(k)N}{24}(k^{n+1}-k)} c_i f_k^i, \ n \text{ even.}$$

Further, the extreme (that is, $c_{[[\frac{\alpha(k)N}{24}(k-\frac{1}{k^n})]]}$ and $c_{\frac{\alpha(k)N}{24}(k^{n+1}-1)}$ ($c_{[[\frac{\alpha(k)N}{24}(1-\frac{1}{k^n})]]}$ and $c_{\frac{\alpha(k)N}{24}(k^{n+1}-k)}$) for odd (even) n) coefficients are nonzero. Thus for odd n (in terms of the local coordinate $x=\exp(2\pi\imath\tau)$ at P_∞ on $\overline{\mathbb{H}^2/\Gamma_o(k)}$,

$$\begin{aligned} &F_{k,n,N}(\tau) \\ &= \prod_{m=1}^{\infty}(1-x^{km})^{\alpha(k)N} \sum_{m=[[\frac{\alpha(k)N}{24}(k-\frac{1}{k^n})]]}^{\infty} P_{\alpha(k)N}\left(k^n m - \frac{\alpha(k)N(k^{n+1}-1)}{24}\right)x^m \\ &= \sum_{i=[[\frac{\alpha(k)N}{24}(k-\frac{1}{k^n})]]}^{\frac{\alpha(k)N}{24}(k^{n+1}-1)} c_i \left[x\frac{\prod_{m=1}^{\infty}(1-x^{km})^{\frac{24}{k-1}}}{\prod_{n=1}^{\infty}(1-x^m)^{\frac{24}{k-1}}}\right]^i, \end{aligned}$$

$$c_{[[\frac{\alpha(k)N}{24}(k-\frac{1}{k^n})]]} = P_{\alpha(k)N}\left(k^n - \frac{\alpha(k)N(k^{n+1}-1)}{24}\right).$$

For even n,

$$\begin{aligned} F_{k,n,N}(\tau) &= \prod_{m=1}^{\infty}(1-x^m)^{\alpha(k)N} \sum_{m=[[\frac{\alpha(k)N}{24}(1-\frac{1}{k^n})]]}^{\infty} P_{\alpha(k)N}\left(k^n m - \frac{\alpha(k)N(k^n-1)}{24}\right)x^m \\ &= \sum_{i=[[\frac{\alpha(k)N}{24}(1-\frac{1}{k^n})]]}^{\frac{\alpha(k)Nk}{24}(k^n-1)} c_i \left[x\frac{\prod_{m=1}^{\infty}(1-x^{km})^{\frac{24}{k-1}}}{\prod_{m=1}^{\infty}(1-x^m)^{\frac{24}{k-1}}}\right]^i, \end{aligned}$$

$$c_{[[\frac{\alpha(k)N}{24}(1-\frac{1}{k^n})]]} = P_{\alpha(k)N}\left(k^n - \frac{\alpha(k)N(k^n-1)}{24}\right).$$

To explain the connection between partition congruences and the last corollary, it is convenient to define three integers. Assume. as before, that we restrict the prime k to the values 2, 3, 5, 7 and 13. Let us fix $n \in \mathbb{Z}^+ \cup \{0\}$ and $N \in \mathbb{Z}^+$. We start with an indexing set

$$I = \{I_1,\ I_1 + 1,\ ...,\ I_2\},$$

where

$$I_1 = \begin{cases} \left[\left[\alpha(k)\frac{N}{24}\left(k - \frac{1}{k^n}\right)\right]\right] & \text{for } n \text{ odd} \\ \left[\left[\alpha(k)\frac{N}{24}\left(1 - \frac{1}{k^n}\right)\right]\right] & \text{for } n \text{ even} \end{cases} \text{ and } I_2 = \begin{cases} \alpha(k)\frac{N}{24}(k^{n+1} - 1) & \text{for } n \text{ odd} \\ \alpha(k)\frac{Nk}{24}(k^n - 1) & \text{for } n \text{ even} \end{cases}.$$

THEOREM 2. *Under the hypothesis (and using the notation) of the last corollary,* $c_i \in \mathbb{Z}$ *and*

$$c = d_1 = d,$$

where

$$c = c(k,n,N) = \gcd\{c_i; i \in I\},$$

$$d_1 = d_1(k,n,N) = \gcd\left\{P_{\alpha(k)N}\left(k^n m - \alpha(k)N\frac{k^{2\left[\frac{n+1}{2}\right]} - 1}{24}\right); m \in I\right\}$$

and

$$d = d(k,n,N) = \gcd\left\{P_{\alpha(k)N}\left(k^n m - \alpha(k)N\frac{k^{2\left[\frac{n+1}{2}\right]} - 1}{24}\right); m \geq I_1\right\}.$$

REMARK 1. The above theorem is based on the properties of the functions $F_{k,n,N}$ with $N \in \mathbb{Z}^+$. A similar result holds for the functions $F_{k,N}$ (that is for $n = 1$ only) with $-N \in \mathbb{Z}^+$. It is not known whether such results can be extended to the case $n > 1$.

In principle, the last two theorems tell us much information about many partition congruences for the primes 2, 3, 5, 7 and 13. Any particular case (a choice of k, n and N) can, of course be resolved by a finite calculation (of $d_1(k,n,N)$). Plenty of work remains to obtain general theorems. The known methods involve a calculation of $F_{k,1} = F_{k,1,1}$ (or more generally $F_{k,N} = F_{k,1,N}$ for small $|N|$) followed by an induction argument on n (n and N) to obtain information on divisibility properties of $c(k,n,N)$. The known general results yield only information about the divisiblity of $d(k,n,N)$ by powers of the prime k. The induction argument has three ingredients: linear operators acting on linear subspaces of $\mathcal{K}(\Gamma_o(k))$, relations obtained by viewing certain $\Gamma_o(k)$-invariant functions as $\Gamma(k,k)$-functions and delicate combinatorial analysis of congruence properties of the integers c_i. The first two of these ingredients are discussed in the next two sections.

4. Linear operators

Fix a prime k. For each nonnegative integer n, we have encountered a linear operator

$$U_{k,n} : \mathcal{K}(\Gamma(k^n,k)) \to \mathcal{K}(\Gamma_o(k))$$

defined by

$$U_{k,n}(f)(\tau) = \frac{1}{k^n}\sum_{l=0}^{k^n-1} f(\tau + l).$$

We abbreviate $U_{k,1} = U_k$. Observe that for $n \in \mathbb{Z}^+ \cup \{0\}$,

$$U_{k,2n+1}\left(\left(\frac{\eta(k\cdot)}{\eta\left(\frac{\cdot}{k^{2n+1}}\right)}\right)^{\alpha(k)N}\right) = F_{k,2n+1,N}$$

$$\text{and } U_{k,2n}\left(\left(\frac{\eta}{\eta\left(\frac{\cdot}{k^{2n}}\right)}\right)^{\alpha(k)N}\right) = F_{k,2n,N}.$$

In particular,

$$U_k\left(\left(\frac{\eta(k\cdot)}{\eta\left(\frac{\cdot}{k}\right)}\right)^{\alpha(k)N}\right) = F_{k,N} \text{ and } \left(\frac{\eta(k\cdot)}{\eta\left(\frac{\cdot}{k}\right)}\right)^{\alpha(k)} = h_k \text{ for } k = 2,\ 3 \text{ and } 5.$$

More important for our work is the $\mathbb{C}$-linear operator

$$V_k : \mathcal{K}(\Gamma_o(k)) \to \mathcal{K}(\Gamma_o(k)),\ V_k(f)(\tau) = \frac{1}{k}\sum_{l=0}^{k-1} f\left(\frac{\tau+l}{k}\right),$$

that is part of a family of operators $V_{k,M}$, $M \in \mathbb{Z}$ ($V_k = V_{k,0}$), defined by

$$V_{k,M}(f)(\tau) = \frac{1}{k}\sum_{l=0}^{k-1}\left(\frac{\eta(k(\tau+l))}{\eta\left(\frac{\tau+l}{k}\right)}\right)^{\alpha(k)M} f\left(\frac{\tau+l}{k}\right),\ \tau \in \mathbb{H}^2.$$

If $k = 2,\ 3,\ 5,\ 7$ or 13, then $f_k^N \in \mathcal{K}(\Gamma_o(k))_0$ ($f_k^N \in \mathcal{K}(\Gamma_o(k))_\infty$) provided $N > 0$ ($N < 0$). Calculations tell us that[6] in these cases

$$V_{k,M}(f_k^N) = f_k^{-N} F_{k,M+\frac{k+1}{\beta(k)}N}.$$

For all primes k, all $N \in \mathbb{Z}$ and all $n \in \mathbb{Z}^+$,

$$V_k(F_{k,2n-1,N}) = F_{k,2n,N} \text{ and } V_{k,N}(F_{k,2n,N}) = F_{k,2n+1,N}.$$

The last two formulae are the basis for one of the two ingredients in induction arguments on n (the level) for congruences for the partition functions P_N. We need a tool for induction on N. It turns out that the same tool is the second ingredient for induction on n. It is discussed in the next section.

5. Modular equations

Every $\Gamma_o(k)$-invariant function is of course $\Gamma(k,k)$-invariant. In particular, for $k = 2,\ 3,\ 5,\ 7$ and 13, $f_k^{-1}(\tau) = \left(\frac{\eta(\tau)}{\eta(k\tau)}\right)^{\frac{24}{k-1}}$ is a $\Gamma_o(k)$-invariant function with divisor $P_0 P_\infty^{-1}$. Its divisor on $\overline{\mathbb{H}^2/\Gamma(k,k)}$ is $P_0 P_1 \ldots P_{k-1} P_\infty^{-k}$. For $k = 2,\ 3$ and 5, $h_k^{-1}(\tau) = \left(\frac{\eta\left(\frac{\tau}{k}\right)}{\eta(k\tau)}\right)^{\frac{24}{k^2-1}}$ is a $\Gamma(k,k)$-invariant function with divisor $P_0 P_\infty^{-1}$. Hence for these k, $\left(\frac{\eta\left(\frac{\tau}{k}\right)}{\eta(k\tau)}\right)^{\frac{24}{k^2-1}}$ is a polynomial of degree k in $\left(\frac{\eta(\tau)}{\eta(k\tau)}\right)^{\frac{24}{k-1}}$ with zero constant term. We call these and similar relations for functions on $\Gamma(k,k)$, *modular equations for the prime k*. It is of interest to also consider the $\Gamma^o(k)$-invariant

---

[6]Note that

$$\frac{k+1}{\beta(k)} = \begin{cases} k+1 \text{ for } k = 2,\ 3,\ 5 \\ \frac{24}{k-1} \text{ for } k = 5,\ 7,\ 13 \end{cases}.$$

function $\left(\frac{\eta(\tau)}{\eta(\frac{\tau}{k})}\right)^{\frac{24}{k-1}}$ with divisor $P_\infty P_0^{-1}$; its projection to $\overline{\mathbb{H}^2/\Gamma(k,k)}$ has divisor $P_\infty P_1 ... P_{\frac{1}{k-1}} P_0^{-k}$. For $k = 2$, 3 and 5 it is a polynomial of degree k in $\left(\frac{\eta(k\tau)}{\eta(\frac{\tau}{k})}\right)^{\frac{24}{k^2-1}}$ with zero constant term.

5.1. $k = 2$. A modular equation in this case is

$$f_2^{-1} = 2^4 h_2^{-1} + h_2^{-2} \text{ or } h_2^{2+N} = f_2(2^4 h_2^{1+N} + h_2^N);$$

the first of these translates, after some elementary algebraic manipulation, to the power series identity

$$\prod_{n=1}^{\infty}(1+x^n)^{16} = 2^4 x \prod_{n=1}^{\infty}(1+x^n)^8 \prod_{n=1}^{\infty}(1+x^{2n})^{16} + \prod_{n=1}^{\infty}(1+x^{2n})^8.$$

The second identity tell us that

$$F_{2,2+N} = f_2(2^4 F_{2,1+N} + F_{2,N})$$

which along with (the easily computed expressions)

$$F_{2,0} = 1 \text{ and } F_{2,1} = 2^3 f_2$$

gives us a recursive way for computing $F_{2,N}$ (for all $N \in \mathbb{Z}$). It is convenient to introduce the following

DEFINITION 4. For each prime k and each $n \in \mathbb{Z} - \{0\}$, we let $\pi_k(n)$ be the highest power of k that divides n. (We define $\pi_k(0) = +\infty$.)

LEMMA 1. *For each* $N \in \mathbb{Z}^+$,

$$F_{2,N} = \sum_{i=[[\frac{N}{2}]]}^{N} c_{N,i} f_2^i \text{ with } c_{N,i} \in \mathbb{Z}^+,$$

and

$$\pi_2(c_{N,i}) \begin{cases} = 0 \text{ if } N \text{ is even and } i = \frac{N}{2} \\ \geq 7 + 8\left(i - \frac{N}{2} - 1\right) \text{ if } N \text{ is even and } i > \frac{N}{2} \\ = 3 \text{ if } N \text{ is odd and } i = \frac{N+1}{2} \\ \geq 3 + 8\left(i - \frac{N+1}{2}\right) \text{ if } N \text{ is odd and } i > \frac{N+1}{2} \end{cases}.$$

In particular,

$$\pi_2(c_{N,i}) \geq 3 + 8\left(i - \frac{N+1}{2}\right).$$

5.2. $k = 3$. The basic equation is now

$$f_3^{-1} = 3^3 h_3^{-1} + 3^2 h_3^{-2} + h_3^{-3} \text{ or } h_3^{3+N} = f_3(3^3 h_3^{2+N} + 3^2 h_3^{1+N} + h_3^N).$$

The first of these leads to the power series identity

$$\prod_{n=1}^{\infty}(1-x^{3n})^{12} = 3^3 x^2 \prod_{n=1}^{\infty}(1-x^n)^3 \prod_{n=1}^{\infty}(1-x^{9n})^9$$
$$+3^2 x \prod_{n=1}^{\infty}(1-x^n)^6 \prod_{n=1}^{\infty}(1-x^{9n})^6 + \prod_{n=1}^{\infty}(1-x^n)^9 \prod_{n=1}^{\infty}(1-x^{9n})^3;$$

the second to the recursion formula ($N \in \mathbb{Z}$)

$$F_{3,3+N} = f_3(3^3 F_{3,2+N} + 3^2 F_{3,1+N} + F_{3,N}).$$

5.3. $k = 5$. In this case, the basic functional equation is

$$f_5^{-1} = 5^2 h_5^{-1} + 5^2 h_5^{-2} + 3 \cdot 5 h_5^{-3} + 5 h_5^{-4} + h_5^{-5}; \tag{6}$$

it is equivalent to the identity between infinite products

$$\prod_{n=1}^{\infty}(1-x^{5n})^6 = 5^2 x^4 \prod_{n=1}^{\infty}(1-x^n)\prod_{n=1}^{\infty}(1-x^{25n})^5 +$$

$$5^2 x^3 \prod_{n=1}^{\infty}(1-x^n)^2 \prod_{n=1}^{\infty}(1-x^{25n})^4 + 3\cdot 5x^2 \prod_{n=1}^{\infty}(1-x^n)^3 \prod_{n=1}^{\infty}(1-x^{25n})^3 +$$

$$5x \prod_{n=1}^{\infty}(1-x^n)^4 \prod_{n=1}^{\infty}(1-x^{25n})^2 + \prod_{n=1}^{\infty}(1-x^n)^5 \prod_{n=1}^{\infty}(1-x^{25n}).$$

We rewrite the modular equations in more general form. For each integer N,

$$h_5^{5+N} = f_5 h_5^N (5^2 h_5^4 + 5^2 h_5^3 + 3\cdot 5 h_5^2 + 5h_5 + 1)$$

and hence (applying the operator U_5)

$$F_{5,5+N} = f_5(5^2 F_{5,4+N} + 5^2 F_{5,3+N} + 3\cdot 5F_{5,2+N} + 5F_{5,1+N} + F_{5,N}). \tag{7}$$

We know that

$$F_{5,0} = 1,\ F_{5,1} = 5f_5,\ F_{5,2} = 5^3 f_5^2 + 2\cdot 5f_5,$$

$$F_{5,3} = 5^5 f_5^3 + 3\cdot 5^3 f_5^2 + 3^2 f_5 \text{ and } F_{5,4} = 5^7 f_5^4 + 2^2 5^5 f_5^3 + 2\cdot 5^2 11 f_5^2 + 2^2 f_5.$$

We have previously seen that for all $N \in \mathbb{Z}^+$,

$$F_{5,N} = \sum_{i=[[\frac{N}{5}]]}^{N} c_{N,i} f_5^i$$

with $c_{N,i} \in \mathbb{Z}$ (Theorem 2). As a consequence of (7) we have an alternate proof of this fact, and we also have a recursive way for computing these constants. We need the following stronger version of the last formula.

LEMMA 2. *For each $N \in \mathbb{Z}^+ \cup \{0\}$,*

$$F_{5,N} = \sum_{i=[[\frac{N}{5}]]}^{N} c_{N,i} f_5^i \text{ with } c_{N,i} \in \mathbb{Z}^+ \text{ and } 5^{\left[\frac{5i-N-1}{2}\right]} | c_{N,i}.$$

In particular, $5|c_{N,i}$ for all i provided $N \equiv 1$ or $2 \mod 5$, and $5|c_{N,i}$ for all $i > [[\frac{N}{5}]]$.

The sum is to be interpreted as 1 for $N = 0$ and $5^{\left[\frac{5i-N-1}{2}\right]}$ is to be replaced by 1 for $5i - N \leq 2$.

PROOF. Direct computations establish the claim for $0 \leq N \leq 4$. For an induction (on N) argument assume that the index i in the sum runs from 1 to N (with $c_{N,i} = 0$ for $i < [[\frac{N}{5}]]$). From the basic recursion formula (7),

$$c_{N+5,i} = 5^2 c_{N+4,i-1} + 5^2 c_{N+3,i-1} + 3\cdot 5c_{N+2,i-1} + 5c_{N+1,i-1} + c_{N,i-1}.$$

Thus by induction $5^{\left[\frac{5i-(N+5)-1}{2}\right]} | c_{N+5,i}$. □

The group $\mathrm{PSL}(2,\mathbb{Z})$ acts on functions by composition. Applying the operator A to (6), we obtain another modular equation

$$f_{1,5}^6 = h_5 + 5h_5^2 + 3\cdot 5h_5^3 + 5^2h_5^4 + 5^2h_5^5.$$

5.4. $k = 7$. This is a more complicated case because h_7 is not $\Gamma(7,7)$-invariant, but h_7^2 is[7]. On $\overline{\mathbb{H}^2/\Gamma(7,7)}$,

$$(f_7^{-1}) = P_0\ P_1\ ...\ P_6\ P_\infty^{-7} \text{ and } (h_7^{-2}) = P_0^2\ P_\infty^{-2}.$$

We can certainly choose constants c_i and d_j so that

$$H = f_7^{-2} - f_7^{-1}\sum_{i=0}^{3} c_i h_7^{-2i} - \sum_{j=1}^{7} d_j h_7^{-2j}$$

has a pole of order ≤ 5 at P_∞ and its Laurent series expansion at this point in terms of the local coordinate $x = \exp\left(\frac{2\pi\imath\tau}{7}\right)$ has zero coefficients for the exponents -4 and -2. It turns out that with this choice of constants, the coefficients for the exponents -5, -3, -1 and 0 also vanish. So the function H is identically zero for this choice of coefficients. We thus have obtained the identity

$$f_7^{-2} = f_7^{-1}\left(7h_7^{-6} + 5\cdot 7h_7^{-4} + 7^2h_7^{-2}\right) +$$

$$h_7^{-14} + 7h_7^{-12} + 3\cdot 7h_7^{-10} + 7^2h_7^{-8} + 3\cdot 7^2h_7^{-6} + 7^3h_7^{-4} + 7^3h_7^{-2}.$$

Applying the operator U_7 yields

$$F_{7,7+N} = f_7\left(7F_{7,4+N} + 5\cdot 7F_{7,5+N} + 7^2F_{7,6+N}\right) + f_7^2\times$$

$$\left(F_{7,N} + 7F_{7,1+N} + 3\cdot 7F_{7,2+N} + 7^2F_{7,3+N} + 3\cdot 7^2F_{7,4+N} + 7^3F_{7,5+N} + 7^3F_{7,6+N}\right).$$

REMARK 2. Not all the coefficients in the recursion formulae for $F_{k,N}$ are divisible by the prime k; this is a main reason why congruence relations are not easily obtained.

5.5. $k = 13$. On $\overline{\mathbb{H}^2/\Gamma(13,13)}$

$$(f_{13}^{-1}) = P_0\ P_1\ ...\ P_{12}\ P_\infty^{-13} \text{ and } (h_{13}^{-7}) = P_0^7\ P_\infty^{-7}.$$

Any two meromorphic functions on the same Riemann surface are, of course, algebraically related. It is not clear how to use the complicated identity that connects the above two functions.

6. Constant functions

Fix a prime k. The functions $F_{k,N}$ are never constant if $N \in \mathbb{Z}^+$. Of course, $F_{k,0} = 1$. Assume that $-N \in \mathbb{Z}^+$. The function $F_{k,N}$ is constant for many choices of k and N. Since the only possible pole of this function is at P_∞, it will be constant if and only if it is regular there. The condition that the function $F_{k,N}$ be regular at P_∞ is that the coefficients for negative indices of an appropriate Laurent series expansion vanish. This is our next

[7] The surface $\overline{\mathbb{H}^2/\Gamma(7,7)}$ has genus one.

THEOREM 3. *Let k be a positive prime and let $N \in \mathbb{Z}^+$. The following conditions are equivalent:*
(a) *$F_{k,-N}$ is a constant function.*
(b) *$P_{-\alpha(k)N}(mk + \beta(k)N) = 0$ for $m = -\left[\frac{\beta(k)N}{k}\right], \ldots, -2, -1$.*
(c) *the function $F_{k,-N}$ is invariant under the map $A_k : \tau \mapsto -\frac{1}{k\tau}$.*
Under these conditions, the constant value of $F_{k,-N}(\tau)$ is $P_{-\alpha(k)N}(\beta(k)N)$.

PROOF. Only that (c) implies (a) requires some comment. If $F_{k,-N}$ is invariant under the automorphism A_k of $\overline{\mathbb{H}^2/\Gamma_o(k)}$, then (since this function is holomorphic at P_0) it is holomorphic at P_∞; hence constant. □

We illustrate in one case. For $k = 5$, the functions $F_{5,-N}$ for $N = 1, 2, 3$ and 4, are constant for trivial reasons (this function has a pole of order at most $\left[\frac{N}{5}\right]$ at P_∞ and is regular elsewhere on $\overline{\mathbb{H}^2/\Gamma_o(5)}$). In addition, only $F_{5,-8}$ and $F_{5,-14}$ are constant. This follows from two facts. The first of these are the general formulae $P_{-N}(1) = -N$ (used for $N \equiv 1 \mod 5$), $P_{-N}(2) = \frac{N(N-3)}{2}$ (used for $N \equiv 2 \mod 5$), $P_{-N}(3) = -\frac{N(N-1)(N-8)}{3!}$ (used for $N \equiv 3 \mod 5$), $P_{-N}(4) = \frac{N(N-1)(N-3)(N-14)}{4!}$ (used for $N \equiv 4 \mod 5$) and $P_{-N}(0) = 1$ (used for $N \equiv 0 \mod 5$). The second fact we need is the calculation $0 = P_{-14}(4) = P_{-14}(9)$. We also use that $P_{-8}(8) = -5^3$ and $P_{-14}(14) = 5^6$.

It is obvious that $F_{k,n,0} = 1$ for all k and $n \in \mathbb{Z}^+$. If for some odd n, $F_{k,m,N} = \text{const}$[8] (thus $N \leq 0$), then also $F_{k,n+1,N} = \text{const}$; while if for some even n, $F_{k,n,N} = \text{const}$, then $F_{k,n+1,N} = \text{const}\ F_{k,N}$. This observation is the basis for the next theorem and provides an alternative way to establish higher level congruences for P_{-N}, $N \in \mathbb{Z}^+$.

THEOREM 4. *If $F_{k,-N}$ is constant $(= P_{-\alpha(k)N}(\beta(k)N))$ for the prime k and $N \in \mathbb{Z}^+$, then so is $F_{k,n,-N}$ $(= P_{-\alpha(k)N}(\beta(k)N)^{\left[\frac{n+1}{2}\right]})$ for all $n \in \mathbb{Z}^+$.*

PROOF. Let $c = P_{-\alpha(k)N}(\beta(k)N)$. Assume by induction that $F_{k,n,-N}$ is constant for $n \in \mathbb{Z}^+$. If n is odd, then

$$F_{k,n+1,-N} = V_k(c^{\left[\frac{n+1}{2}\right]}) = c^{\left[\frac{n+2}{2}\right]}.$$

For even n,

$$F_{k,n+1,-N} = V_k(c^{\left[\frac{n+1}{2}\right]} f_{k^2,1}^{-\alpha(k)N}) = c^{\left[\frac{n}{2}\right]} F_{k,-N} = c^{\left[\frac{n+2}{2}\right]}.$$

□

As a consequence we have, for example, for $k = 5$, $N = -4$ and $n \in \mathbb{Z}^+$,

$$\prod_{m=1}^{\infty} (1 - x^{5m})^{-4} \sum_{m=0}^{\infty} P_{-4}\left(5^n m + \frac{5^{2\left[\frac{n+1}{2}\right]} - 1}{6}\right) x^m = (-5)^{\left[\frac{n+1}{2}\right]}.$$

[8]The symbol const will be used to represent a (usually nonzero) constant whose value might change with multiple uses in a single equation.

7. Partition identities, positive N

There are quite general results about $\pi_k(d(k,n,N))$ for $k = 2, 3, 5, 7$ and 13 (see, for example, [4]; these are the cases with $\overline{\mathbb{H}^2/\Gamma_o(k)}$ of genus 0); the corresponding results ([3]) for $k = 11$, for example, are not as general. These divisibility properties of d are partition congruences. As a sample, we illustrate for the prime 5.

THEOREM 5. *For all n and $m \in \mathbb{Z}^+$,* (a)

$$P\left(5^n m - \frac{5^{2\left[\frac{n+1}{2}\right]} - 1}{24}\right) \equiv 0 \mod 5^n$$

and (b)

$$P_2\left(5^n n - \frac{5^{2\left[\frac{n+1}{2}\right]} - 1}{12}\right) \equiv 0 \mod 5^{1+\left[\frac{n}{2}\right]}.$$

Similar congruences exist for P_N for all $N \in \mathbb{Z}^+$. The results for the primes 2, 3 and 7 are similar. Those for the prime 13, are much more restrictive. Among them are general results of the type

$$P_{10}(13m + 8) \equiv 0 \text{ mod } 13, \text{ all } m \in \mathbb{Z}^+ \cup \{0\},$$

and more specialized, perhaps accidental, formulae such as

$$P_{-15}(13m + 1) \equiv 0 \text{ mod } 5, \text{ all } m \in \mathbb{Z}^+ \cup \{0\}.$$

The congruence ([24], [21])

$$P(13m - 7) \equiv 11P_{-11}(m - 1) \text{ mod } 13, \; m \in \mathbb{Z}^+,$$

hints at possible deeper undiscovered results.

8. Partition identities, negative N

Although Theorem 1 holds for all $N \in \mathbb{Z}$, the consequences (Corollary 1 and Theorem 2) needed for applications hold for $N \in \mathbb{Z}^+$. For negative N, these two facts hold for $n = 1$. We must, in this case, replace I_2 (and all relevant upper limit for sums involving the coefficients c_i) by 0. In some real sense, the case of negative N is easier to handle because of the appearance of constant (and, more generally, low degree) functions. It leads to stronger results; replacing congruences by equalities.

Once again, we illustrate for $k = 5$. We already saw that $F_{5,-N}$ is a constant functions for $N = 1, 2, 3, 4, 8$ and 14. Hence, for these values of N, we have nontrivial *two term* level one congruences.[9] They are

$$P_{-1}(5m + 1) = -P_{-1}\left(\frac{m}{5}\right), \; P_{-2}(5m + 2) = -P_{-2}\left(\frac{m}{5}\right),$$

$$P_{-3}(5m + 3) = 5P_{-3}\left(\frac{m}{5}\right), \; P_{-4}(5m + 4) = -5P_{-4}\left(\frac{m}{5}\right),$$

$$P_{-8}(5m + 3) = -5^3 P_{-8}\left(\frac{m-1}{5}\right) \text{ and } P_{-14}(5m + 4) = 5^6 P_{-4}\left(\frac{m-2}{5}\right).$$

[9]Since we have formulae for $P_N(m)$ for $N = -1$ and -3, all $m \in \mathbb{Z}^+ \cup \{0\}$. Congruence for P_{-3} and P_{-1} should follow directly from (3) and (4), respectively. In this approach, they are part of a more general phenomenum.

Each of these has a higher level analogue (for each $n \in \mathbb{Z}^+ \cup \{0\}$):

$$P_{-1}\left(5^{2n}m + \frac{5^{2n}-1}{24}\right) = (-1)^n P_{-1}(m)$$

$$\text{and } P_{-1}\left(5^{2n+1}m + \frac{5^{2n+2}-1}{24}\right) = (-1)^{n+1} P_{-1}\left(\frac{m}{5}\right),$$

$$P_{-2}\left(5^{2n}m + \frac{5^{2n}-1}{12}\right) = (-1)^n P_{-2}(m)$$

$$\text{and } P_{-2}\left(5^{2n+1}m + \frac{5^{2n+2}-1}{12}\right) = (-1)^{n+1} P_{-2}\left(\frac{m}{5}\right),$$

$$P_{-3}\left(5^{2n}m + \frac{5^{2n}-1}{8}\right) = 5^n P_{-3}(m)$$

$$\text{and } P_{-3}\left(5^{2n+1}m + \frac{5^{2n+2}-1}{8}\right) = 5^{n+1} P_{-3}\left(\frac{m}{5}\right),$$

$$P_{-4}\left(5^{2n}m + \frac{5^{2n}-1}{6}\right) = (-5)^n P_{-4}(m)$$

$$\text{and } P_{-4}\left(5^{2n+1}m + \frac{5^{2n+2}-1}{6}\right) = (-5)^{n+1} P_{-4}\left(\frac{m}{5}\right),$$

$$P_{-8}\left(5^{2n}m + \frac{5^{2n}-1}{3}\right) = (-5)^{3n} P_{-8}(m)$$

$$\text{and } P_{-8}\left(5^{2n+1}m + \frac{5^{2n+2}-1}{3}\right) = (-5)^{3(n+1)} P_{-8}\left(\frac{m}{5}\right),$$

$$P_{-14}\left(5^{2n}m + 7\frac{5^{2n}-1}{12}\right) = 5^{6n} P_{-14}(m)$$

$$\text{and } P_{-14}\left(5^{2n+1}m + 7\frac{5^{2n+2}-1}{12}\right) = 5^{6(n+1)} P_{-14}\left(\frac{m}{5}\right).$$

Partition identities for P_N with negative N are explored in more detail in [**8**] and the literature quoted there. As remarked earlier, the above two term identities arise because certain functions are constant. *Three term* identities result from the construction of functions of degree one. For the prime 5, the resulting identity is the recursion formula for P_{-6}

$$P_{-6}(5m+1) = -5^2 P_{-6}\left(\frac{m-1}{5}\right) - 2\cdot 3P_{-6}(m).$$

9. The primes $k = 11$, 17 and 19

The closed surface $\overline{\mathbb{H}^2/\Gamma_o(k)}$ has genus one for $k = 11$, 17 and 19. A good theory for congruences for P has been developed for the first of these cases, [**4**]. For smaller primes, the modular equation is a key tool for deriving higher level congruences from the one for level one. For $k = 11$, this tool is not easily available and substitutes must be found. Level one congruences can be derived many ways (see, for example, [**8**].) Results for the primes 17 and 19 are even less accessible.

THEOREM 6. *For all n and $m \in \mathbb{Z}^+$,*

$$P\left(11^n m - \frac{11^{2[\frac{n+1}{2}]}-1}{24}\right) \equiv 0 \ mod\ 11^n.$$

There are no known comparable results for the higher partition functions P_N, $N \geq 2$. When combined with the congruences for the primes below 11, the last theorem yields the following

COROLLARY 2. (a) *If* $3m \equiv 1 \mod 2^a$, *then*

$$P_8(m) \equiv 0 \mod 2^{3[[\frac{a+1}{2}]]}.$$

(b) *If* $8m \equiv 1 \mod 3^b$, *then*

$$P_3(m) \equiv 0 \mod 3^{1+[\frac{3b}{2}]}.$$

(c) *If* $24m \equiv 1 \mod 5^c 7^d 11^e$, *then*

$$P(m) \equiv 0 \mod 5^c 7^{1+[\frac{d}{2}]} 11^e.$$

This last result should be contrasted with the identity established by Ramanujan for the Klein $\jmath$-function

$$\jmath(\tau) = \sum_{n=-1}^{\infty} c_n x^n = \frac{\left(1 + 240 \sum_{n=1}^{\infty} \sigma_3(n) x^n\right)^3}{x\mathbf{P}^{-24}(x) - 744}, \quad x = \exp(2\pi\imath\tau), \ \tau \in \mathbb{H}^2,$$

where for $N \in \mathbb{Z}^+$,

$$\sigma_N(n) = \sum_{d \in \mathbb{Z}^+;\ d|n} d^N.$$

Ramanujan's result is: if $m \equiv 0 \mod 2^a 3^b 5^c 7^d 11^e$, then

$$c(m) \equiv 0 \mod 2^{3a+2^3} 3^{2b+3} 5^c 7^d 11^e.$$

10. Abelian differentials of the third kind on spheres

Let G be a subgroup of Γ such that $\overline{\mathbb{H}^2/G}$ is conformally equivalent to the Riemann sphere $\mathbb{C} \cup \{\infty\}$. Let f be a meromorphic function of degree one on $\overline{\mathbb{H}^2/G}$. If f has a zero and pole at x and $y \in \overline{\mathbb{H}^2/G}$, respectively, then $\omega = \frac{df}{f}$ is an abelian differential of the third kind on $\overline{\mathbb{H}^2/G}$ regular except for simple poles at x and y with residues $+1$ and -1, respectively. Independent constructions of f and ω will lead to interesting identities connecting partition functions with the σ-function, involving the divisors of integers. See, for example, Theorem 8, below.

11. Combinatorial interpretations of some identities

The purpose of this second part of our paper is to convince the reader that some of the identities discussed in the first part have extremely interesting combinatorial and number theoretic interpretations. It is not clear whether all the results we expose require these identities, but the fact that they follow so easily seems worthy of remark and note. In fact we shall see that some of the results also follow from simpler well known classical identities. A popular title for this section could very well be "Curious properties of integers." We describe these properties in the form of theorems. The theorems all involve in some way partitions of the positive integers.

11.1. Let S_0 denote the set consisting of 8 copies of the positive even integers and S_1 denote the set consisting of 8 copies of the positive odd integers. We shall consider partitions of the integers with parts (summands) taken from the sets S_0 and (or) S_1. It is of course immediate that there are no partitions of an odd integer from the set S_0.

Denote by $L_0^8(2N)$ the number of partitions of the positive even integer $2N$ with parts taken from S_0 and by $L_1^8(N)$ the number of partitions of the positive integer N with parts from S_1.

THEOREM 7. *For all* $N \in \mathbb{Z}^+$,

$$L_1^8(2N+1) = 8L_0^8(2N).$$

PROOF. The theorem is a consequence of the theta identity

$$\eta^8\left(\frac{\tau}{2}\right) + \eta^8\left(\frac{\tau+1}{2}\right) = -2^4\eta^8(2\tau)$$

(part of the theory for the prime $k = 2$) that translates to the infinite product identity (for $x = \exp\{\pi\imath\tau\}$)

$$\prod_{n=1}^{\infty}(1-x^{2n-1})^8 - \prod_{n=1}^{\infty}(1+x^{2n-1})^8 = -2^4x\prod_{n=1}^{\infty}(1+x^{2n})^8. \tag{8}$$

□

We find the above result quite striking but can understand that one could react to it with "so what?" Why are statements about partitions interesting? In order to answer this question we present another example which on the one hand attempts to answer this question and on the other presents another curious property. The above identity is also an immediate consequence of the Jacobi quartic identity.

REMARK 3. The modular equation for $k = 2$ is equivalent to (8). The theory for $k = 4$ leads to the identity

$$2^4\left(\left(\frac{\theta\begin{bmatrix}0\\0\end{bmatrix}(0,2\tau)}{\theta\begin{bmatrix}1\\0\end{bmatrix}(0,2\tau)}\right)^4 - 1\right) = \left(\frac{\eta(\tau)}{\eta(4\tau)}\right)^8$$

that also translates to the product identity (8).

11.2. Let S be the set consisting of 4 copies of the odd positive integers and 4 copies of the even positive integers which are congruent to zero mod 4. A partition of $N \in \mathbb{Z}^+$ taken from the set S is called *even (odd)* provided it contains an even (odd) number of even parts (summands); $d(N) = e(N) - o(N)$ denotes the difference between the number of even and odd partitions of N. Denote by $T_4(N)$ the number of representations of N as a sum of four triangular numbers (a number of the form $\frac{m(m+1)}{2}$ with $m \in \mathbb{Z}^+ \cup \{0\}$. Each representation of N is a vector (n_1, n_2, n_3, n_4) with n_j triangular and $\sum_{j=1}^4 n_j = N$ with different vectors corresponding to distinct representations. As usual $\sigma(N)$ is the sum of the positive divisors of N.

THEOREM 8. *For each* $N \in \mathbb{Z}^+$,

$$d(N) = \sigma(2N+1) = T_4(N).$$

PROOF. The $k = 2$ theory leads to the theta identity

$$\text{const}\ \frac{\theta'\begin{bmatrix}1\\1\end{bmatrix}^{\frac{8}{3}}(0,2\tau)}{\theta'\begin{bmatrix}1\\1\end{bmatrix}^{\frac{4}{3}}(0,\tau)} = \frac{d}{d\tau}\log\frac{\theta^4\begin{bmatrix}0\\0\end{bmatrix}(0,\tau)}{\theta^4\begin{bmatrix}0\\1\end{bmatrix}(0,\tau)}.$$

The identity is a consequence of the fact that both sides of the equality are meromorphic one forms on the Riemann surface which is the compactification of the upper half plane factored by the group $\Gamma(2,2)$ with the same divisors. This identity translates to the infinite product identity

$$x\prod_{n=1}^{\infty}(1-x^{2n})^4(1+x^{2n})^8 = \sum_{n=1}^{\infty}\sigma(2n-1)x^{2n-1}.$$

We explain how the combinatorial results follow.

Simple algebraic manipulations (using $1 - y^2 = (1-y)(1+y)$) followed by a change of variable (replacing x^2 by x) and some more algebraic manipulation leads to

$$\prod_{n=1}^{\infty}(1-x^{4n})^4(1+x^{2n-1})^4 = \sum_{n=1}^{\infty}\sigma(2n-1)x^{n-1};$$

whose combinatorial content is the first equality of the theorem.

It remains to make the connection with $T_4(N)$. This follows from the identity

$$\prod_{n=1}^{\infty}(1-x^{4n})(1+x^{2n-1}) = \sum_{n=0}^{\infty}x^{\frac{n^2+n}{2}} \tag{9}$$

which is a consequence of the Jacobi triple product formula (5). □

The above results are consequences of the $k = 2$ theory and as such are in a sense classical since this theory has been well understood for a long while. We do not know if the above interpretations have been made but they probably have. Before we leave this case however it seems worthwhile to point out that the above relates partition theory to one of the most basic functions of number theory, the σ-function. As a consequence of this relation we have the following characterization of primes.

THEOREM 9. *The odd number $2l + 1$ is a prime if and only if*

$$d(l) = T_4(l) = 2(l+1).$$

We now turn to the case $k = 3$ where we expect some new results. The first of these again involves the difference between the number of even and odd partitions.

11.3. Let S be the set consisting of 4 copies of those positive integers congruent to 0 mod 9, 3 copies of the positive integers which are relatively prime to 9, and 1 copy of those positive integers which are congruent to either 3 or 6 mod 9. As before, the *parity* of a partition is the parity of the number of its parts. We shall let $E(N)$ and $O(N)$ denote the number of even and odd partitions of N.

THEOREM 10. *For all $N \in \mathbb{Z}^+$, $N \geq 2$,*

$$E(N-1) - O(N-1) = \begin{cases} 0 \text{ if } N \equiv 0 \mod 3 \\ \sigma(N) \text{ if } N \equiv 1 \mod 3 \\ -\sigma(N) \text{ if } N \equiv 2 \mod 3 \end{cases}.$$

PROOF. The theorem follows from, once again, a theta constant identity:

$$\text{const}\,\frac{\theta'\begin{bmatrix}1\\1\end{bmatrix}(0,3\tau)\,\theta'\begin{bmatrix}1\\1\end{bmatrix}(0,\frac{\tau}{3})}{\theta'\begin{bmatrix}1\\1\end{bmatrix}^{\frac{2}{3}}(0,\tau)} = \frac{d}{d\tau}\log\frac{\theta\begin{bmatrix}\frac{1}{3}\\\frac{1}{3}\end{bmatrix}(0,\tau)}{\theta\begin{bmatrix}\frac{1}{3}\\\frac{5}{3}\end{bmatrix}(0,\tau)},$$

which in terms of the η-function is

$$\text{const}\,\frac{\eta^3(3\tau)\,\eta^3\left(\frac{\tau}{3}\right)}{\eta^2(\tau)} = \frac{d}{d\tau}\log\frac{\theta\begin{bmatrix}\frac{1}{3}\\\frac{1}{3}\end{bmatrix}^3(0,\tau)}{\theta\begin{bmatrix}\frac{1}{3}\\\frac{5}{3}\end{bmatrix}^3(0,\tau)},$$

whose translation in terms of infinite products and sums (with $x = \exp\left(\frac{2\pi\imath\tau}{3}\right)$) is

$$x\prod_{n=1}^{\infty}\frac{(1-x^{9n})^3(1-x^n)^3}{(1-x^{3n})^2} = \sum_{n=0}^{\infty}\left(\sigma(3n+1)x^{3n+1} - \sigma(3n+2)x^{3n+2}\right).$$

□

We see that 9 plays a special role in the previous theorem. It seems to be the case that 9 is always special in the theory for $k = 3$. We give another example to demonstrate this point.

11.4. Two different constructions of a certain meromorphic differentials on the Riemann surface $\overline{\mathbb{H}^2/\Gamma(3)}$ yield the theta identity

$$\text{const}\,\frac{\theta'\begin{bmatrix}1\\1\end{bmatrix}^{\frac{10}{3}}(0,\tau)}{\theta'\begin{bmatrix}1\\1\end{bmatrix}(0,3\tau)\,\theta'\begin{bmatrix}1\\1\end{bmatrix}(0,\frac{\tau}{3})} = \frac{d}{d\tau}\log\frac{\theta'\begin{bmatrix}1\\1\end{bmatrix}(0,3\tau)}{\theta'\begin{bmatrix}1\\1\end{bmatrix}(0,\frac{\tau}{3})};$$

which in terms of the η-function can be written

$$\text{const}\,\frac{\eta^{10}(\tau)}{\eta^3(3\tau)\,\eta^3\left(\frac{\tau}{3}\right)} = \frac{d}{d\tau}\log\frac{\eta^3(3\tau)}{\eta^3\left(\frac{\tau}{3}\right)},$$

that translates in terms of the variable $x = \exp(\frac{2\pi\imath\tau}{3})$ to the identity

$$\text{(10)}\quad \prod_{n=1}^{\infty}\frac{(1-x^{3n})^{10}}{(1-x^n)^3(1-x^{9n})^3} = 1+3\sum_{n=1}^{\infty}\sigma(n)(x^n - 9x^{9n}) = 1+3\sum_{n=1}^{\infty}\hat{\sigma}(n)x^n,$$

where $\hat{\sigma}(N)$ is the sum of the divisors of N which are **not** congruent to zero mod 9. A different variant of the σ-function, is obtained by setting $\tilde{\sigma}(N)$ to be the sum of the divisors of N which are **not** congruent to zero mod 3. Averaging the meromorphic one form on $\mathbb{H}^2/\Gamma(3,3) = \mathbb{H}^2/\Gamma(3)$ given above over a group of translations, we obtain a one form on $\mathbb{H}^2/\Gamma_o(3)$ that yields the theta identity

$$\text{const}\,\frac{\theta'\begin{bmatrix}1\\1\end{bmatrix}^{\frac{10}{3}}(0,\tau)}{\theta'\begin{bmatrix}1\\1\end{bmatrix}(0,3\tau)}\sum_{l=0}^{2}\frac{\exp\left(\frac{2\pi\imath l}{24}\right)}{\theta'\begin{bmatrix}1\\1\end{bmatrix}(0,\frac{\tau+l}{3})} = \frac{d}{d\tau}\log\left(\frac{\theta'\begin{bmatrix}1\\1\end{bmatrix}(0,3\tau)}{\theta'\begin{bmatrix}1\\1\end{bmatrix}(0,\tau)}\right)^4.$$

This theta identity translates to

$$\prod_{n=1}^{\infty} \frac{(1-x^n)^{10}}{(1-x^{3n})^3} \left(\sum_{n=0}^{\infty} P_3(3n) x^n \right) = 1 + 12 \sum_{n=1}^{\infty} \tilde{\sigma}(n) x^n.$$

Replacing x by x^3, multiplying by $\prod_{n=1}^{\infty} \frac{1}{(1-x^n)^3}$ and using (10) tells us that

$$\left(1 + 3 \sum_{n=1}^{\infty} \hat{\sigma}(n) x^n \right) \prod_{n=1}^{\infty} (1-x^n)^3 = \frac{1 + 12 \sum_{n=1}^{\infty} \tilde{\sigma}(n) x^{3n}}{\sum_{n=0}^{\infty} P_3(3n) x^{3n}};$$

which as a consequence of the Jacobi identity can be rewritten as

$$\left(1 + 3 \sum_{n=1}^{\infty} \hat{\sigma}(n) x^n \right) \left(\sum_{n=0}^{\infty} (-1)^n (2n+1) x^{\frac{n(n+1)}{2}} \right) = \frac{1 + 12 \sum_{n=1}^{\infty} \tilde{\sigma}(n) x^{3n}}{\sum_{n=0}^{\infty} P_3(3n) x^{3n}}. \tag{11}$$

It is clear from the above identity that the right hand side is a function of x^3 so that the same is true of the left hand side. We have thus established most of

THEOREM 11. *If $n \in \mathbb{Z}^+$ is congruent to either 1 or 2 mod 3, then*

$$3 \sum_{j=0}^{\frac{j^2+j}{2} \le n-1} (-1)^j (2j+1) \hat{\sigma}\left(n - \frac{j(j+1)}{2} \right) + P_{-3}(n) = 0.$$

PROOF. Take the Cauchy product of the two series to the left of the equality in (11) and use the fact that the terms corresponding to exponents which are not congruent to 0 mod 3 vanish. □

The above theorem has an amusing and interesting by-product. Let us begin by recalling that triangular numbers, those of the form $\frac{j(j+1)}{2}$, are always congruent to either 0 or 1 mod 3. They are never congruent to 2 mod 3. Furthermore, if n is congruent to 2 mod 3, then for any j, the difference $n - \frac{j(j+1)}{2}$ is never congruent to 0 mod 3; so surely never congruent to 0 mod 9. It thus follows that

COROLLARY 3. *If N is congruent to 2 mod 3, then*

$$\sum_{j=0}^{\frac{j^2+j}{2} \le N-1} (-1)^j (2j+1) \sigma\left(N - \frac{j(j+1)}{2} \right) = 0.$$

This gives a recursion formula for $\sigma(n)$ in terms of a rather thin set of values.

The above corollary leaves open the question of what happens when N is congruent to 1 mod 3. It turns out that the last formula with a correction for triangular numbers is valid in this case. This was pointed out to one of us by Frank Garvan. Garvan's proof uses the Jacobi identity (3) and proceeds as follows. If we let $f = \mathbf{P}^{-3}$, then

$$-x \frac{f'(x)}{f(x)} = \sum_{n=1}^{\infty} \frac{3x^n}{1-x^n} = 3 \sum_{n=1}^{\infty} \sigma(n) x^n, \ x \in \Delta.$$

Hence it follows that

$$- \sum_{n=1}^{\infty} (-1)^n (2n+1) \frac{n(n+1)}{2} x^{\frac{n(n+1)}{2}} =$$

$$\left(\sum_{n=0}^{\infty}(-1)^n(2n+1)x^{\frac{n(n+1)}{2}}\right)\left(3\sum_{n=1}^{\infty}\sigma(n)x^n\right)$$

and

$$-\sum_{n=1}^{\infty}(-1)^n(2n+1)\frac{n(n+1)}{2}x^{\frac{n(n+1)}{2}}=3\sum_{n=1}^{\infty}a_nx^n,$$

with

$$a_n=\sum_{j=0}^{\frac{j(j+1)}{2}\leq n-1}(-1)^j(2j+1)\sigma\left(n-\frac{j(j+1)}{2}\right).$$

Hence we conclude that if N is not a triangular number, then

$$\sum_{j=0}^{\frac{j(j+1)}{2}\leq N-1}(-1)^j(2j+1)\sigma\left(N-\frac{j(j+1)}{2}\right)=0;$$

while if N is a triangular number $N=\frac{m(m+1)}{2}$, we have

$$\sum_{j=0}^{\frac{j(j+1)}{2}\leq N-1}(-1)^j(2j+1)\sigma\left(N-\frac{j(j+1)}{2}\right)=(-1)^{m+1}(2m+1)\frac{m(m+1)}{6}.$$

We therefore see that the formula of the corollary is also correct for numbers that are congruent to 1 mod 3 provided they are not triangular.

The above observations together with the last theorem therefore give

THEOREM 12. *Assume N is congruent to 1 modulo 3. If N is not a triangular number we have*

$$\sum_{j=0}^{\frac{j(j+1)}{2}\leq N-1}(-1)^j(2j+1)\sigma\left(N-\frac{j(j+1)}{2}\right)=$$

$$\sum_{j=0}^{\frac{j(j+1)}{2}\leq N-1}(-1)^j(2j+1)\widehat{\sigma}\left(N-\frac{j(j+1)}{2}\right)$$

and in particular

$$\sum_{j\in\mathbb{Z}^+,\ j\equiv 1 \mod 3}(-1)^j(2j+1)\sigma\left(N-\frac{j(j+1)}{2}\right)=$$

$$\sum_{j\in\mathbb{Z}^+,\ j\equiv 1 \mod 3}(-1)^j(2j+1)\widehat{\sigma}\left(N-\frac{j(j+1)}{2}\right).$$

If N is a triangular number

$$\sum_{j=0}^{\frac{j(j+1)}{2}\leq N-1}(-1)^j(2j+1)\sigma\left(N-\frac{j(j+1)}{2}\right)=$$

$$N\sum_{j=0}^{\frac{j(j+1)}{2}\leq N-1}(-1)^j(2j+1)\widehat{\sigma}\left(N-\frac{j(j+1)}{2}\right).$$

In the theorem, we interpret both $\sigma(n)$ and $\widehat{\sigma}(n)$ to be zero if $n\notin\mathbb{Z}^+$.

11.5. In the above theorems and examples we have seen that the σ-function or some variant of it enters into the discussion of partitions of the integers. That this is so should not be surprising. It follows from the fact that

$$\frac{x\mathbf{P}'(x)}{\mathbf{P}(x)} = \sum_{n=1}^{\infty} \frac{nx^n}{1-x^n} = \sum_{n=1}^{\infty} \sigma(n)x^n, \ x \in \Delta.$$

Therefore

$$\sum_{n=1}^{\infty} nP(n)x^n = \left(\sum_{n=0}^{\infty} P(n)x^n\right)\left(\sum_{n=1}^{\infty} \sigma(n)x^n\right) = \sum_{n=1}^{\infty} a_n x^n,$$

with

$$nP(n) = a_n = \sum_{j=0}^{n-1} P(j)\sigma(n-j).$$

Hence we see that the σ-function is intimately connected with the partition function. It is well known that $P(n)$ is expressible in terms of previous values $P(m)$, $m < n$. This last fact follows from the observation that $\mathbf{P}\mathbf{P}^{-1} = 1$; that is, from the identity

$$1 = \left(\sum_{n=0}^{\infty} P(n)x^n\right)\left(\sum_{n=-\infty}^{\infty} (-1)^n x^{\frac{3n^2+n}{2}}\right).$$

This last identity says that for all $n \in \mathbb{Z}^+$

$$\sum_{j\in\mathbb{Z};\ \frac{3j^2+j}{2}\le n} (-1)^j P\left(n - \frac{3j^2+j}{2}\right) = 0.$$

In a similar vein, if we work with $f = \mathbf{P}^{-1}$, then,

$$\frac{-xf'(x)}{f(x)} = \sum_{n=1}^{\infty} \sigma(n)x^n, \ x \in \Delta;$$

so that

$$\sum_{n=-\infty}^{\infty} (-1)^{n+1} \frac{3n^2+n}{2} x^{\frac{3n^2+n}{2}} = \left(\sum_{n=1}^{\infty} \sigma(n)x^n\right)\left(\sum_{n=-\infty}^{\infty} (-1)^n x^{\frac{3n^2+n}{2}}\right).$$

Therefore if n is not pentagonal

$$\sum_{j\in\mathbb{Z};\ \frac{3j^2+j}{2}\le n-1} (-1)^j \sigma\left(n - \frac{3j^2+j}{2}\right) = 0$$

and if n is pentagonal, $n = \frac{3m^2+m}{2}$, $m \in \mathbb{Z}$,

$$\sum_{j\in\mathbb{Z};\ \frac{3j^2+j}{2}\le n-1} (-1)^j \sigma\left(n - \frac{3j^2+j}{2}\right) = (-1)^{m+1} \frac{3m^2+m}{2}.$$

11.6. We present another curious property which is related to the results in Theorem 8. The identity (9) that we used there can obviously be rewritten as

$$\prod_{n=1}^{\infty} \frac{1-x^{2n}}{1-x^{2n-1}} = \sum_{n=0}^{\infty} x^{\frac{n(n+1)}{2}}.$$

Denoting the left hand side, as usual, by $f(x)$ we get

$$\frac{xf'(x)}{f(x)} = \sum_{n=1}^{\infty} \left(\frac{(2n-1)x^{2n-1}}{1-x^{2n-1}} - \frac{2nx^{2n}}{1-x^{2n}} \right) = \sum_{n=1}^{\infty} (\sigma_o(n) - \sigma_e(n))x^n,$$

where $\sigma_o(n)$ and $\sigma_e(n)$ represent the sum of the odd and even divisors of n respectively.

It thus follows that

$$\sum_{n-1}^{\infty} \frac{n^2+n}{2} x^{\frac{n^2+n}{2}} = \left(\sum_{n=1}^{\infty} (\sigma_o(n) - \sigma_e(n))x^n \right) \left(\sum_{n=0}^{\infty} x^{\frac{n^2+n}{2}} \right) = \sum_{n=1}^{\infty} a_n x^n,$$

with

$$a_n = \sum_{j=0}^{\frac{j^2+j}{2} \le n-1} \sigma_o\left(n - \frac{j^2+j}{2}\right) - \sigma_e\left(n - \frac{j^2+j}{2}\right).$$

The conclusion is therefore

THEOREM 13. *If N is not a triangular number then*

$$\sum_{j=0}^{\frac{j^2+j}{2} \le N-1} \sigma_o\left(N - \frac{j^2+j}{2}\right) - \sigma_e\left(N - \frac{j^2+j}{2}\right) = 0.$$

11.7. Corollary 3 has an interesting consequence. It implies that $\sigma(3N+2) \equiv 0 \bmod 3$. However this fact follows easily without the corollary. If n is congruent to 2 mod 3, then clearly n must have at least one prime factor congruent to 2 mod 3 appearing with odd multiplicity. Denoting the factor by p and its multiplicity by $2l+1$ we see that $\sigma(p^{2l+1}) = \frac{p^{2l+2}-1}{p-1} \equiv 0 \bmod 3$. Since the σ-function is multiplicative we are done. Once one realizes this fact, one immediately suspects that the σ-function satisfies lots of congruences. It is not hard to see that

1. $\sigma(2N+1) \equiv 0 \bmod 2$ unless $2N+1$ is a square,
2. $\sigma(3N+2) \equiv 0 \bmod 3$,
3. $\sigma(4N+3) \equiv 0 \bmod 4$,
4. $\sigma(6N+5) \equiv 0 \bmod 6$,
5. $\sigma(8N+7) \equiv 0 \bmod 8$,
6. $\sigma(12N+11) \equiv 0 \bmod 12$ and
7. $\sigma(24N+23) \equiv 0 \bmod 24$.

It would be nice to see if these relatively simple congruences for the σ-function give some of the less simple congruences for the partition function.

References

[1] G.E. Andrews, *The Theory of Partitions*, Encyclopedia of Mathematics and its Applications, vol. 2, Addison-Wesley, 1976.
[2] G.E. Andrews and R.A. Askey et al. (editors), *Ramanujan Revisited*, Academic Press, 1988.
[3] A.O.L. Atkin, *Proof of a conjecture of Ramanujan*, Glasgow J. Math. **8** (1967), 14–32.
[4] ———, *Ramanujan congruences for $p_{-k}(n)$*, Can. J. Math. **20** (1968), 67–78.
[5] A.O.L. Atkin and J.N. O'Brien, *Some properties of $p(n)$ and $c(n)$ modulo powers of 13*, Trans. Amer. Math. Soc. **126** (1967), 442–459.

[6] B.C. Berndt, *Ramanujan's modular equations*, Ramanujan Revisited, Academic Press, 1988, pp. 313–333.
[7] F.J. Dyson, *A walk through Ramanujan's garden*, Ramanujan Revisited, Academic Press, 1988, pp. 7–28.
[8] H.M. Farkas and I. Kra, *Ramanujan partition identities*, Proceedings of I Iberoamerican Congress on Geometry, January 1998, Chile, 107–128 and in this volume.
[9] ———, *Theta constants, Riemann surfaces and the modular group*, monograph in preparation.
[10] ———, *Automorphic forms for subgroups of the modular group*, Israel J. Math. **82** (1993), 87–131.
[11] ———, *Automorphic forms for subgroups of the modular group. II: Groups containing congruence subgroups*, J. d'Analyse Math. **70** (1996), 91–156.
[12] ———, *Theta constant identities with applications to combinatorial number theory*, Contemporary Math. **211** (1997), 227–252.
[13] N.J. Fine, *Basic Hypergeometric Series and applications*, Mathematical Surveys and Monographs, vol. 27, American Mathematical Society, 1989.
[14] F. Franklin, *Sur le développement du produit infini* $(1-x)(1-x^2)(1-x^3)\dots$, Comptes Rendus **82** (1881), 448–450.
[15] F.G. Garvan, *Combinatorial interpretations of Ramanujan's partition identities*, Ramanujan Revisited, Academic Press, 1988, pp. 29–45.
[16] G. Gasper and M. Rahman, *Basic Hypergeometric Series*, Encyclopedia of Mathematics and its Applications, vol. 35, Cambridge University Press, 1990.
[17] G.H. Hardy and E.M. Wright, *An Introduction to the Theory of Numbers (fifth edition)*, Clarendon Press, 1995.
[18] M.I. Knopp, *Modular Functions in Analytic Number Theory*, Chelsea Publishing Company, 1993.
[19] J. Lehner, *Lectures on Modular Forms*, Appl. Math. Ser., vol. 61, Nat. Bur. Stand., 1969.
[20] M. Newman, *An identity for the coefficients of certain modular forms*, J. Lond. Math. Soc. **30** (1955), 488–493.
[21] ———, *Congruences for the coefficients of modular forms and some new congruences for the partition function*, Can. J. Math. **9** (1957), 549–552.
[22] S. Ramanujan, *Some properties of $p(n)$, the number of partitions of n*, Proc. Camb. Phil. Soc. **19** (1919), 207–210.
[23] G.N. Watson, *Ramanujans Vermutung über Zerfällungsanzahlen*, J. Reine Angew. Math. **179** (1938), 97–118.
[24] H. Zuckerman, *Identities anologous to Ramanujan's identities involving the partition function*, Duke Math. J. **5** (1939), 88–110.

The Hebrew University of Jerusalem
Jerusalem 91904, Israel
E-mail: farkas@sunset.huji.ac.il
and
State University of New York at Stony Brook
Stony Brook, NY 11794, USA
E-mail: irwin@math.sunysb.edu

Contemporary Mathematics
Volume **240**, 1999

Holomorphic Motions and Quasifuchsian Manifolds

Frederick Gardiner and Linda Keen

ABSTRACT. We use holomorphic motions to show the Hölder exponent $1/K$ of a $K-$quasiconformal mapping and to show that any sufficiently small quakebend deformation of a quasifuchsian group remains quasifuchsian.

Dedicated to the memory of Lipman Bers

Introduction

This article is an expository account of why mappings in a holomorphic motion $f_t(z)$ of the $z-$plane parameterized by a complex variable t with $|t| < 1$ are necessarily quasiconformal maps and an account of some geometric applications of this fact. Our approach is different from the one ordinarily given. In the first section we demonstrate how the generalized Schwarz lemma implies f_t is quasiconformal with dilatation less than or equal to $\frac{1+|t|}{1-|t|}$ and Hölder continuous with Hölder exponent $\frac{1-|t|}{1+|t|}$. Extremal length methods are not used to obtain the precise Hölder exponent. In the second section, we describe Thurston's construction of a quasifuchsian manifold with prescribed geometry as a holomorphic motion.

Our exposition is stimulated by lectures, given by Lipman Bers near the end of his mathematical career, on a paper he wrote jointly with Halsey Royden. Bers first learned about holomorphic motions in a lecture by Sullivan in the mid-eighties, (see the reference to Màñé, Sad and Sullivan, [**23**]). The ensuing theorems of Sullivan and Thurston [**30**], of Bers and Royden [**7**] and, finally, of Slodkowski [**29**] and of Astala [**4**], clarified the subject of quasiconformal mapping as a tool for deforming conformal dynamical systems.

A holomorphic motion of a closed set J in the extended complex plane $\bar{\mathbb{C}}$ is a curve $f_t(z)$, defined for every z in J and for every t in the unit disk, such that:

i) $f_0(z) = z$ for all z in J,
ii) $z \mapsto f_t(z)$ is injective as a function from J into $\overline{\mathbb{C}}$ and
iii) $t \mapsto f_t(z)$ is holomorphic for $|t| < 1$ and for each fixed z in J.

The variable t is the complex time-parameter and as t changes the set $J_t = f_t(J)$ moves in $\overline{\mathbb{C}}$. Although J may start out to be as smooth as a circle and although the points of J are moving holomorphically, for every $t \neq 0$, J_t can be an interesting fractal set with fractional Hausdorff dimension.

1991 *Mathematics Subject Classification.* Primary 32G15, 30C62, 30F40, 30F60; Secondary 30C70, 30C75.

Key words and phrases. Cantor sets, bounded geometric type, non-conical limit sets, Fuchsian groups, hyperbolic distortion lemma, fundamental domains.

Research partially supported by PSC-CUNY and by NSF.

Any holomorphic motion of a closed subset of $\overline{\mathbb{C}}$ defined for $|t| < 1$ automatically extends to a holomorphic motion of $\overline{\mathbb{C}}$. Moreover, for each fixed t, the mapping $z \mapsto f_t(z)$ is quasiconformal with dilatation K less than or equal to $\frac{1+|t|}{1-|t|}$.

Thus injectivity and holomorphic dependence automatically lead to quasiconformality. Viewed from first principles, this is surprising because quasiconformality is a geometric idea about distortion. By definition, an arbitrary orientation preserving homeomorphism is quasiconformal if it distorts standard shapes by a bounded amount. It may distort sizes by a large amount; no assumption is made on this point. A beautiful first result of the theory is that quasiconformal mappings are necessarily Hölder continuous with Hölder exponent $1/K$. (This result is due to Lavrientiev or to Ahlfors, although it is difficult to know who deserves credit because each attributed it to the other [**6**].) Thus, quasiconformal control of distortion of shape implies some control on distortion of size. Moreover, a normalized family of quasiconformal mappings with a common bound on the dilatation of every member is equicontinuous.

The realization that any holomorphic motion defined for $|t| < 1$ and for z in some closed set could be extended to all z in $\overline{\mathbb{C}}$ and for all $|t| < 1$ was the combined result of several research papers. The first one by Màñé, Sad and Sullivan presented the idea of a holomorphic motion and showed that mappings in the motion necessarily have quasiconformal extensions. Then the papers by Sullivan and Thurston [**30**] and by Bers and Royden [**7**] showed the local holomorphic extendibility to a domain $|t| < \epsilon$ where ϵ is universal, not depending on J or the particular holomorphic motion. (Bers and Royden showed that $\epsilon \geq 1/3$.) Finally, Slodkowski [**29**] showed that any such motion could be extended to a holomorphic motion of the whole complex plane defined on the full unit disk, $|t| < 1$.

Slodkowski's result complements nicely the 1961 paper by Ahlfors and Bers [**3**] in which singular operators are used to solve the global Beltrami equation

$$f_{\overline{z}}(z) = t\mu(z)f_z,$$

where μ is an arbitrary L_∞ complex-valued function defined on $\overline{\mathbb{C}}$ with $||\mu||_\infty = 1$ and $|t| < 1$. The use of singular integral operators to solve the Beltrami equation had been carried through earlier by Bojarski in [**8**]. The solution $f^{t\mu}$ is a quasiconformal homeomorphism of $\overline{\mathbb{C}}$ uniquely determined up to post-composition by a Möbius transformation. A major part of the emphasis of the paper by Ahlfors and Bers was that $f^{t\mu}(z)$ depends holomorphically on t for $|t| < 1$. When $t = 0$, $f^{t\mu}$ is the identity and, for each fixed $|t| < 1$, $f^{t\mu}(z)$ is an injective function of z. In other words, their paper showed the existence of many holomorphic motions of $\overline{\mathbb{C}}$. The new theory of holomorphic motions showed their emphasis on the holomorphic dependence had further significance: holomorphic dependence combined with injectivity of $z \mapsto f^{t\mu}(z)$ implies $z \mapsto f^{t\mu}(z)$ is quasiconformal; the holomorphic dependence of the quasfuchsian groups defined by $G_0 \mapsto f^{t\mu}(z) \circ G \circ (f^{t\mu}(z))^{-1}$ for fixed quasifuchsian $G_0 \subset PSL(2, C)$ and quasiconformal $f^{t\mu}(z)$ satisfying $\mu(g(z))\frac{\overline{g'(z)}}{g'(z)} = \mu(z)$ for all $g \in G_0$ has been used to develop the complex structure on Teichmüller space.

In Section 1 of this paper we give an analytic application of holomorphic motions. We show why a holomorphic motion of injective mappings of the plane starting at the identity necessarily consists of mappings which are Hölder continuous and which are quasiconformal. The method of proof of Hölder continuity shows how the $(|\epsilon \log \epsilon|)^{-1}$ growth of the infinitesimal form of the Poincaré metric near a puncture is related, after integration of an ordinary differential equation, to the Hölder continuity of the mappings.

In Section 2 we give a geometric application. In his Princeton lecture notes [**31**], Thurston described a method to construct hyperbolic 3-manifolds. The main step is to construct a quasifuchsian manifold. Such a manifold is the quotient of hyperbolic 3-space by a quasifuchsian group. The construction, which is a generalization of topological methods of Dehn, is geometric. It involves "bending" the surface along geodesics; the resulting manifold not only is quasifuchsian, but it has prescribed geometric properties. We describe Thurston's construction and indicate how to prove, as a consequence of the main theorems on holomorphic motions, that it yields quasifuchsian manifolds with the prescribed geometry.

1. Quasiconformal mapping

The goal of this section is to explain why a holomorphic motion of $\mathbb{C}$ necessarily moves through quasiconformal mappings. Our first step is to show that f_t is Hölder continuous with Hölder exponent $\frac{1-|t|}{1+|t|}$.

We show this exponent by analyzing the tangent vectors to the curve f_t. Let $V_t(z)\frac{\partial}{\partial z}$ be the vector field defined by the equation

$$f_{s+t} \circ (f_t)^{-1}(z) = z + sV_t(z) + o(t).$$

Since $f_t(0) = 0, f_t(1) = 1$ and $f_t(\infty) = \infty$, for each $|t| < 1$, the vector field $V_t(z)\frac{\partial}{\partial z}$ vanishes at $0, 1$ and ∞. Note that a continuous vector field $V(z)\frac{\partial}{\partial z}$ vanishes at ∞ if, and only if, $V(z)/z^2$ approaches 0 as $z \to \infty$.

A cross-ratio Q of four distinct points cannot equal $0, 1$ or ∞. Since f_t is injective, the same cross-ratio Q_t of f_t applied to these four distinct points is also not equal to $0, 1$ or ∞. Therefore, by applying Schwarz's lemma to the map $t \mapsto Q_t$, from $|t| < 1$ into $\Sigma = \overline{\mathbb{C}} \setminus \{0, 1, \infty\}$, we find that the speed of Q_t at time t measured with the Poincaré metric on Σ does not exceed $\frac{2}{1-|t|^2}$.

Let ρ be the infinitesimal form of the Poincaré metric on Σ normalized to have constant negative curvature equal to -4. Assume z_1, z_2, z_3 and z_4 are four distinct points and let $A = z_2 - z_1$, $B = z_3 - z_2$, $C = z_4 - z_3$, and $D = z_1 - z_4$. We call z_1 and z_2 the initial and terminal points of A, and use the notation

$$V'(A) = \frac{V(\textit{terminal point}) - V(\textit{initial point})}{\textit{terminal point} \; - \; \textit{initial point}}.$$

If we choose Q to be the cross-ratio $\frac{AC}{BD}$, a calculation shows that the speed of Q_t at $t = 0$ is

$$\text{(1)} \qquad |\rho(Q)Q\{V'(A) - V'(B) + V'(C) - V'(D)\}|.$$

Since $V(z)/z^2$ approaches 0 as $z \to \infty$, it follows that if $z_4 = \infty$ then the term $V'(C) - V'(D)$ in (1) vanishes. By letting A be the interval from z to 0 and B be the interval from 0 to 1, we find $Q = z$, $V'(A) = -V(z)/z$ and $V'(B) = 0$. Thus,

$$|\rho(z)V_t(z)| \leq \frac{2}{1-|t|^2}.$$

Consequently, by the well-known estimate for the Poincaré metric in a neighborhood of a puncture [**2**], for every $\epsilon > 0$, there is a $\delta > 0$, such that if $|z| < \delta$ then

$$|V_t(z)| \leq (\frac{1+\epsilon}{1-\epsilon})|z|(\log\frac{1}{|z|})\frac{2}{1-|t|^2}. \tag{2}$$

In fact, if $\epsilon = \exp(-d_n)$ where $d_n = \log\log 2^n$, one can choose $\delta = \frac{1}{2^n}$.

LEMMA 1. *Suppose f_t is a holomorphic motion of $\mathbb{C}$ such that $f_t(0) = 0, f_t(1) = 1$ and $f_t(\infty) = \infty$, for each $|t| < 1$. Then there exists a constant C depending only on R such that if $|z| < R$ and $|w| < R$ and if $|z - w| < \frac{1}{2}$, then*

$$|f_t(z) - f_t(w)| \leq C|z-w|^{\frac{1-|t|}{1+|t|}}.$$

Proof: Let $\Delta(t) = |f_t(z) - f_t(w)|$.

Step 1. We first consider the case $w = 0$ and $|z| < 1/2$ so $\Delta(t) = |f_t(z)|$. Since one of the cross ratios of the quadruple $f_t(z), 0, 1, \infty$ is $f_t(z)$, by the bound on the velocity given in (2), we find, for $|z| \leq \frac{1}{2^n}$,

$$\frac{d}{dt}\Delta(t) \leq B\frac{2}{1-|t|^2}\Delta(t)\log\frac{1}{\Delta(t)},$$

where $B = \frac{1+\epsilon_n}{1-\epsilon_n}$ and $\epsilon_n = 1/\log 2^n$. The solution to this differential inequality with the inequality replaced by equality is

$$\Delta(t) = \Delta(0)^{(\frac{1-|t|}{1+|t|})^B}.$$

By the inequality principle for ordinary differential equations, we find

$$|f_t(z)| \leq |z|^{(\frac{1-|t|}{1+|t|})^B}.$$

Step 2. The last term in the previous inequality can be replaced by $C|z|^{\frac{1-|t|}{1+|t|}}$. To prove this assume that $\frac{1}{2^{n+1}} \leq |z| \leq \frac{1}{2^n}$ and let $\alpha = \frac{1-|t|}{1+|t|}$. We wish to find a constant C such that

$$|z|^{\alpha^B} \leq C|z|^\alpha.$$

It is sufficient to find C such that

$$|1/2^n|^{\alpha^B} \leq C|1/(2^{n+1})|^\alpha.$$

This is equivalent to finding an upper bound for

$$(\alpha - \alpha^B)(n\log 2) = \alpha(1-\alpha^{(B-1)})n\log 2.$$

Since $0 < \alpha < 1$, this is equivalent to finding an upper bound for

$$(B-1)n(\log 2)\alpha(\log(\frac{1}{\alpha})). \tag{3}$$

But by (2), $B - 1 = \frac{2\epsilon_n}{1-\epsilon_n}$ and $\epsilon_n = \frac{1}{n\log 2}$ and thus the quantity displayed in (3) is bounded independently of n.

Step 3. Now suppose g_t is a holomorphic motion fixing three points $0, z$ and ∞ and assume $|z - w| \leq \min\{1/2, |z|/2\}$. Then the same arguments as above yield the similar inequality:

$$|g_t(w) - z| \leq C|w-z|^{\frac{1-|t|}{1+|t|}}.$$

Step 4. Apply Step 3 to $g_t(w) = \frac{f_t(w)}{f_t(z)}$, a motion which fixes $0, z$ and ∞. Since $f_t(z)$ is bounded for $|t| \leq |t_0| < 1$, we obtain

$$|f_t(w) - f_t(z)| \leq C|w-z|^{\frac{1-|t|}{1+|t|}},$$

where C is a new constant equal to the product of the constant in Step 3 and the bound on $f_t(z)$. □

The following result is given on page 60 of [**1**].

LEMMA 2. *If* $|t| < 2 - \sqrt{3}$, *equivalently, if the hyperbolic distance from* 0 *to* t *is less than* $\frac{1}{2}\log 3$, *and if* $f_t(\infty) = \infty$, *then the image of oriented vertices of any equilateral triangle under* $z \mapsto f_t(z)$ *are mapped to vertices of a triangle with the same orientation.*

Proof: Let z_1, z_2, z_3 be the vertices of such a triangle labeled in counterclockwise order and let A be the segment from z_1 to z_2, B be the segment from z_2 to ∞, C be the segment from ∞ to z_3 and D the segment from z_3 to z_1. Then

$$Q = \frac{AC}{BD} = \frac{A}{-D} = \frac{z_2 - z_1}{z_3 - z_1},$$

which for an equilateral triangle is equal to

$$\frac{1}{2} + i\frac{1}{2}\sqrt{3}.$$

The mapping $z \mapsto f_t(z)$ maps the vertices z_1, z_2, z_3 to vertices w_1, w_2, w_3. Let Q_t be a cross ratio of w_1, w_2, w_3, ∞. As the points move, the orientation of the triangle remains counterclockwise until the points become collinear. This happens precisely when Q_t is real and cannot happen if $d_\Sigma(Q, Q_t) < d_\Sigma(Q, \mathbb{R})$. By Schwarz's lemma

$$d_\Sigma(Q, Q_t) \leq \log\frac{1+|t|}{1-|t|}.$$

In order to complete the proof of the lemma it suffices to show the Poincaré distance from $\omega = \frac{1}{2}+\frac{1}{2}\sqrt{3}$ to the real axis is $\log\sqrt{3}$. Consider $S_1(z) = 1-z, S_2(z) = 1/z$ and $B(z) = \bar{z}$. Then $B \circ S_1$ and $B \circ S_2$ are anticonformal automorphisms of Σ. Moreover, the intersection of the locus of fixed points of $B \circ S_1$ and $B \circ S_2$ consists of two points, $\frac{1}{2} \pm i\frac{1}{2}\sqrt{3}$, each of which is equidistant from the real axis. In the standard universal covering of Σ by the upper half plane, the upper half of Σ is covered by part of a fundamental domain lying to the right of the y-axis, to the left of the vertical line with x-coordinate equal to 1, and above the semicircle with diameter coinciding with the unit interval $[0, 1]$. The point ω in Σ is covered by the point $\frac{1}{2} + i\frac{1}{2}\sqrt{3}$. Using the infinitesimal form $|d\zeta|/\mathrm{Im}(\zeta)$ for the Poincaré metric in the universal covering, we see that the Poincaré distance from ω to the real axis is equal to

$$\int_{1/2}^{\sqrt{3}/2} \frac{dt}{t} = \log\sqrt{3}.$$

□

LEMMA 3. *Let* $f_t(z)$ *be a holomorphic motion of the whole plane fixing* $0, 1$ *and* ∞, *defined for* $|t| < 1$. *If* $|t| < 2-\sqrt{3}$, *then the mapping* $z \mapsto f_t(z)$ *is quasiconformal.*

Proof: To show $z \mapsto f_t(z)$ is quasiconformal we construct a normalized sequence f_n of quasiconformal mappings such that $K(f_n) \leq K_0$ which converges pointwise to $f_t(z)$. Let Γ_n be a grid of equilateral triangles which fill the plane. Assume each triangle of Γ_n has side-length equal to $\frac{1}{2^n}$ and one of the triangles of Γ_n, which we call the initial triangle, lies in the first quadrant, has a vertex at the origin and one side on the positive real axis. The mapping f_n is defined by declaring that it coincides with f_t on the vertices of Γ_n and is affine in each triangle of Γ_n.

To calculate the complex dilatation of f_n on a triangle Δ of Γ_n, note that that this dilatation is the same as the dilatation of $A \circ f_n \circ B$ where A is affine and $B(z) = cz + d$, where c is real. Select B so that it maps the triangle with vertices at $0, 1$ and $\tau_0 = \frac{1}{2} + i\frac{1}{2}\sqrt{3}$ to a triangle Δ of Γ_n and select A so that $A \circ f_n \circ B$ fixes 0 and 1. Let $\tau = A \circ f_n \circ B(\tau_0)$. Then the complex dilatation μ_n of f_n in the triangle Δ satisfies:

$$\mu_n = -\frac{\tau - \tau_0}{\tau - \overline{\tau_0}}.$$

Assuming $|t| \leq 2 - \sqrt{3} - \epsilon_0$, the argument of the previous lemma implies $|\frac{\tau - \tau_0}{\tau - \overline{\tau_0}}| \leq k_0$, where $k_0 < 1$ and k_0 depends only on $\epsilon_0 > 0$.

We conclude that the normalized family of mappings $\{f_n\}$ satisfies $K(f_n) \leq K_0 = \frac{1+k_0}{1-k_0}$. From the previous lemma, this implies that, restricted to any bounded set in the plane, the family $\{f_n\}$ is equicontinuous. Therefore f_n converges pointwise to f_t.

At this point we adopt the geometric definition of quasiconformality, which says a mapping is quasiconformal if it distorts standard shapes by a bounded amount. What we take as a model for a standard shape is somewhat arbitrary but in this case, by a standard shape we mean a square with horizontal and vertical sides. The image $f_n(S)$ of the square S is a topological rectangle and there is a conformal map g_n from $f_n(S)$ to a rectangle R_n such that $g_n \circ f_n$ preserves horizontal and vertical sides. If a_n is the length of the rectangle and b_n is its height, then by Grötzsch's argument ([**1**], page 7), $1/K_0 \leq a_n/b_n \leq K_0$. This is true for each n and, since f_n converges locally uniformly to f_t, the same inequality is also true for the side lengths of the rectangle $R = g \circ f_t(S)$. Therefore, $K(f_t) \leq \frac{1+|t|}{1-|t|}$. □

LEMMA 4. *Let $f_t(z)$ be a holomorphic motion of the whole plane fixing* $0, 1$ *and* ∞, *defined for* $|t| < 1$. *Then* $z \mapsto f_t(z)$ *is quasiconformal for every* $|t| < 1$.

Proof: Assume $|t| < 1$. On the hyperbolic geodesic joining 0 to t choose points $0 = t_0, t_1, ..., t_n = t$ so that the hyperbolic distance from t_{k-1} to t_k is equal to $1/n$ times the hyperbolic distance from 0 to t. We choose n large enough so that this distance is less than $\frac{1}{2}\log 3$. Let $g_k = f_{t_k} \circ (f_{t_{k-1}})^{-1}$ so that

$$f_t = g_n \circ g_{n-1} \circ ... \circ g_1.$$

Also, let $F_s = f_t \circ (f_{t_{k-1}})^{-1}$ where $t = \frac{s + t_{k-1}}{1 + \overline{t_{k-1}}t}$. Then the dependence of F_s on s for $|s| < 1$ is holomorphic, when $s = 0$, F_s is the identity, and when $t = t_k$, $F_s = g_k$. Since for this value of t, $|s| = |\frac{t_k - t_{k-1}}{1 - \overline{t_{k-1}}t_k}| < 2 - \sqrt{3}$, it follows from the previous lemma g_k is quasiconformal. Therefore, f_t is quasiconformal because it is a composition of quasiconformal mappings. □

It is known that if a mapping f satisfies the geometric definition of quasiconformality, then it has first partial derivatives in the sense of distributions which are locally L_2 and, moreover, the Beltrami coefficient $\mu_f = f_{\bar{z}}/f_z$ is an L_∞–function with $||\mu||_\infty < 1$.

LEMMA 5. *The map*

$$t \mapsto \mu_t = \text{the Beltrami coefficient of } f_t,$$

is a holomorphic mapping from $|t| < 1$ *into the open unit ball of the Banach space* $L_\infty(\mathcal{C})$.

Proof: We must show that $t \mapsto \mu_t$ is differentiable in the complex sense. We may assume f_t is normalized to fix $0, 1$ and ∞ because, if it does not, we may select Möbius transformations A_t so that $A_t \circ f_t$ is normalized this way. But the Beltrami coefficient of $A_t \circ f_t$ is is equal to the Beltrami coefficient of f_t. Let

$$g_s(z) = f_{t+s} \circ (f_t)^{-1}(z) = z + sV_t(z) + o(s).$$

Let z_1, z_2 and z_3 be the vertices of an equilateral triangle and consider the motion of the quadruple z_1, z_2, z_3, ∞ induced by g_s, which depends holomorphically on s. Let $\omega_1 = z_2 - z_1$, $\omega_2 = z_3 - z_2$ and $\omega_3 = z_1 - z_3$ and let W be an affine function whose values at z_1, z_2 and z_3 coincide with the values v_1, v_2 and v_3 of V_t at these points. On the one hand,

$$\bar{\partial} W = \frac{\omega_2 v_1 + \omega_3 v_2 + \omega_1 v_3}{\overline{\omega}_2 \omega_1 - \overline{\omega}_1 \omega_2}.$$

On the other hand, we know from (1) that

$$\bar{\partial} W = \frac{\omega_2 v_1 + \omega_3 v_2 + \omega_1 v_3}{\omega_1^2}$$

is bounded. Approximating V_t by a sequence affine functions V_n which coincide with V_t at the vertices of a hexagonal grid whose equilateral triangles have side length $1/2^n$, we find that $\frac{\partial}{\partial \bar{z}} V_t(z) = \nu(z)$, exists as a generalized partial derivative and $||\nu||_\infty < \infty$. Then

$$[\frac{\nu}{1 - |\mu|^2}](\frac{1}{\theta}) \circ f^{-1},$$

where $f = f_t$, $\theta = \frac{\bar{p}}{p}$ and $p = \frac{\partial}{\partial z} f_t(z)$, is the complex derivative of this mapping with respect to s at $s = 0$. □

Combining the previous lemmas and using Schwarz's lemma we obtain the following theorem.

THEOREM 6. *Let $f_t(z)$ be a holomorphic motion of $\overline{\mathbb{C}}$ defined for $|t| < 1$ where $f_0 =$ identity. Then, for each t with $|t| < 1$, $z \mapsto f_t(z)$ is quasiconformal with dilatation less than $\frac{1+|t|}{1-|t|}$ and is Hölder continuous with Hölder exponent $\frac{1-|t|}{1+|t|}$.*

2. Quasifuchsian groups

2.1. Quakebends. In his work on simultaneous uniformization, Bers defined a *quasifuchsian group* analytically (see [**1**], [**5**]). Any hyperbolic Riemann surface S_0 can be represented as the quotient of the upper half plane U by a Fuchsian group G_0. We assume here for simplicity that S_0 is a compact surface of genus $g \geq 1$ from which at most n, $0 \leq n < \infty$, points have been removed and if $g = 1$, $n > 0$. A *Beltrami differential* μ defined on U, is an L_∞ function on U with $||\mu|| < 1$; it is *compatible* with G_0 if it satisfies the condition $\mu(g(z))\frac{\overline{g'(z)}}{g'(z)} = \mu(z)$ for all $g \in G_0$. If μ is extended to the lower half plane L so that the compatibility condition is satisfied, the differential equation $f_{\bar{z}} = \mu f_z$ has a solution f^μ that is a quasiconformal homeomorphism; the compatibility condition implies that $g^\mu = f^\mu \circ g \circ (f^\mu)^{-1}$ for any $g \in G_0$ is a Möbius transformation. The group of Möbius transformations $G^\mu = f^\mu \circ G_0 \circ (f^\mu)^{-1}$ is called *quasifuchsian*. It is discrete and in general, its limit set $\Lambda(G^\mu) = f^\mu(\Lambda(G_0))$ is not a subset of a circle so that it is not Fuchsian; in fact it is usually a fractal with Hausdorff dimension greater than 1.

The group G^μ represents *two* Riemann surfaces, $f^\mu(U)/G^\mu$ and $f^\mu(L)/G^\mu$ each topologically equivalent to S_0. Note however, that any finitely generated Kleinian group G whose domain of discontinuity Ω has exactly two invariant connected components is a quasifuchsian group representing two Riemann surfaces of the same topological type (see [**26**]).

In the sixties, Fenchel circulated a manuscript that was never published. In it, following on ideas of Poincaré from the turn of the century, he discussed Kleinian groups as discrete groups of isometries of hyperbolic three space $\mathbb{H}^3$. The quotient of $\mathbb{H}^3$ by a Kleinian group is thus a three manifold with a hyperbolic structure. In the mid-seventies, the first paper ([**24**]) discussing the geometry of the hyperbolic three manifolds and orbifolds arising from Kleinian groups of was published. A few years later, Thurston revolutionized the whole subject of three dimensional geometry and topology using Kleinian groups as a major tool ([**31**]). Among other things he showed how to construct general hyperbolic manifolds by gluing together special ones of the form $S_0 \times (0,1)$; these building blocks are the manifolds represented by quasifuchsian groups.

In developing his theory, Thurston first considered deformations of the hyperbolic structure of a Riemann surface, and hence of the Fuchsian group representing it. He generalized Dehn twisting along a simple closed curve to a process he called *earthquaking along a measured lamination.* He then extended the earthquake deformation to a deformation from a Fuchsian group into a quasifuchsian group by bending and called the combination of earthquaking and bending *quakebending.* We describe these deformations below and see that the quakebend deformation is actually a holomorphic motion.

A *geodesic lamination* on a Riemann surface S is a generalization of a closed geodesic. It is a closed set that is a union of pairwise disjoint simple geodesics called leaves. A *measured lamination* is a geodesic lamination with a transverse measure invariant under isotopies preserving its leaves.

If the geodesic lamination consists of k simple closed geodesics $\gamma_1, \dots \gamma_k$ and β is a transverse arc, the measure is $\sum_{p \in \beta \cap \gamma_j} t_j \delta_j$ where δ_j is atomic measure and $(t_1, \dots t_k) \in \mathbb{R}^+$. Such a lamination is called *rational.* The weak topology on measures gives a natural topology on the space of measured laminations and the rational laminations are dense in this space (see [**10**]).

Two measured laminations λ_1 and λ_2 are projectively equivalent if they are supported on the same geodesic lamination and if there exists some $t > 0$ such that for any transverse arc β, $\lambda_1(\beta) = t\lambda_2(\beta)$. The set of projective classes of measured laminations on S is denoted $PML(S)$.

An earthquake is a generalization of a *partial Dehn twist* along an oriented simple closed geodesic γ. A partial Dehn twist may be described roughly as follows: mark a point p on γ, cut the surface along γ and then reglue so that the point p on the right side is glued to a point q on the left, at distance t (the twist parameter) from p measured along γ. For an earthquake γ is a geodesic lamination with a marked point p. One would like to cut along γ and reglue so that the point p on the right side is glued to a point q on the left at a distance t (now called the earthquake parameter) from p. To make this meaningful one has to introduce the measure and work in the universal covering space realized as the hyperbolic plane. See [**10**] for a detailed discussion of these ideas. This process changes the lengths of geodesics transverse to the lamination and thus the hyperbolic structure of the resulting surface. In the case γ is a closed curve, these lengths can actually be computed from the parameter t (see [**15**]). In the general case, it is not clear how to compute the new lengths, but the variation $d\ell(\eta)/dt$ with respect to the earthquake parameter where $\ell(\eta)$ is the length of a closed geodesic intersecting γ can be computed (see [**32, 19**]). In [**20**] it is shown that the new structure depends real analytically on the earthquake parameter.

Below we describe the earthquake and its extension by bending to a quakebend in the case where the geodesic lamination γ on the base surface S_0 consists of a single simple closed oriented geodesic. The quakebend process is easier to understand in this case and the proof that a small bend results in a quasifuchsian group contains all the geometric ideas of the general proof. After we sketch the proof for this case we indicate how to proceed in the general case. We refer the reader to [**13**] for a detailed description of quakebends. An alternate approach may be found in [**22**].

Fix a simple closed geodesic β intersecting γ and let $\tau = t + i\theta$ be a complex number. We define the quakebend as the composition of an earthquake by t and a (pure) bend by θ relative to the choice of β. Represent S_0 by G_0 acting on the disk $\mathbb{D}$ (instead of U) and embed $\mathbb{D}$ as the equatorial plane in the ball model of $\mathbb{H}^3$. Let p be an intersection point of γ and β; let $\tilde{p}$ be a lift of p to $\mathbb{D}$ and let $\tilde{\gamma}$ be a lift of γ to $\mathbb{D}$ with endpoints $\tilde{p}$ and $\tilde{p}_1$. Let V be the element in G_0 such that $V(\tilde{p}) = \tilde{p}_1$. Since γ is geodesic, $\tilde{\gamma}$ is contained in the axis A_V of V. Set $\mathcal{V} = \cup_{g \in G_0} g(A_V)$. These axes divide $\mathbb{D}$ into pieces $\{P_k\}$ that are infinite sided polygons with vertices on $\partial\mathbb{D}$. Let P_0 be a piece that has A_V as one of its boundaries.

Let $\tilde{q}$ be another point on A_V; we determine its signed distance s from $\tilde{p}$ as follows: if, standing on P_0, $\tilde{q}$ is to the left of $\tilde{p}$, we take $s > 0$, whereas if it is to the right, we take $s < 0$.

Let $\tilde{\beta}$ be the lift of β intersecting P_0 with one endpoint at $\tilde{p}$ and the other at a point $\tilde{p}_2$ and let $W(\tilde{p}_2) = \tilde{p}$. Now let P_j be the piece adjacent to P_0 along A_V so that $W(P_0) = P_j$. The group G_0 acts equivariantly on the decomposition of $\mathbb{D}$ into these pieces. In fact, if H_0 is the stabilizer in G_0 of P_0, then G_0 is generated by H_0 and W.

To construct the earthquake given t, let $\tilde{q}$ be the point on A_V at signed distance t from $\tilde{p}$. Define W_t as the Möbius transformation taking the fixed points of $A_{V'}$, where $V' = W^{-1}VW$, to the fixed points of A_V and taking $\tilde{p}_2$ to $\tilde{q}$. Note that $P_j(t) = W_t(P_0)$ is the same set P_j but sheared in the direction of A_V by a distance t. This construction defines an isomorphism $\Phi_t : G_0 \to G_t$ as follows. Define $\Phi_t(h_0) = h_0$ for all h_0 in the stabilizer H_0 of P_0 and define $\Phi_t(W) = W_t$. Since G_0 is generated by H_0 and W, Φ_t is an isomorphism. Moreover we have $g_t\Phi_t(g_0) = \Phi_t^{-1}(g_t)g_0$ for all $g_0 \in G_0$ and $g_t \in G_t$.

We can form the developing map $\phi_t : \mathbb{D} \to \mathbb{D}$ as the identity on P_0 and for every piece P_k, if $P_k = g_0(P_0)$ for $g_0 \in G_0$, set $\phi_t(P_k) = \Phi_t(g_0)(P_0)$. To determine g_0 uniquely, choose it as W for $P_k = P_j$ as above and develop. It is proved in [**15**] that the image of the developing map ϕ_t again gives an embedding of $\mathbb{D}$ into $\mathbb{H}^3$ on which G acts discontinuously.

Now to make a pure bend, let $\bar{v}$ be a unit normal vector to A_V at $\tilde{p}$ pointing from P_0 into P_j. Bend P_j along A_V so that the bent vector $\bar{w}$ pointing into the bent P_j makes an angle θ with $\bar{v}$. The tangent $\bar{t}$ to A_V pointing towards its attracting fixed point is orthogonal to both $\bar{v}$ and $\bar{w}$. Make the sign convention that $\theta > 0$ if $(\bar{v}, \bar{w}, \bar{t})$ is a right handed system.

We can define W_θ as the Möbius transformation taking the fixed points of $A_{V'}$, where $V' = W^{-1}VW$, to the fixed points of A_V and taking $W^{-1}(\tilde{p})$ to $\tilde{p}$ and

$W^{-1}(\bar{v})$ to $\bar{w}$; W_θ is W followed by a rotation of angle θ with fixed axis A_V. We can define an isomorphism $\Phi_\theta : G_0 \to G_\theta$ as above so that it is the identity on H_0 and sends W to W_θ. We also have an analogous developing map Φ_θ defined on $\mathbb{D}$ whose image $\mathbb{D}_\theta \subset \mathbb{H}^3$ is invariant under G_θ and consists of planar pieces joined at angle θ, along *bending lines* that are the images in $\mathbb{H}^3$ of the axes in $\mathcal{V}$. Again Φ_θ intertwines the actions of G_0 on $\mathbb{D}$ and G_θ on $\mathbb{D}_\theta$. Note that since P_0 and $\phi(P_j)$ are not coplanar, $\Lambda(G_\theta)$ is not necessarily contained in a circle. It is not too hard to see that if $\theta = \pi$, $\Lambda(G_\theta)$ is again a circle but the set of discontinuity of G_θ cannot have two distinct invariant components.

The quakebend by τ is a composition of an earthquake and a pure bend. We have an isomorphism $\Phi_\tau(\gamma) : G_0 \to G_\tau$ and a developing map $\phi_\tau : \mathbb{D} \to \mathbb{D}_\tau$. Unlike the earthquake, however, the group G_τ thus obtained is not necessarily discrete.

We have

THEOREM 7. *Let G be a Fuchsian group and let γ be a simple closed geodesic on $S = \mathbb{D}/G$. Let $\tau = t + i\theta$ and let $\Phi_\tau(\gamma)$ be the associated quakebend homomorphism. There exists $\epsilon > 0$, depending on the hyperbolic length of γ, such that if $|\theta| < \epsilon$, then $\mathbb{D}_\tau$ is embedded and G_τ is quasifuchsian.*

Proof: We sketch a proof of this by showing by purely geometric means that the quakebend defines a holomorphic motion of the fixed points of the elements of G_0. A slightly different proof of the theorem, also based on the geometric lemmas below, is carried out in full detail in [**17**] where it is applied to the case of the punctured torus.

The proof that the quakebend defines a holomorphic motion of the fixed points divides into two parts. First we show that the fixed points of G_τ depend holomorphically on τ; this will be true whether or not G_τ is discrete and so does not use the hypothesis that the bend is small. Second we show that the map ϕ_τ is an embedding and extends to an equivariant map on the fixed points and moreover, that the extension is injective on the fixed points; this part does require the small bending hypothesis.

1. Holomorphicity

Using the method developed in [**28**], we shall find a formula for the trace $\mathrm{Tr}(W_\tau)$ that shows it depends holomorphically on τ. Since the entries of W_τ can be recovered from information about H_0 and $\mathrm{Tr}(W_\tau)$, the entries and fixed points are holomorphic in τ. Since any element in G_τ is a word in the fixed elements of H_0 and $W_\tau^{\pm n}$, its entries and fixed points will also be holomorphic in τ.

The trace $\mathrm{Tr}(W_\tau)$ and the complex translation length $2\lambda_{W_\tau}$ are related by $\mathrm{Tr}(W_\tau) = 2\cosh\lambda_{W_\tau}$; we always take $\Re\lambda_{W_\tau} > 0$. Following the construction in [**28**], we can form a *skew hexagon* in hyperbolic space by traversing the geodesic arcs l_i, $i = 1, \dots, 6$ defined as follows:

$l_1 = A_V$ and let l_2 be the the common orthogonal to A_V and $A_{V'}$ with respective endpoints ξ, ξ' on $A_V, A_{V'}$; let l_3 be the orthogonal to P_0 through the midpoint of the line (ξ, ξ').

Let $l_5 = W_\tau(l_3)$, $l_6 = W_\tau(l_2)$ and let l_4 be the common orthogonal to l_3 and l_5.

We claim that l_4 is the axis A_{W_τ} and that we can find its complex translation length using the hyperbolic cosine rule applied to the hexagon formed by the lines $l_1, \dots l_6$. To prove the claim, recall that any loxodromic element can be written as the product of reflections in a pair of disjoint hyperbolic lines perpendicular to its axis. Denote the reflection in l_k by i_k (see e.g. [**14**]). Let η be the midpoint of the segment of l_1 joining ξ and $W_\tau(\xi')$ and let l_0 be the geodesic normal to l_1 at η such that $i_0(l_2) = l_6$. Observe that the real hyperbolic length of the segment (ξ, η) is $t/2$ and that l_0 makes an angle of $\theta/2$ with P_0 and with $W_\tau(P_0)$.

Note that $i_0(l_3) = l_5$ so $i_5 = i_0 i_3 i_0$; $i_3(A_V) = A_{V'}$ and if $V'' = W_\tau V W_\tau^{-1}$, then $i_5(A_V) = A_{V''}$. Thus

$$i_0 i_3(A_{V'}) = i_0(A_V) = A_V = W_\tau(A_{V'})$$

$$i_0 i_3(A_V) = i_0(A_{V'}) = A_{V''} = W_\tau(A_V).$$

Therefore $i_0 i_3 = i_5 i_0 = W_\tau$ and A_{W_τ} is perpendicular to l_0, l_3 and l_5; since l_4 was defined as this perpendicular, $l_4 = A_{W_\tau}$ as claimed.

Denote the complex length of the segment of l_2 from l_1 to l_3 by $d_0 + i\psi$. Since d_0 is half the distance between A_V and $A_{V'}$ on P_0 it depends only on G_0. Since l_3 is perpendicular to P_0, $\psi = \pm\pi/2$ where the sign depends on the sign of t and appropriate conventions. These are discussed in detail in [**28**]. Applying these conventions again, the complex length of the segment $(W_\tau(\xi), \xi)$ is $\pm\tau$ and the complex length of l_4 from l_3 to l_5 is $2\lambda_{W_\tau} + i\pi$. Thus, by the hyperbolic cosine rule applied to the sides $l_1 \ldots l_6$ we have

$$\cosh l_4 = \cosh l_1 \cosh l_2 \cosh l_6 - \sinh l_2 \sinh l_6$$

so that

$$\cosh 2\lambda_{W_\tau} = \cosh \tau \cosh^2(d_0) - \sinh^2(d_0).$$

Rewriting, this becomes

$$\cosh \lambda_{W_\tau} = \cosh d_0 \sqrt{\frac{\cosh \tau + 1}{2}}$$

where the root is chosen with sign conventions of [**28**].

2. Injectivity

The proof of the injectivity is based on two geometric lemmas. Let $\eta : [0, \infty) \to \mathbb{D}$ be a geodesic parametrized by distance; assume it is transverse to a subset of the axes of $\mathcal{V}$ and set $\eta_\tau(s) = \phi_\tau(s)$. If $\eta(s_0)$ intersects an axis, its image $\eta_\tau(s_0)$ is called a *bending point*; denote by ψ the angle between the right and left limits of the forward pointing tangents at $\eta_\tau(s_0)$.

The first lemma says $|\psi| < |\theta|$ and is proved by elementary geometry.

For the second lemma we need the following definitions:

Let $X \in \mathbb{H}^3$ and let $\bar{v}$ be a vector in $\mathbb{H}^3$ at X. The *forward cone* of angle α at X in the direction $\bar{v}$ is the closed solid cone of angle α with vertex at X whose central axis is the $\mathbb{H}^3$ geodesic in the direction of $\bar{v}$. The *shadow* of a cone at X in direction $\bar{v}$ on $\partial\mathbb{H}^3$ is the set of endpoints on $\partial\mathbb{H}^3$ of the semi-infinite geodesics emanating from X and contained inside the forward cone at X.

The second lemma, called the cone lemma, says that given $\alpha \in (0, \pi/2)$, there is an $\epsilon > 0$ depending only on α and the hyperbolic length of γ such that if $|\theta| < \epsilon$, then the developed geodesic $\eta_\tau(s + t)$ is contained inside the forward cone at any point $\eta_\tau(s)$ — whichever of the two directions for the cone is used at the bending points — and moreover, there is a constant d_γ such that the forward cones along η_τ at points d_γ apart are nested. The proof of this lemma is based on estimates using elementary hyperbolic geometry.

From the cone lemma it follows that the developed image η_τ of any oriented geodesic $\eta \subset \mathbb{D}$ is embedded since the forward and backward cones at any point separate the forward and backward trajectories of the geodesic.

Next, assume $\eta(s)$ has as endpoint the fixed point of some element of G_0. The cone lemma applied to a sequence $\eta(s_n) \to \partial\mathbb{D}$ implies that the accumulation set of $\eta_\tau(s_n) \in \partial\mathbb{H}^3$. Since $\{s_n\}$ may be chosen so that the cones at $\eta_\tau(s_n)$ are nested, it follows that the diameters of their shadows go to zero. Using the group invariance, it follows that ϕ_τ extends to the fixed points of elements of G_0, taking them to corresponding fixed points of elements of G_τ.

Now suppose that $x, y \in \mathbb{D}$ are fixed points — of perhaps different elements of G_0 — and let z be a point on the geodesic η joining x and y. As above, the forward and backward cones at $\phi_\tau(z)$ separate the endpoints of η_τ so that $\phi_\tau(x) \neq \phi_\tau(y)$ and ϕ_τ is injective.

3. Application of the λ-lemma

Finally, if Λ_0 is the set of fixed points of elements of G, because the fixed points of all elements of G_τ are holomorphic functions of τ and they move injectively, ϕ_τ defines a holomorphic motion on $\Lambda_0 \times \Delta$ where $\Delta = \{\tau = t + i\theta,\, t \in \mathbb{R}, |\theta| < \epsilon\}$. Moreover, the motion acts equivariantly with respect to G_0 on a dense subset of its limit set. By Slodkowski's version of the λ-lemma, the holomorphic motion can be extended to a quasiconformal homeomorphism f of $\hat{\mathcal{C}}$ and by Theorem 1 of [**12**] the continuous extension can be made to act equivariantly. Thus, $G_\tau = fG_0f^{-1}$ and G_τ is quasifuchsian. □

If μ is an any rational measured lamination on S with support $\{\gamma_1, \dots, \gamma_k\}$, $0 < k \leq 3g-3$ and transverse measure $\{t_1\delta_{\gamma_1}, \dots, t_k\delta_{\gamma_k}\}$, we define the quakebend $\Phi_\tau(\gamma)$ as a composition of quakebends $\Phi_{t_j\tau}(\gamma_j)$, $j = 1, \dots, k$. Clearly the theorem holds in this case where the condition is that $|\Im(t_j\tau)| < \epsilon$ for all $j = 1, \dots, k$.

4. Arbitrary laminations

Now suppose γ is an arbitrary geodesic lamination on $S = \mathbb{D}/G$ with a complex valued measure ω supported on γ. In this case, the essential geometric ideas are the same, but carrying out the full details involves a discussion of how to take limits of quakebends along rational laminations. Here we only sketch what has to be done.

In chapter 3 of [**13**] there is a full description of how to define the *quakebend cocycle* B_ω and the corresponding developed surface $\phi_\omega : \mathbb{D} \to \mathbb{D}_\omega \subset \mathbb{H}^3$ that determine the *quakebend* along ω by a limiting procedure. It follows immediately from section 3.7 of [**13**] that B_ω determines a group $G_\omega \subset PSL(2, \mathcal{C})$ such that ϕ_ω intertwines the actions of G_0 on $\mathbb{D}$ and G_ω on $\mathbb{H}^3$.

The theorem in the general case is

THEOREM 8. *Let G be a Fuchsian group and let γ be a measured lamination on $S = \mathbb{D}/G$. Let $\tau \in \mathcal{C}$ be given and let $B_{\tau\mu}(\gamma)$ be the associated quakebend cocycle. There exists $\epsilon > 0$, depending on γ such that if $|\Im\tau| < \epsilon$, then $\mathbb{D}_{\tau\mu} = \phi_{\tau\mu}(\mathbb{D})$ is embedded and $G_{\tau\mu}$ is quasifuchsian.*

Proof: The proof follows the same outline as the sketched proof above. Again the holomorphicity does not depend on small bending but follows from the general quakebend construction applied to complex measures of the form $\omega = \tau\mu$ where μ is a transverse measure supported on γ in the usual sense and $\tau \in \mathcal{C}$. Theorem 3.9.1 of [**13**] can be carried out for this measure and it says that the trace (or equivalently the complex translation length) of any geodesic on S transverse to the lamination γ is a holomorphic function of τ. It follows that the fixed points of G_τ also depend holomorphically on τ.

The main difference in the proof of the extendibility and injectivity of ϕ_τ is that the cone lemma needs to be generalized. In the rational case, because there is a minimum to the lengths of geodesic arcs joining leaves of the lamination, elementary hyperbolic geometric arguments can be used to prove the cone lemma. In the general case however, there is no minimum and a new argument is required. This is given in detail for punctured tori in **[17]**; the arguments there extend without difficulty. We outline the steps here.

First, one needs to get a uniform control of the approximations to the quakebend cocycle. This is done using area estimates on S_0: one proves that there are constants K_μ and d_μ such that if β is any geodesic segment on S of length less than d_μ, then $\mu(\beta) < K_\mu$.

Second, the cones have to be properly defined. To do this, an appropriate metric D_T on the tangent space to $\mathbb{H}^3$ is introduced. Next the geodesics $\eta(s) \subset \mathbb{D}$ and $\eta_\tau(s) \subset \mathbb{D}_\tau$ are approximated. Using tangents to the approximations and taking limits in D_T, one can prove that tangent vectors to $\eta_\tau(s)$ can be properly defined even though the bending points form a Cantor set in $\eta_\tau(s)$,

Finally, using the constant d_μ one can prove that the cone lemma still holds; that is, $\eta_\tau(s+t)$ is entirely contained in forward cone at $\eta_\tau(s)$ and that the cones at $\eta_\tau(s+t)$ are nested for $t > d_\mu$.

The same argument as in the rational case shows that the map ϕ_τ extends injectively to Λ_0 and defines a holomorphic motion. Again, extending to obtain an equivariant quasiconformal homeomorphism of $\hat{\mathcal{C}}$, we conclude that G_τ is quasi-fuchsian. □

2.2. The convex hull boundary. One of the most elegant aspects of the study of Fuchsian and Kleinian groups is the interplay between the analytical structure of the Riemann surfaces and their parameter spaces and the geometric structure of the quotient spaces. For a Fuchsian group G_0, the conformal structure and the hyperbolic structure of the disk descends to the quotient $S_0 = \mathbb{D}/G_0$. Traces of elements of G_0 can be used as parameters and they also have a geometric meaning in terms of the lengths of closed geodesics on S_0.

If S_0 is a Riemann surface of finite type but not finite volume — a surface of finite genus from which finitely many points and disks have been removed — the limit set $\Lambda(G_0)$ is a Cantor set in the boundary of the hyperbolic plane. Nielsen realized that if one takes the hyperbolic convex hull of $\Lambda(G_0)$ (the Nielsen convex region) and forms the quotient, the result is a surface $\bar{S}_0$ of finite volume that contains all of the interesting geometric information; in particular, it is completely determined by trace parameters for G_0 or by the lengths of geodesics on S_0.

For quasifuchsian groups and more generally for Kleinian groups, the relation between the conformal structure of the quotient surfaces and the traces of group elements is very mysterious. By contrast however, the traces can be interpreted in terms of the hyperbolic structure of the quotient three manifold. Thurston pointed out that again, the convex hull of the limit set $\mathcal{C}(G)$, and its quotient $\mathcal{C}/G$, the convex core, contain all of the relevant geometric information.

A *pleated surface* is a pair S, ϕ where S is a complete hyperbolic surface and $\phi : S \to M$ is an isometry into a hyperbolic 3-manifold such that at every point

$p \in S$ there is at least one geodesic segment σ such that $\phi(\sigma)$ is geodesic and such that ϕ induces an injective map on fundamental groups. The set of $p \in S$ such that there is only one such σ is called the *pleating locus* (see [**13**] for a full discussion). For example, the pair $(\mathbb{D}, \phi_\tau)$, where ϕ_τ is the developing map defined above, is a pleated surface, and the bending lines are the pleating locus. For simplicity we also call the immersed image surface $\mathbb{D}_\tau = \phi_\tau(\mathbb{D})$ pleated.

The quotient surfaces $\partial\mathcal{C}/G$ have a hyperbolic structure and are pleated surfaces. Thurston conjectured that the pleating locus of the convex hull boundary with its natural bending measure should completely determine the group. This has been proved for Schottky groups in the case that the pleating locus is rational by Otal [**27**] and for arbitrary laminations by Bonahon [**9**]. These proofs use the deep machinery of Kerckhoff-Hodgson and Thurston-Morgan-Shalen.

Using the quakebend construction and Theorem 8 it follows that

THEOREM 9. *For a quasifuchsian group G, the boundary surface components of $\partial\mathcal{C}/G$ are pleated and the pleating locus for one component can be prescribed arbitrarily.*

Proof: Again we begin with the rational case. To prove the boundary surfaces are pleated, we observe that by convexity, every bi-infinite geodesic in a component of $\partial\mathcal{C}$ extends to Λ. We can thus define an isometry from each component of $\partial\mathcal{C}$ to $\mathbb{D}$ by unfolding.

That the pleating locus of one side can be prescribed follows from

PROPOSITION 1. *Suppose μ is a rational measured lamination on the surface S with support $\{\gamma_1, \dots, \gamma_k\}$ and measure $\{t_1\delta_{\gamma_1}, \dots, t_k\delta_{\gamma_k}\}$. Let $\tau \in \mathcal{C}$ and let $\mathbb{D}_\tau$ be the developed surface associated to the quakebend $\Phi_\tau(\gamma)$. If ϵ is as in the cone lemma and if $0 \leq \Im(t_j\tau) < \epsilon$ for all j, then one of the half spaces cut out of $\mathbb{H}^3 \cup \hat{\mathcal{C}}$ by $\overline{\mathbb{D}_\tau}$ is convex.*

Proof: Since $\partial\mathbb{D}_\tau$ is a Jordan curve on $\hat{\mathcal{C}}$, $\mathbb{D}_\tau$ does indeed divide $\mathbb{H}^3 \cup \hat{\mathcal{C}}$ into two half spaces, H_i, $i = 1, 2$. Suppose H_1 is not convex so there are points $x_1, y_1 \in H_1$ and a geodesic σ_1 joining them that leaves H_1. Let Π_1 be a plane through σ_1 intersecting H_1. Then $\Pi_1 \cap (\sigma_1 \cup \partial H_1)$ is a hyperbolic n-gon, the sum of whose interior angles is less than $(n-2)\pi$. By hypothesis, the angles from ∂H_1 all have the same sign and are less than π in absolute value. If H_2 is also not convex, we form another polygon by taking points $x_2, y_2 \in H_2$, an arc σ_2, and a plane Π_2 through σ_2 and H_2; since this time the angles are measured on the opposite side of $\mathbb{D}_\tau$, they again all have the same sign, but their absolute values are all greater than π. Such a polygon cannot exist and this gives a contradiction. □

COROLLARY 10. *The pleated surface $\mathbb{D}_\tau$ is a component of $\partial\mathcal{C}(G_\tau)$.*

Proof: On one hand, H_2 is convex so $H_2 \supset \mathcal{C}$. On the other, $\mathbb{D}_\tau$ is contained in the convex hull of its bending lines so that $\mathbb{D}_\tau \supset \partial\mathcal{C}$ and $\mathbb{D}_\tau = \partial H_2$ by definition. □

Since the projective rational laminations are dense in $PML(S)$, we can extend these results to arbitrary laminations by approximation. In [**18**] it is proved that the map from quasifuchsian space to $PML(S)$ defined by sending a group to the pleating locus of one component of the convex hull boundary is continuous. This coupled with the observation that the limit of convex sets is convex completes the proof of theorem 9. □

References

[1] L. V. Ahlfors, *Lectures on Quasiconformal Mappings*, Van Nostrand Studies **10**, 1966.

[2] _____ *Conformal Invariants; Topics in Geometric Function Theory*, McGraw-Hill, 1973.

[3] L. V. Ahlfors and L. Bers, Riemann's mapping theorem for variable metrics, *Annals Math.*, 72:345–404, 1961.

[4] K. Astala, Area distortion of quasiconformal mappings, *Acta Math.*, 173:37-60, 1994.

[5] L. Bers, Simultaneous uniformization, *Bull AMS*, 54:311–315, 1948.

[6] _____ Videotaped lecture in Sullivan's CUNY seminar, 1983 Available for viewing at CUNY, Graduate Center

[7] L. Bers and H. L. Royden, Holomorphic families of injections, *Acta Math.*, 157:259–286, 1986.

[8] B. Bojarski, Generalized solutions of a system of differential equations of first order and elliptic type with discontinuous coefficients, *Math. Sb.*, **85**:451–503, 1957.

[9] F. Bonahon, Private communication.

[10] R. D. Canary and D. B. A. Epstein and P. Green, Notes on notes of Thurston In D. B. A. Epstein, editor, *Analytical and Geometric Aspects of Hyperbolic Space*, LMS Lecture Notes 111, pages 3–92. Cambridge University Press, 1987.

[11] C. J. Earle, Some maximal holomorphic motions, In *Lipa's Legacy: Proceedings of the Bers Colloquium, 1995*, AMS Contemp. Math. Ser. 211, 1997.

[12] C. J. Earle, I. Kra and S. I. Krushkal, Holomorphic Motions and Teichmüller Spaces, *Trans. A.M.S.*, 343(2):927–948, 1994.

[13] D. B. A. Epstein and A. Marden, Convex hulls in hyperbolic space, a theorem of Sullivan, and measured pleated surfaces, In D. B. A. Epstein, editor, *Analytical and Geometric Aspects of Hyperbolic Space*, LMS Lecture Notes 111, pages 112–253. Cambridge University Press, 1987.

[14] W. Fenchel, *Elementary Geometry in Hyperbolic Space,* de Gruyter, 1989.

[15] W. Fenchel and J. Nielsen, Discontinuous groups of non-Euclidean motions, Unpublished manuscript

[16] F. Gehring and C. Pommerenke, On the Nehari univalence criterion and quasicircles, *Comm. Math. Helv.*, 59: 226-242, 1984.

[17] L. Keen and C. Series, How to bend punctured tori, In *Lipa's Legacy: Proceedings of the Bers Colloquium, 1995*, AMS Contemp. Math. Ser. 211: 359–387, 1997.

[18] L. Keen and C. Series, Pleating coordinates for the Maskit embedding of the Teichmüller space of punctured tori, *Topology*, 32(4):719–749, 1993.

[19] S. Kerckhoff, The Nielsen realization problem, *Annals Math.*, 117:235–265, 1983.

[20] S. Kerckhoff Earthquakes are analytic, , *Comm. Mat. Helv.*, 60:17–30, 1985.

[21] N. Lakic, Infinitesimal Teichmüller geometry, *Complex Variables*, 30:1-17, 1996.

[22] C. McMullen, Complex earthquakes and Teichmüller theory, preprint

[23] R. Màñé, P. Sad, and D. Sullivan, On the dynamics of rational maps, *Ann. Èc. Norm. Sup.*, 96:193–217, 1983.

[24] A. Marden, The geometry of finitely generated Kleinian groups, *Annals Math.*, 99:383–462, 1974.

[25] G. J. Martin, The distortion theorem for quasiconformal mappings, Schottky's theorem & holomorphic motions, to appear, *Proc. Amer. Math. Soc.* 1997.

[26] B. Maskit, *Kleinian Groups*, Springer, New York, 1987.

[27] J.-P. Otal. Sur le coeur convexe d'une variété hyperbolique de dimension 3, to appear, *Invent. Math.*

[28] J. R. Parker and C. Series, Bending formulae for convex hull boundaries, *J.d'Analyse Math.*, 67:165–198, 1995.

[29] Z. Slodkowski, Holomorphic motions and polynomial hulls, *Proc. Amer. Math. Soc.*, 111:347-355, 1991.

[30] D. P. Sullivan and W. P. Thurston, Extending holomorphic motions, *Acta Math.*, 157:243–257, 1986.

[31] W. Thurston, The geometry and topology of three manifolds, Lecture notes, Princeton 1979.

[32] S. Wolpert, The Fenchel-Nielsen deformation, *Annals Math.*, 115:501–528, 1982.

FREDERICK P. GARDINER, DEPT. OF MATH., BROOKLYN COLLEGE, CUNY, 11210
E-mail address: `fgardine@email.gc.cuny.edu`

MATHEMATICS DEPT. OF MATH., LEHMAN COLLEGE, CUNY, 10468
E-mail address: `linda@alpha.lehman.cuny.edu`

Contemporary Mathematics
Volume **240**, 1999

Computation of Topological Numbers Via Linear Algebra: Hypersurfaces, Vector Fields and Vector Fields on Hypersurfaces

Luis Giraldo, Xavier Gómez-Mont, and Pavao Mardešić

The pattern we develop is that for problems over the complex numbers the dimension of a finite dimensional vector space computes a topologically defined number, while over the real numbers one needs to introduce a natural symmetric non-degenerate bilinear form in the real part of this finite dimensional vector space and then its signature is the topological number searched for. This is carried out for hypersurfaces in section 1, for vector fields in section 2 and for vector fields tangent to a hypersurface in section 3. A significant role is played by Algebra structures in finite dimensional vector spaces and non-degenerate dualities defined in them due to the existence of a 'minimal ideal' which is generated by a natural element: The Hessian determinant of a function, the Jacobian determinant of a vector field or the relative Jacobian determinant, for a vector field restricted to a hypersurface.

Let f be a germ of a holomorphic function at 0 in $\mathbb{C}^{n+1}$, having 0 as an isolated critical point. We can associate to f the finite dimensional algebras obtained by considering the ring of germs of holomorphic functions $\mathcal{O}_{\mathbb{C}^{n+1},0}$ modulo the ideal generated by the partial derivatives $(f_0, \dots, f_n)$ of f

$$\mathbf{A} := \frac{\mathcal{O}_{\mathbb{C}^{n+1},0}}{(f_0, \dots, f_n)} \qquad , \qquad f_j := \frac{\partial f}{\partial z_j}$$

The dimension of this algebra corresponds to the number of independent 'vanishing cycles' ([**8**]), i.e.

$$dim H_n(V_\varepsilon, \mathbb{C}) \qquad , \qquad V_\varepsilon = f^{-1}(\varepsilon) \cap \{|z| < \varepsilon'\}.$$

The algebra structure in $\mathbf{A}$ is remarkable: The class of the Hessian determinant of f

$$Hess(f) := det \begin{pmatrix} \frac{\partial^2 f}{\partial z_1^2} & \cdots & \frac{\partial^2 f}{\partial z_1 \partial z_n} \\ & \cdots & \\ \frac{\partial^2 f}{\partial z_n z_1} & \cdots & \frac{\partial^2 f}{\partial z_n z_n} \end{pmatrix} \tag{1}$$

in $\mathbf{A}$ (that is, the class of the Hessian determinant of f modulo the partials of f) generates an ideal $(Hess(f)) \subset \mathbf{A}$ which has dimension 1 as a $\mathbb{C}$-vector space, and further it is contained in every non-zero ideal of $\mathbf{A}$ ([**2**],[**7**]). It is a 'minimal ideal' and plays a dual role to the maximal ideal in the local algebra $\mathbf{A}$. We can

---

1991 *Mathematics Subject Classification.* Primary: 14A05; Secondary: 14M10.
Supported by Conacyt.

make this duality precise as follows: Choose any linear map $L : \mathbf{A} \longrightarrow \mathbb{C}$ with $L(Hess(f)) \neq 0$, then the symmetric bilinear form obtained by the composition of multiplication in A with the linear map L

$$<\ ,\ >_L : \mathbf{A} \times \mathbf{A} \stackrel{\cdot}{\longrightarrow} \mathbf{A} \stackrel{L}{\longrightarrow} \mathbb{C}$$

is a non-degenerate bilinear form ([**2**],[**7**]). Under this duality, the orthogonal of an ideal corresponds to the annihilator ideal, and in particular, the orthogonal to the maximal ideal of $\mathbf{A}$ is the ideal $(Hess(f))$. The dual property to the maximality of the maximal ideal is that every non-zero ideal in $\mathbf{A}$ contains the 1-dimensional ideal $(Hess(f))$.

Doing the same thing over the real numbers, and asking the linear map $\mathbf{A}_{\mathbb{R}} \stackrel{L}{\longrightarrow} \mathbb{R}$ to satisfy $L(Hess(f)) > 0$, one obtains that the symmetric bilinear form

$$<\ ,\ >_L : \mathbf{A}_{\mathbb{R}} \times \mathbf{A}_{\mathbb{R}} \stackrel{\cdot}{\longrightarrow} \mathbf{A}_{\mathbb{R}} \stackrel{L}{\longrightarrow} \mathbb{R} \qquad , \qquad \mathbf{A}_{\mathbb{R}} := \frac{\mathcal{A}_{\mathbb{R}^{n+1},0}}{(f_0, \dots , f_n)}$$

is non-degenerate, where $\mathcal{A}_{\mathbb{C}^{n+1},0}$ is the ring of germs of real analytic functions. In the real case, one has in addition the invariant provided by the signature σ of the bilinear form $<\ ,\ >_L$, number which is independent of the choice of L. The geometric meaning of σ is that $1 \pm \sigma$ is the Euler characteristic of $V_{\pm\varepsilon}^{\mathbb{R}} := V_{\pm\varepsilon} \cap \mathbb{R}$ ([**1**, Theorem 1.2]).

One may proceed further: For every $h \in \mathbf{A}_{\mathbb{R}}$ one may repeat the above method applied to the algebra $\frac{\mathbf{A}_{\mathbb{R}}}{Ann_{\mathbf{A}_{\mathbb{R}}}(h)}$ with the relative Hessian determinant $\frac{Hess(f)}{h} \in \frac{\mathbf{A}_{\mathbb{R}}}{Ann_{\mathbf{A}_{\mathbb{R}}}(h)}$ as generator of the 'minimal' ideal of $\frac{\mathbf{A}_{\mathbb{R}}}{Ann_{\mathbf{A}_{\mathbb{R}}}(h)}$ (see [**5**])

$$<\ ,\ >_L : \frac{\mathbf{A}_{\mathbb{R}}}{Ann_{\mathbf{A}_{\mathbb{R}}}(h)} \times \frac{\mathbf{A}_{\mathbb{R}}}{Ann_{\mathbf{A}_{\mathbb{R}}}(h)} \stackrel{\cdot}{\longrightarrow} \frac{\mathbf{A}_{\mathbb{R}}}{Ann_{\mathbf{A}_{\mathbb{R}}}(h)} \stackrel{L}{\longrightarrow} \mathbb{R} \quad , \quad L(\frac{Hess(f)}{h}) > 0$$

The signature of the associated non-degenerate bilinear forms defines an integer valued function

$$Sign_{\mathbf{A}_{\mathbb{R}}} : \mathbf{A}_{\mathbb{R}} \longrightarrow \mathbb{Z} \tag{2}$$

Since all the forms obtained are non-degenerate, then $Sign_{\mathbf{A}_{\mathbb{R}}}$ is constant as long as the ideal $(h) \subset \mathbf{A}_{\mathbb{R}}$ has constant dimension. That is, we may stratify $\mathbf{A}_{\mathbb{R}}$ into a finite number of real semi-algebraic subsets, defined by the rank of the multiplication map $h : \mathbf{A}_{\mathbb{R}} \longrightarrow \mathbf{A}_{\mathbb{R}}$, and (2) is the association to each strata an integer. The integer corresponding to the 0 strata is 0. The exact geometric meaning of this signature function has still elluded us. Nevertheless, we will see in section 3 that it appears in a formula for the index of vector fields on $V_0^{\mathbb{R}}$ and n an odd positive integer.

Similarly, let $\tilde{X} := \sum_{j=0}^{n} X^j \frac{\partial}{\partial z_j}$ be a germ of a holomorphic vector field at 0 in $\mathbb{C}^{n+1}$ with an isolated zero at 0. We can associate to X the finite dimensional algebras obtained by considering the ring of germs of holomorphic functions $\mathcal{O}_{\mathbb{C}^{n+1},0}$ modulo the ideal generated by the entries of the vector field $(X^0, \dots , X^n)$:

$$\mathbf{B} := \frac{\mathcal{O}_{B,0}}{(X^0, \dots , X^n)}$$

This algebra is as remarkable as $\mathbf{A}$, with the the role of the Hessian determinant of f taken by the Jacobian determinant of X:

$$J(X) := det \begin{pmatrix} \frac{\partial X^0}{\partial z_1} & \cdots & \frac{\partial X^0}{\partial z_n} \\ & \cdots & \\ \frac{\partial X^n}{\partial z_1} & \cdots & \frac{\partial X^n}{\partial z_n} \end{pmatrix}. \tag{3}$$

In this case, the dimension of $\mathbf{B}$ and the signature of its natural bilinear form in $\mathbf{B}_{\mathbb{R}}$ correspond to the Poincaré-Hopf index of the holomorphic or real analytic vector field, respectively (see section 2).

Now assume that X is tangent to the hypersurface $V := f^{-1}(0)$ with an isolated singularity at 0. The tangency condition is equivalent to the existence of a germ of a holomorphic function h satisfying

$$df(\tilde{X}) = \sum_{j=0}^{n} f_j X^j = f \cdot h \qquad h := \frac{df}{f}(\tilde{X}) \in \mathcal{O}_{\mathbb{C}^{n+1},0} \tag{4}$$

Observe that the relation (4) implies that $fh = 0$ in both rings $\mathbf{A}$ and $\mathbf{B}$, so that in particular we have

$$(f) \subset Ann_{\mathbf{A}}(h) \subset \mathbf{A} \quad , \quad (f) \subset Ann_{\mathbf{B}}(h) \subset \mathbf{B}$$

and dually

$$(h) \subset Ann_{\mathbf{A}}(f) \subset \mathbf{A} \quad , \quad (h) \subset Ann_{\mathbf{B}}(f) \subset \mathbf{B}$$

The index of the vector field $X := \tilde{X}|_V$ on the hypersurface V at 0 may be computed algebraically by ([**3**]):

$$Ind_{V,0}(X) = \begin{cases} dim_{\mathbb{C}} \frac{\mathbf{B}}{Ann_{\mathbf{B}}(h)} + dim_{\mathbb{C}} \frac{\mathbf{A}}{(f)} & \text{for even n} \\ dim_{\mathbb{C}} \frac{\mathbf{B}}{(f)} - dim_{\mathbb{C}} \frac{\mathbf{A}}{(f)} & \text{for odd n} \end{cases}$$

This then gives a procedure using Linear Algebra to compute a topologically defined number. We describe in section 3 how one introduces bilinear forms in certain algebras to solve the computation of the index of a real analytic vector field restricted to an algebraically isolated singularity. We use the for this the fact that the following modules are equidimensional as vector spaces ([**3**]):

$$\frac{Ann_{\mathbf{A}}(f)}{(h)} \quad , \quad \frac{Ann_{\mathbf{A}}(h)}{(f)} \quad , \quad \frac{Ann_{\mathbf{B}}(f)}{(h)} \quad , \quad \frac{Ann_{\mathbf{B}}(h)}{(f)}.$$

1. Real and Complex Hypersurfaces

Let $n \geq 1$ be an integer. Throughout the paper B will denote the ball of radius $r > 0$ in $\mathbb{C}^{n+1}$ and $B_{\mathbb{R}}$ will denote the ball of radius r in $\mathbb{R}^{n+1}$.

1.1. The Holomorphic Case. Let $f : \bar{B} \longrightarrow \mathbb{C}$ be a holomorphic function with 0 as its only critical point and denote $f^{-1}(\delta)$ by V_δ. Assume further that V_0 is a contractible topological space. Denote by

$$\mathbf{A} := \frac{\mathcal{O}_{B,0}}{(f_0, \dots, f_n)}$$

the finite dimensional algebra obtained by considering the ring of germs of holomorphic functions $\mathcal{O}_{B,0}$ modulo the ideal generated by the partial derivatives $(f_0, \dots, f_n)$ of f. It is a local commutative algebra. If we call the Hessian determinant of f

the determinant of the Hessian matrix (1) and we consider its class in **A**, then it generates a 1 dimensional ideal, called the **socle** of **A**. It plays a dual role to the maximal ideal: It is contained in every non-zero ideal of **A** ([**2**],[**7**]). With the help of this fundamental property, one associates a family of non-degenerate complex-valued bilinear forms defined on **A** as follows: Choose any $\mathbb{C}$-linear map $L : \mathbf{A} \longrightarrow \mathbb{C}$ which sends $Hess(f) \in \mathbf{A}$ to a non-zero complex number, and define the symmetric non-degenerate bilinear form as

$$< , >_L : \mathbf{A} \times \mathbf{A} \stackrel{\cdot}{\longrightarrow} \mathbf{A} \stackrel{L}{\longrightarrow} \mathbb{C}$$

These bilinear forms have the property that for every ideal $J \subset \mathbf{A}$ its L-orthogonal is the annihilator $Ann_{\mathbf{A}}(J)$ (and hence independent of the choice of L).

The geometric meaning of the algebra **A** appears as:

THEOREM 1.1. *([**8**])The dimension of* **A** *is equal to the dimension of the homology group* $H_{n-1}(V_\epsilon, \mathbb{C})$*, with* $|\epsilon|$ *small and non-zero.*

They are the only non-vanishing reduced homology groups of V_ϵ and are called 'vanishing cycles'. The dimension of A is then related to the 'hidden topology' inside the singularity V_0.

1.2. The Real Analytic Case. Let $f_{\mathbb{R}} : \bar{B}_{\mathbb{R}} \longrightarrow \mathbb{R}$ be a real analytic function that accepts an analytic extension f to a holomorphic function on the closed unit ball $\bar{B}$ in $\mathbb{C}^n$. Assume that 0 is the only critical point of f on $\bar{B}$ (i.e. 0 is an algebraically isolated singularity of $f_{\mathbb{R}}$) and denote the real hypersurfaces $f_{\mathbb{R}}^{-1}(\delta)$ by $V_\delta^{\mathbb{R}}$, $\delta \in \mathbb{R}$. Assume further that $V_0^{\mathbb{R}}$ is a contractible topological space and denote by $V_+^{\mathbb{R}}$ and $V_-^{\mathbb{R}}$ the differentiable type of the fibers of $f_{\mathbb{R}}$ for small positive and negative ϵ. Denote by

$$\mathbf{A}_{\mathbb{R}} := \frac{\mathcal{A}_{\mathbb{R}^{n+1},0}}{(f_0, \cdots, f_n)}$$

the finite dimensional algebra obtained by considering the ring of germs of real analytic functions $\mathcal{A}_{B_{\mathbb{R}^{n+1},0},0}$ modulo the ideal generated by the partial derivatives of f. It is a local commutative algebra. The class of the Hessian determinant (1) of f in $\mathbf{A}_{\mathbb{R}}$, generates the socle: A 1 dimensional ideal that is contained in every non-zero ideal of $\mathbf{A}_{\mathbb{R}}$ ([**2**],[**7**]). One associates a family of non-degenerate real-valued bilinear forms defined in $\mathbf{A}_{\mathbb{R}}$ as before:

$$L : \mathbf{A} \longrightarrow \mathbb{R} \quad , \quad L(Hess(f)) > 0 \quad , \quad < , >_L : \mathbf{A}_{\mathbb{R}} \times \mathbf{A}_{\mathbb{R}} \stackrel{\cdot}{\longrightarrow} \mathbf{A}_{\mathbb{R}} \stackrel{L}{\longrightarrow} \mathbb{R}$$

Call σ the signature of any of the associated non-degenerate bilinear forms. The geometric meaning of the signature σ appears since it determines the Euler characteristic $\chi(V_\pm^{\mathbb{R}})$ of $V_\pm^{\mathbb{R}}$:

THEOREM 1.2. *([**2**])Under the above hypothesis, we have*

$$\chi(V_\pm^{\mathbb{R}}) := \begin{cases} 1 - \sigma & n \text{ is odd} \\ 1 \mp \sigma & n \text{ is even} \end{cases}$$

2. The Index of Real and Complex Vector Fields

2.1. The Holomorphic Case. Let $X = \sum_{j=0}^{n} X^j \frac{\partial}{\partial z_j}$ be a holomorphic vector field on $\bar{B} \subset \mathbb{C}^{n+1}$ vanishing only at 0. Denote by

$$\mathbf{B} := \frac{\mathcal{O}_{B,0}}{(X^0, \dots, X^n)}$$

the finite dimensional algebra obtained by considering germs of holomorphic functions modulo the ideal generated by the components of the vector field X. It is a local commutative algebra. Define the Jacobian determinant of X as (3) and we consider its class in $\mathbf{B}$. It is shown in [**2**],[**7**], that it generates the socle of $\mathbf{B}$: A 1 dimensional ideal, which is contained in every non-zero ideal of $\mathbf{B}$. One associates a family of non-degenerate complex-valued bilinear forms defined in $\mathbf{B}$ as before:

$$L : \mathbf{B} \longrightarrow \mathbb{C} \quad , \quad L(Hess(f)) \neq 0 \quad , \quad <\,,\,>_L : \mathbf{B} \times \mathbf{B} \stackrel{\cdot}{\longrightarrow} \mathbf{B} \stackrel{L}{\longrightarrow} \mathbb{C}$$

The geometric meaning of the algebra $\mathbf{B}$ is:

THEOREM 2.1. *([**6**])The dimension of* $\mathbf{B}$ *(over* $\mathbb{C}$*) is the Poincaré-Hopf index of the vector field* X *at the isolated singular point* $0 \in \mathbb{C}^{n+1}$.

2.2. The Real Analytic Case. Let $B_{\mathbb{R}}$ be the ball of radius 1 in $\mathbb{R}^{n+1}$, $n \geq 1$, and let $X = \sum_{j=0}^{n} X^j \frac{\partial}{\partial x_j}$ be a real analytic vector field on $\bar{B}_{\mathbb{R}}$ vanishing only at 0 which extends analytically to a holomorphic vector field X on $\bar{B}$ vanishing only at 0. Such points are called algebraically isolated singular points of the vector field X. Denote by

$$\mathbf{B}_{\mathbb{R}} := \frac{\mathcal{A}_{\mathbb{R}^{n+1},0}}{(X^0, \dots, X^n)} \tag{2.1}$$

the finite dimensional algebra obtained by considering the ring $\mathcal{A}_{B_{\mathbb{R}},0}$ of germs of real analytic functions modulo the ideal generated by the components of the vector field X. It is a local commutative algebra. The class in $\mathbf{B}_{\mathbb{R}}$ of the Jacobian determinant of X as in (3) generates the socle of $\mathbf{B}_{\mathbb{R}}$. One associates a family of non-degenerate real-valued bilinear forms defined in $\mathbf{B}_{\mathbb{R}}$ as before:

$$L : \mathbf{B} \longrightarrow \mathbb{R} \quad , \quad L(Hess(f)) > 0 \quad , \quad <\,,\,>_L : \mathbf{B}_{\mathbb{R}} \times \mathbf{B}_{\mathbb{R}} \stackrel{\cdot}{\longrightarrow} \mathbf{B}_{\mathbb{R}} \stackrel{L}{\longrightarrow} \mathbb{R}$$

Let $\sigma(X)$ be the signature of $<\,,\,>_L$. The geometric meaning of this signature is:

THEOREM 2.2. *([**2**],[**7**])Under the above hypothesis, the signature* $\sigma(X)$ *of the above family of non-degenerate bilinear forms in* $\mathbf{B}_{\mathbb{R}}$ *is the Poincaré-Hopf index of the vector field* X *at the algebraically isolated critical point* $0 \in \mathbb{R}^{n+1}$.

3. The Index of Vector Fields Tangent to a Hypersurface

3.1. The Holomorphic Case. Let B be the ball of radius 1 in $\mathbb{C}^{n+1}$, $n \geq 1$, and $f : \bar{B} \longrightarrow \mathbb{C}$ a holomorphic function with 0 as its only critical point and denote $f^{-1}(\delta)$ by V_δ. Let $\tilde{X} = \sum_{j=0}^{n} X^j \frac{\partial}{\partial z_j}$ be a holomorphic vector field on $\bar{B}$ vanishing only at 0 and tangent to $V := V_0$. Denote by X the vector field induced on V by $\tilde{X}$. The condition that $\tilde{X}$ is tangent to V is equivalent to having $df(\tilde{X})$ a multiple of f, so that $\frac{df}{f}(\tilde{X})$ is a holomorphic function on B, called the **cofactor**:

$$df(\tilde{X}) = \sum_{j=0}^{n} f_j X^j = f \cdot h \qquad h := \frac{df}{f}(\tilde{X}) \in \mathcal{O}_{B,0} \tag{3.1}$$

Topologically, one has the Poincaré-Hopf index, $Ind_{V_\epsilon,0}(X)$, defined by transporting the vector field X to vector fields X_ϵ on $V_\epsilon \cap \partial B_\mathbb{R}$, extending then continuously to all V_ϵ with isolated critical points, and adding up the usual Poincaré-Hopf indices of X_ϵ at the critical points:

$$Ind_{V,0}(X) = \sum_{\substack{q \in V_\epsilon \\ X_\epsilon(q)=0}} Ind_{V_\epsilon,q}(X_\epsilon)$$

The index measures the number of equilibrium points with sign of the vector field X_ϵ in the smoothening (V_ϵ, X_ϵ) before it degenerated to (V, X).

THEOREM 3.1. *([**3**]) Under the above hypothesis, the index of the vector field X on the hypersurface V at 0 is:*

$$Ind_{V,0}(X) = \begin{cases} dim_\mathbb{C} \frac{\mathbf{B}}{Ann_\mathbf{B}(h)} + dim_\mathbb{C} \frac{\mathbf{A}}{(f)} & \text{for even } n \\ dim_\mathbb{C} \frac{\mathbf{B}}{(f)} - dim_\mathbb{C} \frac{\mathbf{A}}{(f)} & \text{for odd } n \end{cases} \tag{3.2}$$

Formula (3.2) may be interpreted geometrically as follows. The number of singular points in the ambient B of $\tilde{X}$ at 0 can be measured by the $\mathbb{C}$-dimension of $\mathbf{B}$ and it may be pictured geometrically by the point *Spec*$\mathbf{B}$, consisting of the point $\{0\}$ and the algebra of functions $\mathbf{B}$. There are 2 subpoints of *Spec*$\mathbf{B}$ associated to the tangency with the hypersurface V: The relation (3.1) in the ring $\mathbf{B}$ says that f and h annihilate each other, and so we have the inclusions

$$0 \subset (f) \subset Ann_\mathbf{B}(h) \subset \mathbf{B}$$

which give rise to the inclusions

$$Spec \frac{\mathbf{B}}{Ann_\mathbf{B}(h)} \subset Spec \frac{\mathbf{B}}{(f)} \subset Spec\mathbf{B}. \tag{3.3}$$

That is, the middle point in (3.3) is obtained by imposing on the zero set *Spec*$\mathbf{B}$ of $\tilde{X}$ the condition $f = 0$. One obtains the smaller point by imposing the further conditions contained in $Ann_\mathbf{B}(h)$. This smaller point is the dual (complementary or orthogonal) point to $Spec\frac{\mathbf{B}}{(h)}$. Formula (3.2) then is saying thay fon n even we should measure the index with the smaller point, while for n odd we should measure with the bigger point.

3.2. The Real Analytic Case. Let $f_\mathbb{R} : \bar{B}_\mathbb{R} \longrightarrow \mathbb{R}$ be a real analytic function which has an analytic extension f to $\bar{B}$ with 0 as its only critical point. Denote $f_\mathbb{R}^{-1}(\delta)$ by $V_\delta^\mathbb{R}$. Let $\tilde{X}_\mathbb{R} = \sum_{j=0}^n X^j \frac{\partial}{\partial x_j}$ be a real analytic vector field on $\bar{B}_\mathbb{R}$ which extends analytically to a holomorphic vector field $\tilde{X}_\mathbb{R}$ on $\bar{B}$ vanishing only at 0. Let $\mathbf{B}_\mathbb{R}$ be the finite dimensional algebra (2.1). As we saw in section 2.2. the class of the Jacobian determinant $J\tilde{X}_\mathbb{R}$ of $\tilde{X}_\mathbb{R}$ in $\mathbf{B}_\mathbb{R}$ is a generator of the 1 dimensional socle.

Assume further that $\tilde{X}_{\mathbb{R}}$ is tangent to $V^{\mathbb{R}} := V_0^{\mathbb{R}}$. Denote by X the vector field induced on $V^{\mathbb{R}}$ by $\tilde{X}_{\mathbb{R}}$. The condition that $\tilde{X}_{\mathbb{R}}$ is tangent to $V^{\mathbb{R}}$ is equivalent to (3.1).

Topologically, one has the Poincaré-Hopf index, $Ind_{\pm,0}(X)$, defined by transporting the vector field $X|_{V\cap\partial B_{\mathbb{R}}}$ to vector fields $X_\pm$ on $V_\pm \cap \partial B_{\mathbb{R}}$, extending them continuously to all $V_\pm$ with isolated critical points, and adding up the usual Poincaré-Hopf indices of $X_\pm$ at the critical points:

$$Ind_{V_\pm,0}(X) = \sum_{\substack{q \in V_\pm \\ X_\pm(q)=0}} Ind_{V_\pm,q}(X_\pm)$$

The index measures the number of equilibrium points with sign of the vector field $X_\pm$ in the smoothening $(V_\pm, X_\pm)$ before it degenerated to (V, X).

Algebraically, we can now consider the non-degenerate bilinear form in $\frac{\mathbf{B}_{\mathbb{R}}}{Ann_{\mathbf{B}_{\mathbb{R}}}(h)}$ induced by the choice of $\frac{J(\tilde{X}_{\mathbb{R}})}{h}$ as generator of its socle:

$$<\,,\,>_L : \frac{\mathbf{B}_{\mathbb{R}}}{Ann_{\mathbf{B}_{\mathbb{R}}}(h)} \times \frac{\mathbf{B}_{\mathbb{R}}}{Ann_{\mathbf{B}_{\mathbb{R}}}(h)} \xrightarrow{\cdot} \frac{\mathbf{B}_{\mathbb{R}}}{Ann_{\mathbf{B}_{\mathbb{R}}}(h)} \xrightarrow{L} \mathbb{R}$$

$$L : \frac{\mathbf{B}_{\mathbb{R}}}{Ann_{\mathbf{B}_{\mathbb{R}}}(h)} \longrightarrow \mathbb{R} \quad , \quad L\left(\frac{J(\tilde{X}_{\mathbb{R}})}{h}\right) > 0$$

Denote by $Sign_{f,0}(\tilde{X}_{\mathbb{R}})$ the common signature of these bilinear forms.

THEOREM 3.2. *([**4**]) Let $n \in \mathbb{N}$ be an even positive number, and let f be a real analytic function with an algebraically isolated singularity as above. There are integers $K_\pm$ that depend only on f such that*

$$Ind_{\pm,0}(X) = Sgn_{f,0}(\tilde{X}_{\mathbb{R}}) + K_\pm \tag{3.4}$$

for every real analytic vector field $\tilde{X}_{\mathbb{R}}$ with an algebraically isolated singularity at 0 and tangent to V, $X := \tilde{X}_{\mathbb{R}}|_{V^{\mathbb{R}}}$.

Hence, for n even in formula (3.2), we have the same pattern as the previous results for the smooth case in sections 1 and 2, providing an algebraic method to compute a topological number. For n odd, formula (3.2) does not give at first hand a procedure to define a non-degenerate bilinear form so that its signature is related to the index. We must first show that formula (3.2) may be written for n odd as:

$$Ind_{f,0}(\tilde{X}) = dim_{\mathbb{C}} \frac{\mathbf{B}}{Ann_{\mathbf{B}}(h)} - dim_{\mathbb{C}} \frac{\mathbf{A}}{Ann_{\mathbf{A}}(h)} \tag{3.5}$$

Again under the same pattern as in the smooth case, we obtain:

THEOREM 3.3. *([**5**])Let $n \geq 1$ be an odd positive integer and $f_{\mathbb{R}}$ and $\tilde{X}_{\mathbb{R}}$ as above, then the index of the vector field $X = \tilde{X}|_{V^{\mathbb{R}}}$ on $V^{\mathbb{R}}$ at 0 is:*

$$Ind_{\pm,0}(X) = Sgn_{f,0}(\tilde{X}_{\mathbb{R}}) - Sgn_{\mathbf{A}}\left(\frac{df}{f}(\tilde{X}_{\mathbb{R}})\right), \tag{3.6}$$

A geometric interpretation compatible with formula (3.6) is the following: There are 2 kinds of singular points of X at 0:

a) The singular points captured in $\frac{\mathbf{B}}{Ann_{\mathbf{B}}(h)}$ and measured by the relative Jacobian.

b) The singular points of $Spec\mathbf{A}$ which are not in $\{X = 0\}$, and which are measured by the relative Hessian determinant in $\frac{\mathbf{A}}{Ann_{\mathbf{A}}(h)}$.

One then counts the first type of points minus the second kind of points. This way of counting is particular to the case when n is odd. Fon n even, one only counts the first kind of points, and adds a constant $K_{\pm}$, as in (3.4).

The validity of formula (3.5) follows directly from the isomorphism of modules

$$\frac{Ann_{\mathbf{A}}(h)}{(f)} \simeq \frac{Ann_{\mathbf{B}}(h)}{(f)} \tag{3.7}$$

since

$$dim\frac{\mathbf{B}}{(f)} - dim\frac{\mathbf{A}}{(f)} = dim\frac{\mathbf{B}}{Ann_{\mathbf{B}}(h)} + dim\frac{Ann_{\mathbf{B}}(h)}{(f)} - dim\frac{\mathbf{A}}{Ann_{\mathbf{A}}(h)} - dim\frac{Ann_{\mathbf{A}}(h)}{(f)}$$
$$= dim\frac{\mathbf{B}}{Ann_{\mathbf{B}}(h)} - dim\frac{\mathbf{A}}{Ann_{\mathbf{A}}(h)}$$

The isomorphism (3.7) is somewhat surprising, since the modules are formed with very different rings **A** and **B**. Nevertheless, the relation (3.1) supplies a certain duality.

References

[1] V. Arnold *The index of a singular point of a vector field, the Petrovski-Oleinik inequality and mixed Hodge Structures,* Funct. Anal. Appl., 12, 1978, 1 - 14

[2] D. Eisenbud, H. Levine *An algebraic formula for the degree of a C^∞ map germ.* Ann. Math., 106, 1977, 19 - 38.

[3] X. Gómez–Mont *An Algebraic formula for the index of a vector field on a hypersurface with an isolated singularity.* J. Alg. Geom., 7 (1998), 731-752

[4] X. Gómez–Mont, P. Mardešić *The index of a vector field tangent to an even dimensional hypersurface and the signature of the relative Jacobian.* Ann. Inst. Fourier, 47, 1997, 1523-1539.

[5] X. Gómez–Mont, P. Mardešić *The index of a vector field tangent to an odd dimensional hypersurface and the signature of the relative Hessian.* Functional Analysis and Applic. (to appear)

[6] Ph. Griffiths, J. Harris, *Principles of Algebraic Geometry,* J. Wiley, 1978.

[7] Khimishishiashvili, G.N., *On the local degree of a smooth map.* Trudi Tbilisi math. Inst, 1980, 105-124.

[8] J. Milnor *Singular points of Complex Hypersurfaces.* Princeton U. P., 1968.

Luis Giraldo, Departamento de Algebra, Universidad Complutense, Madrid, España
E-mail address: `giraldo@ eucmax.sim.ucm.es`

Xavier Gómez-Mont, CIMAT, A.P. 402, Guanajuato, 36000, México
E-mail address: `gmont@ fractal.cimat.mx`

Pavao Mardešić, U. Dijon, France., B.P 400, Dijon, 21011, France
E-mail address: `mardesic @ satie.u-bourgogne.fr`

Contemporary Mathematics
Volume **240**, 1999

Cyclic Coverings of a Smooth Curve and Branch Locus of the Moduli Space of Smooth Curves

Esteban Gómez González

ABSTRACT. We give a geometric description of cyclic coverings of a complete irreducible smooth curve. From this description, we obtain a necessary and sufficient condition for there exists a smooth curve of genus g that has an automorphism of order n with s fixed points. Furthermore, we construct irreducible closed subschemes among which are the irreducible components of the locus of smooth curves of given genus with an automorphism of given order.

INTRODUCTION

The aim of this paper is to study cyclic coverings of degree n of a complete irreducible smooth curve C over an algebraically closed field k with g. c. d.$(n, \text{ch}(k)) = 1$. We associate with each cyclic covering the data (D, L, χ), where D is an effective divisor on C, L is an invertible sheaf on C such that $L^{\otimes n} \simeq \mathcal{O}_C(D)$ and χ is a generator character of the dual Cartier group of $\mathbb{Z}/n\mathbb{Z}$. Moreover, we recover the covering from these data and prove the existence of a bijective map between the set of cyclic coverings of degree n of C and the quotient set of data on C by a equivalence relation. In section §3, we study the properties of the covering (branch points, ramification indexes, irreducibility) in terms of the construction data. This theory of cyclic coverings can be considered as a geometric formulation of Kummer theory for the cyclic field extensions of a function field of one variable (see [**D**], [**P**]).

If $\mathcal{M}_g$ is the moduli space of complete irreducible smooth curves of genus g over k, the branch locus of $\mathcal{M}_g$ is defined as:

$$\mathcal{B}_g = \Big\{ [X] \in \mathcal{M}_g \ : \ \text{Aut}(X) \neq \{\text{Id}\} \Big\}$$

It is known that for $g \geq 4$, $\mathcal{B}_g$ coincides with the singular locus of $\mathcal{M}_g$ ([**P**], [**L**]). This definition is not valid for genus 2 because all the curves are hyperelliptic; $\mathcal{B}_2$ is defined as the locus of curves of genus 2 whose group of automorphisms is different from $\mathbb{Z}/2\mathbb{Z}$.

1991 *Mathematics Subject Classification.* Primary 14H10, 14H30; Secondary 14E09.

This research was partly supported by the Spanish DGES through the research project PB96-1305 and by the "Junta de Castilla y León" through the research project SA27/98.

Every curve with non-trivial automorphisms can be recovered as a cyclic covering of a smooth curve. Therefore, this theory of coverings allows one to construct all curves with non-trivial automorphisms. In particular, if we fix natural numbers $\{g, n, s\}$, the necessary and sufficient condition for there exists a complete irreducible smooth curve of genus g that has an automorphism of order n with s fixed points is that the numbers $\{g, n, s\}$ should fulfill the Riemann-Hurwitz formula (Corollary 3.12).

Furthermore, we generalize this construction of cyclic coverings to the relative case; that is, when the base is a scheme. In consequence, this theory is very useful for the study the problems about the moduli space of curves with non-trivial automorphisms (see [**G**]). An example of this is the construction of irreducible closed subschemes, among which are the irreducible components of the locus $\mathcal{M}_g^{\mathbb{Z}/n\mathbb{Z}}$ of complete irreducible smooth curves of genus g with an automorphism of order n (Theorem 4.1). This problem was suggested by Ries in [**R**]. When one studies the moduli space of curves we must bear in mind that one works with irreducible curves; therefore, it is necessary to have a criterion about the irreducibility of the covering associated in terms of the data (Corollary 3.9).

The existence of Riemann surfaces with non-trivial automorphisms is studied via Fuchsian group theory ([**H1**], [**H2**], [**Gu**], [**RR**],...). The divisor of the construction data of coverings that we show is equivalent to a convenient presentation of the Fuchsian group for studying the Riemann surface with automorphisms defined by that group (see Corollary 3.3).

From now on, let C be a complete irreducible smooth curve over an algebraically closed field k and let n be a natural number which is not divisible by the characteristic of k. If α is a real number, we use $[\alpha]$ to denote the greatest integer less than or equal to α.

I wish to thank Prof. C. Sancho de Salas for many useful discussions and suggestions.

1. Preliminaries

Let $\overline{C}$ be a complete smooth curve over k and G a finite group of order n.

DEFINITION 1.1. A G-covering is a finite morphism $\pi\colon \overline{C} \to C$ of degree n and an immersion $G \hookrightarrow \operatorname{Aut}(\overline{C}/C)$ such that the quotient scheme $\overline{C}/G$ is isomorphic to C. A cyclic covering is a G-covering where G is a cyclic group.

A morphism of G-coverings $\overline{C} \to \overline{C}'$ is a morphism of C-schemes and is compatible with the actions of G on $\overline{C}$ and $\overline{C}'$.

Let $\overline{C} \to C$ be a G-covering. If $\overline{C}$ is not irreducible, $\overline{C}$ is not connected because $\overline{C}$ is smooth. Since G acts transitively on the fibres and since C is connected, the connected components of $\overline{C}$ are isomorphic. Let $\widetilde{C}$ be a irreducible component of $\overline{C}$ and let H be the subgroup of G that acts on $\widetilde{C}$. By the natural inclusion $\widetilde{C} \hookrightarrow \overline{C}$, we have that $\widetilde{\pi}\colon \widetilde{C} \to C$ is an H-covering. Hence, there exists an isomorphism $\varphi\colon \widetilde{G} \xrightarrow{\sim} H \subset G$ where $\widetilde{G} = \operatorname{Aut}(\widetilde{C}/C)$.

We consider the curve $G \times \widetilde{C}$ with the following actions of G and $\widetilde{G}$:

$$\begin{aligned} G \times (G \times \widetilde{C}) &\longrightarrow G \times \widetilde{C} \\ (g, (g', x)) &\longmapsto (g \cdot g', x) \\ \widetilde{G} \times (G \times \widetilde{C}) &\longrightarrow G \times \widetilde{C} \\ (\widetilde{g}, (g', x)) &\longmapsto (g' \cdot \varphi(\widetilde{g}^{-1}), \widetilde{g}(x)) \end{aligned}$$

One has that the curve $(G \times \widetilde{C})/\widetilde{G}$ is isomorphic to $\overline{C}$ as G-coverings. Therefore, in this way it is possible to recover the covering $\overline{C} \to C$ and the action of G on $\overline{C}$ from the immersion $\varphi\colon \widetilde{G} \hookrightarrow G$ or, equivalently, from the action of H on the irreducible component $\widetilde{C}$ of $\overline{C}$.

From this one deduces that there exists a canonical bijective map from the set of the different structures of G-covering on $\overline{C} \to C$ onto the quotient set of the immersions $\widetilde{G} \hookrightarrow G$ by the inner automorphisms of G. In particular,

COROLLARY 1.2. *If $G = \mathbb{Z}/n\mathbb{Z}$ is a cyclic group, the different structures of G-covering on $\overline{C} \to C$ are identified to* $\mathrm{Aut}(\mathbb{Z}/r\mathbb{Z})$, *where n/r is the number of irreducible components of $\overline{C}$.*

2. Construction data associated with a cyclic covering

We shall now associate construction data on the curve C with a cyclic G-covering $\pi\colon \overline{C} \to C$ of degree n. Since the group G is semisimple, the G-module $\pi_*\mathcal{O}_{\overline{C}}$ decomposes into a direct sum of sub-G-modules in which the group acts through a character:

$$\pi_*\mathcal{O}_{\overline{C}} \simeq \bigoplus_{\chi \in \widehat{G}} \mathcal{O}_\chi$$

where $\widehat{G} = \mathrm{Hom}_{\text{groups}}(G, k^*)$ is the Cartier dual group.

LEMMA 2.1. *For every $\chi \in \widehat{G}$, the $\mathcal{O}_C$-module $\mathcal{O}_\chi$ is non-null and is an invertible sheaf on C. In particular, the above decomposition is the decomposition of $\pi_*\mathcal{O}_{\overline{C}}$ into simple G-modules.*

PROOF. Since π is a finite morphism, the sheaves $\mathcal{O}_\chi$ are flat $\mathcal{O}_C$-modules and are hence locally free. The sections of the sheaves $\mathcal{O}_\chi$ on the complementary open subset of the branch points of π are non-null because π is a Galois covering over this open subset. Therefore, $\mathcal{O}_\chi$ is non-null. Bearing in mind that $\pi_*\mathcal{O}_{\overline{C}}$ is a $\mathcal{O}_C$-module of rank n and that we have n-summands in the decomposition, each one is necessarily of rank one. □

For every $\chi \in \widehat{G}$, we have that $\mathcal{O}_\chi^{\otimes n}$ is invariant by G. Hence, there exists an immersion $\mathcal{O}_\chi^{\otimes n} \hookrightarrow \mathcal{O}_C$ that induces an effective divisor, D_χ, associated with the invertible sheaf $(\mathcal{O}_\chi^{-1})^{\otimes n}$.

LEMMA 2.2. *For every character $\chi \in \widehat{G}$, the coefficient a of every point $q \in C$ in the divisor D_χ fulfills $0 \leq a < n$.*

PROOF. We assume that $a > n$; hence, $a = \lambda n + b$ with $b < n$ and $\lambda > 0$. If t is a local generator of $\mathcal{O}_\chi$ at the point q and w is a local coordinate of C at q and v_q is the valuation defined by the point q, then:

$$v_q\left(\frac{t^n}{w^{\lambda n}}\right) = a - \lambda n = b \geq 0$$

Therefore, $t/w^{\lambda n} \in (\mathcal{O}_\chi)_q$; that is, $t/w^{\lambda n}$ is a multiple of t, but since $\lambda > 0$, w^λ is not invertible in $\mathcal{O}_{C,q}$. □

DEFINITION 2.3. Given a cyclic G-covering $\pi\colon \overline{C} \to C$, we define the data associated with π as the triplets $(D_\chi, \mathcal{O}_\chi^{-1}, \chi)$ with $\chi \in \widehat{G}$.

EXAMPLE 2.4. Let

$$\pi\colon \overline{C} = C \coprod \overset{(n}{\cdots} \coprod C \longrightarrow C$$

be the trivial covering of degree n. One has that the data associated with this covering are equal to the datum defined by the null divisor and the trivial invertible sheaf.

The following theorem gives the relationship between the different data associated with a cyclic covering. In particular, if we know the data defined by a generator character of dual group, we know all the data associated with the covering.

THEOREM 2.5. *If $\chi_1, \chi_2 \in \widehat{G}$, then*

$$\mathcal{O}_{\chi_1\cdot\chi_2} \simeq \mathcal{O}_{\chi_1} \otimes \mathcal{O}_{\chi_2} \otimes \mathcal{O}_C(\sum_i \left[\frac{a_i^1 + a_i^2}{n}\right] \cdot q_i)$$

$$-D_{\chi_1\cdot\chi_2} = -(D_{\chi_1} + D_{\chi_2}) + n \sum_i \left[\frac{a_i^1 + a_i^2}{n}\right] \cdot q_i$$

where $D_{\chi_j} = \sum_i a_i^j \cdot q_i$, for $j = 1, 2$.

PROOF. Let t_1, t_2, t be local generators at a point $q \in C$ of the invertible sheaves $\mathcal{O}_{\chi_1}, \mathcal{O}_{\chi_2}, \mathcal{O}_{\chi_1\cdot\chi_2}$, respectively. We have that $t_1 \cdot t_2 = f \cdot t$ for some $f \in \mathcal{O}_{C,q}$. Then:

$$a^1 + a^2 = v_q(t_1^n \cdot t_2^n) = v_q(f^n \cdot t^n) = a + n \cdot v_q(f)$$

where a^1, a^2, a are the coefficients of q in the divisors $D_{\chi_1}, D_{\chi_2}, D_{\chi_1\cdot\chi_2}$, respectively. Therefore, $a \equiv a^1 + a^2 \pmod n$. Since $a < n$, $v_q(f) = \left[\frac{a^1+a^2}{n}\right]$, which is all that we need to know. □

We have seen that given a cyclic covering we can associate data on the base curve. Let us now reconstruct the cyclic covering from the data defined by the generator characters.

Let C be a complete irreducible smooth curve over k and let n be a natural number such that $\text{g.c.d.}(n, \text{ch}(k)) = 1$. Let $D = \sum_{i=1}^s a_i q_i$ be an effective divisor on C with the degree of D a multiple of n, and let L be an invertible sheaf on C such that $L^{\otimes n} \simeq \mathcal{O}_C(D)$; we fix an isomorphism between $(L^{-1})^{\otimes n}$ and the sheaf of ideals $\mathcal{O}_C(-D) \hookrightarrow \mathcal{O}_C$. Let χ be a generator character of the dual group of $G = \mathbb{Z}/n\mathbb{Z}$.

Let us consider the sheaf of $\mathcal{O}_C$-algebras:

$$\mathcal{A} = \mathcal{O}_C \oplus L^{-1} \oplus (L^{-1})^{\otimes 2} \oplus \cdots \oplus (L^{-1})^{\otimes(n-1)}$$

where the product is given by the product of sections and by the inclusion defined by the divisor D: $(L^{-1})^{\otimes n} \hookrightarrow \mathcal{O}_C$. We define the curve $C'(n, C; D, L, \chi)$ as the spectrum of this sheaf of $\mathcal{O}_C$-algebras and the natural inclusion of $\mathcal{O}_C$ defines a morphism

$$\pi'\colon C'(n, C; D, L, \chi) \longrightarrow C$$

which is finite of degree n. We define the following action of G on the sheaf of rings of $C'(n,C;D,L,\chi)$:

$$\begin{array}{ccc} G \times \mathcal{A} & \longrightarrow & \mathcal{A} \\ (g,(s,s_1,\dots,s_{n-1})) & \longmapsto & (s,\chi(g)\cdot s_1,\dots,\chi^{n-1}(g)\cdot s_{n-1}) \end{array}$$

Since χ is a generator character, the invariants by this action are isomorphic to $\mathcal{O}_C$; that is, $C'(n,C;D,L,\chi)/G \simeq C$. We define the curve $\overline{C}(n,C;D,L,\chi)$ as the desingularization of $C'(n,C;D,L,\chi)$; if $C'(n,C;D,L,\chi)$ is not irreducible, its desingularization is the disjoint union of the desingularizations of each irreducible component.

Since G acts on $C'(n,C;D,L,\chi)$ and the quotient scheme is irreducible, G acts transitively on the irreducible components of the curve $C'(n,C;D,L,\chi)$. Therefore, G acts on the desingularization. Thus,

$$\pi\colon \overline{C}(n,C;D,L,\chi) \longrightarrow C$$

is a cyclic covering of degree n. The construction does not depend on the isomorphism $(L^{-1})^{\otimes n} \simeq \mathcal{O}_C(-D)$ since all elements of k have a n-th root.

When there is no confusion on the curve C and the number n, we shall write

$$\overline{C}(n,C;D,L,\chi) = \overline{C}(D,L,\chi) \quad \text{and} \quad C'(n,C;D,L,\chi) = C'(D,L,\chi)$$

DEFINITION 2.6. We shall say that the triplets (D,L,χ) fulfilling the previous conditions are construction data of degree n on the curve C.

LEMMA 2.7. *If (D,L,χ) is a construction datum of degree n on C, one has that:*

$$\overline{C}\left(\sum_{i=1}^{s}(a_i - \left[\frac{a_i}{n}\right] n)q_i \ , \ L(-\sum_{i=1}^{s}\left[\frac{a_i}{n}\right] q_i),\chi\right) \simeq \overline{C}(D,L,\chi)$$

PROOF. There exists a surjective G-morphism between the singular curves and therefore there exists a G-morphism between their desingularizations which is an isomorphism because these G-coverings have the same degree. □

Consequently, from now on, we assume that construction data fulfill the condition that the coefficients in divisors are less than n.

EXAMPLE 2.8. Given a cyclic covering $\pi\colon \overline{C} \to C$ of degree n, for every generator character $\chi \in \widehat{G}$ we have the datum $(D_\chi, \mathcal{O}_\chi^{-1}, \chi)$ defined by this character. In particular, the triplet $(D_\chi, \mathcal{O}_\chi^{-1}, \chi)$ is a construction datum of degree n on C fulfilling the condition that the coefficients in the divisor D_χ are less than n (Lemma 2.2)

The next step is to recover the construction datum (D,L,χ) from the covering $\overline{C}(D,L,\chi) \to C$:

LEMMA 2.9. *If (D,L,χ) is a construction datum of degree n on the curve C, then $L^{-1} \simeq \mathcal{O}_\chi$ and $D = D_\chi$, where $(D_\chi, \mathcal{O}_\chi^{-1}, \chi)$ is the datum associated with the covering $\pi\colon \overline{C}(D,L,\chi) \to C$ defined by the character χ.*

PROOF. Since the sheaf of rings of the curve $\overline{C}(D,L,\chi)$ is the integral closure of the sheaf of rings of $C'(D,L,\chi)$, we have the morphisms of G-modules and $\mathcal{O}_C$-algebras:

$$L^{-1} \hookrightarrow \bigoplus_{r=0}^{n-1}(L^{-1})^{\otimes r} \hookrightarrow \pi_*\mathcal{O}_{\overline{C}(D,L,\chi)} = \bigoplus_{r=0}^{n-1}\mathcal{O}_{\chi^r}$$

By definition of the action of G, we have that G acts on L^{-1} through the character χ. Since L^{-1} is an invertible sheaf,

$$L^{-1} \hookrightarrow \mathcal{O}_\chi$$

Bearing in mind that the coefficients in the divisors D and D_χ are less than n, we deduce the result as in the proof of Theorem 2.5. □

THEOREM 2.10. *Every cyclic covering $\pi\colon \overline{C} \to C$ of degree n is defined by construction data of degree n on C, which are the data $(D_\chi, \mathcal{O}_\chi^{-1}, \chi)$ associated with the covering π defined by the different generator characters of $\mathbb{Z}/n\mathbb{Z}$.*

PROOF. By previous Lemma, we must proof that

$$\overline{C} \simeq \overline{C}(n, C; D_\chi, \mathcal{O}_\chi^{-1}, \chi)$$

as G-coverings for every generator character χ. We have that:

$$\pi_* \mathcal{O}_{\overline{C}} \simeq \mathcal{O}_C \oplus \mathcal{O}_\chi \oplus \mathcal{O}_{\chi^2} \oplus \cdots \oplus \mathcal{O}_{\chi^{n-1}}$$

Furthermore,

$$\pi'_* \mathcal{O}_{C'(D_\chi, \mathcal{O}_\chi^{-1}, \chi)} = \mathcal{O}_C \oplus \mathcal{O}_\chi \oplus \mathcal{O}_\chi^{\otimes 2} \oplus \cdots \oplus \mathcal{O}_\chi^{\otimes (n-1)}$$

For every $r \leq n$, there exists a morphism $(\mathcal{O}_\chi)^{\otimes r} \hookrightarrow \mathcal{O}_{\chi^r}$ of G-modules that induces an injective morphism $\pi'_* \mathcal{O}_{C'(D_\chi, \mathcal{O}_\chi^{-1}, \chi)} \hookrightarrow \pi_* \mathcal{O}_{\overline{C}}$ of $\mathcal{O}_C$-algebras and G-modules. Therefore, there exists a surjective morphism $\overline{C} \to C'(D_\chi, \mathcal{O}_\chi^{-1}, \chi)$ of G-coverings. Since $\overline{C}$ is non-singular, the above morphism factors through the desingularization of $C'(D_\chi, \mathcal{O}_\chi^{-1}, \chi)$:

$$\overline{C} \longrightarrow \overline{C}(D_\chi, \mathcal{O}_\chi^{-1}, \chi)$$

which is an isomorphism because these G-coverings have the same degree. □

COROLLARY 2.11. *Construction data of degree n on C, $(D = \sum a_i q_i, L, \chi)$ and $(D' = \sum a'_i q_i, L', \chi')$, define isomorphic G-coverings if and only if there exists $r \in \mathbb{N}$ with $1 \leq r < n$ such that:*

a) $\chi^r = \chi'$. In particular, $\mathrm{g.c.d.}(r, n) = 1$.

b) $r \cdot a_i \equiv a'_i \pmod n$ for every i. In particular, $\sup D = \sup D'$.

c) $L^{\otimes r} \simeq L'(\sum \left[\frac{r a_i}{n}\right] q_i)$.

PROOF. If $\phi\colon \overline{C}(D, L, \chi) \xrightarrow{\sim} \overline{C}(D', L', \chi')$ as G-coverings, then the submodule L'^{-1} of the sheaf of rings of $\overline{C}(D', L', \chi')$, where G acts through the character χ', is isomorphic to the submodule $\mathcal{O}_{\chi'}$ of the sheaf of rings of $\overline{C}(D, L, \chi)$ where G acts through the character χ'. Therefore, $L \simeq \mathcal{O}_\chi^{-1}$ and $L' \simeq \mathcal{O}_{\chi'}^{-1}$ for the characters $\chi, \chi' \in \widehat{G}$. In particular, $\chi' = \chi^r$ for some $r \in \mathbb{N}$ with $1 \leq r < n$ and $\mathrm{g.c.d.}(r, n) = 1$. By the relations of Theorem 2.5 between the invertible sheaves $\mathcal{O}_{\chi^r}$ and $\mathcal{O}_\chi^{\otimes r}$ and between their associated divisors, one concludes.

Conversely, we have the covering $\overline{C}(D, L, \chi)$ with $L^{-1} \simeq \mathcal{O}_\chi$. From conditions b) and c), one has that $L'^{-1} \simeq \mathcal{O}_{\chi^r}$ and $D' = D_{\chi^r}$. By Theorem 2.10, $\overline{C}(D, L, \chi)$ and $\overline{C}(D', L', \chi')$ are isomorphic as G-coverings. □

In consequence, we can define a bijective map between the set of cyclic coverings of degree n of C and the quotient set of construction data of degree n on C by the equivalence relation defined by the above corollary.

REMARK 2.12. If (D, L, χ) is a construction datum of degree n on C, by Corollary 1.2 the different structures of $\mathbb{Z}/n\mathbb{Z}$-covering of the covering $\overline{C}(D, L, \chi) \to C$ are obtained by changing the generator character χ, because the natural morphism

$$(\mathbb{Z}/n\mathbb{Z})^* \longrightarrow (\mathbb{Z}/r\mathbb{Z})^* \simeq \mathrm{Aut}(\mathbb{Z}/r\mathbb{Z})$$

where n/r is the number of irreducible components of $\overline{C}(D, L, \chi)$, is surjective. Hence, when we work with the curve $\overline{C}(D, L, \chi)$ as C-scheme and we forget the $\mathbb{Z}/n\mathbb{Z}$-structure, it is not necessary to fix the generator character χ.

The following theorem gives the behavior of construction data of degree n by base changes.

THEOREM 2.13. *Let X be a complete irreducible smooth curve over k and $f\colon X \to C$ a finite morphism. Given a cyclic covering $\pi\colon \overline{C} = \overline{C}(n, C; D, L, \chi) \to C$, if we denote the desingularization of $\overline{C} \times_C X$ by $\overline{C}_X$, we have that $\pi_X : \overline{C}_X \to X$ is a cyclic covering of degree n. Then,*

$$\overline{C}_X \simeq \overline{C}(n, X; f^*D, f^*L, \chi)$$

PROOF. Since (f^*D, f^*L, χ) is a construction datum of degree n on X, the theorem of base change of cohomology gives us a morphism

$$\overline{C} \times_C X \longrightarrow C'(n, X; f^*D, f^*L, \chi)$$

of G-coverings over X. Hence, by degrees, their desingularizations are isomorphic coverings. □

We must bear in mind that the coefficients in the divisor f^*D may be greater than n; hence f^*L^{-1} can be non-isomorphic to the submodule of $(\pi_X)_*\mathcal{O}_{\overline{C}_X}$, where G acts through the character χ. If we consider the equivalent datum to (f^*D, f^*L, χ) that results by making modulo n (Lemma 2.7), then this datum indeed fulfills the above condition.

By Theorem 2.5, we have that the decomposition into simple G-modules of the sheaf of rings of the covering $\overline{C}(D, L, \chi)$ is:

$$\pi_*\mathcal{O}_{\overline{C}(D,L,\chi)} \simeq \bigoplus_{r=0}^{n-1} (L^{-1})^{\otimes r}(\sum_{i=1}^{n} \left[\frac{ra_i}{n}\right] q_i)$$

Since the morphism $\pi\colon \overline{C}(D, L, \chi) \to C$ is affine, we have an explicit description of the covering associated with the datum (D, L, χ).

This description allows us to generalize the construction of cyclic coverings to the case of a relative curve: Let $f\colon C \to S$ be a flat smooth curve of genus g over the scheme S; let $\{q_1, \dots, q_s, q_{s+1}\}$ be rational disjoint points of C over S and $\{a_1, \dots, a_s\}$ natural numbers fulfilling $1 \le a_i < n$ and $a_1 + \dots + a_s \equiv 0 \pmod{n}$. Let D be the effective Cartier divisor on C defined by $D = \sum_{i=1}^{s} a_i q_i$ and let $\mathcal{L} \in \mathrm{Pic}(C/S)$ a relative invertible sheaf on C such that $\mathcal{L}^{\otimes n} \simeq \mathcal{O}_C(D)$ in the relative Picard scheme $\mathrm{Pic}(C/S)$. Let L be the unique invertible sheaf on C representative of $\mathcal{L}$ such that its restriction to the rational point q_{s+1} is trivial. Thus, $L^{\otimes n} \simeq \mathcal{O}_C(D)$ as absolute invertible sheaves on C. In particular, $(L^{-1})^{\otimes n} \hookrightarrow \mathcal{O}_C$ through the section given by the divisor D. Let χ be a generator character of the dual group of $G = \mathbb{Z}/n\mathbb{Z}$.

Copying the construction carried out above in the case of $S = \operatorname{Spec} k$, we define:

$$\overline{C} = \operatorname{Spec}\left(\bigoplus_{r=0}^{n-1} (L^{-1})^{\otimes r}(\sum_{i=1}^{s}\left[\frac{ra_i}{n}\right] q_i)\right) \longrightarrow C$$

which is a cyclic covering of degree n of C. We have that $\overline{C} \longrightarrow S$ is a relative smooth curve over S.

It should be noted that the point q_{s+1} is only necessary in order to determine the choice of the invertible sheaf L and the isomorphism $L^{\otimes n} \simeq \mathcal{O}_C(D)$.

3. Properties of the cyclic covering defined by a construction datum

In this section, we shall study the relationship between the construction data and the branch points and the ramification indexes of the associated covering.

Let us first assume that the cyclic covering $\pi\colon \overline{C} \to C$ is irreducible.

If $q \in C$, we denote the number of distinct points in the fibre of q by d, the ramification index of each point in the fibre by m and the coefficient of the point q in the divisor D_χ associated with a generator character by a.

LEMMA 3.1. *Let $q \in C$ be a branch point of π. There exists a construction datum (D, L, χ_q) associated with π such that the coefficient of the point q in the divisor D is the number d of distinct points in the fibre of the point q.*

PROOF. If $K \subseteq G = \mathbb{Z}/n\mathbb{Z} \subseteq \operatorname{Aut}(\overline{C}/C)$ is the isotropy group of each point in $\pi^{-1}(q)$, then $\#K = m$ and $\#(G/K) = d = n/m$. Let $\pi^{-1}(q) = \{p_1, \dots, p_d\}$ and we denote the ideal of the point p_i by $\mathfrak{m}_i$, $B = (\pi_*\mathcal{O}_{\overline{C}})_q$ and $I = \mathfrak{m}_1 \cdot \ldots \cdot \mathfrak{m}_d \subset B$. If we denote by $k_\chi = \mathcal{O}_{\chi,q}/\mathfrak{m}_q \cdot \mathcal{O}_{\chi,q}$ for each $\chi \in \widehat{G}$, the decomposition into simple G-modules of $\pi_*\mathcal{O}_{\overline{C}}$ induces the following decomposition into G-modules:

$$B/I^m = B/\mathfrak{m}_q \cdot B = \bigoplus_{\chi\in\widehat{G}} k_\chi$$

Furthermore, since G acts on B and I, we have the exact sequence of G-modules for every $0 < r \leq m$

$$0 \longrightarrow I^r/I^{r+1} \longrightarrow B/I^{r+1} \longrightarrow B/I^r \longrightarrow 0$$

which splits as G-modules because G is a semisimple group. Thus

$$B/I^m \simeq B/I \oplus I/I^2 \oplus \cdots \oplus I^{m-1}/I^m$$

For every $\chi \in \widehat{G}$, $k_\chi \subset I^r/I^{r+1}$ for some r, because k_χ is a simple G-module. We shall compute such r for some generator character. Since K is the isotropy group of the points p_i and since $\overline{C}$ is irreducible, the natural morphism:

$$\chi^K : K \hookrightarrow \operatorname{Aut}_k(\mathfrak{m}_1/\mathfrak{m}_1^2) \simeq k^*$$

is injective; that is, χ^K is a generator character of $\widehat{K}$. Since G is abelian, K acts on $\mathfrak{m}_i/\mathfrak{m}_i^2$ through the character χ^K for every i. Therefore, K acts on I/I^2 through the character χ^K.

Let χ_q be a generator character of $\widehat{G}$ such that $\chi_{q|_K} = \chi^K$. Since K acts on I/I^2 through the character χ^K, it follows that $k_{\chi_q} \subset I/I^2$. Hence, a local generator t of $\mathcal{O}_{\chi_q,q}$ at the point q belongs to I/I^2; that is, for some point p_i, $t \in \mathfrak{m}_i$ and $t \notin \mathfrak{m}_i^2$. If we denote the coefficient of the point q in the divisor D_{χ_q} by a, then

$$n = v_{p_i}(t^n) = m \cdot v_q(t^n) = m \cdot a$$

Therefore, $a = n/m = d$. □

THEOREM 3.2. *For every construction datum (D, L, χ) associated with the irreducible cyclic covering $\pi\colon \overline{C} \to C$ one has that the points of the support of D are the branch points of π. Moreover, if d is the number of distinct points in the fibre of a point $q \in C$ and a is the coefficient of q in the divisor D, then* $\mathrm{g.c.d.}(a, n) = d$.

PROOF. Let $q \in \sup D$ and let t be a local generator of L^{-1} at the point q and $p \in \pi^{-1}(q)$. We have that $v_p(t) = (m \cdot v_q(t^n))/n = (m \cdot a)/n > 0$. Since $v_p(t)$ is a natural number and $a < n$, then $m > 1$; that is, q is a branch point of π.

Conversely, let $q \in C$ be a branch point; since χ is a generator character, $\chi = \chi_q^r$ for some natural number $r < n$ with $\mathrm{g.c.d.}(r, n) = 1$. By Theorem 2.5,

$$r \cdot d = a + n \cdot \left[\frac{r \cdot d}{n}\right]$$

Hence, by the Euclides algorithm,

$$\mathrm{g.c.d.}(a, n) = \mathrm{g.c.d.}(n, r \cdot d) = d \cdot \mathrm{g.c.d.}(n/d, r) = d$$

Moreover $d < n$, thus $a > 0$; that is, $q \in \sup D$. □

COROLLARY 3.3. *For every branch point $q \in C$ of the irreducible cyclic covering $\pi\colon \overline{C} \to C$, there exists a generator character $\chi^K \in \widehat{K}$, where K is the isotropy group of the fibre of the point q, such that for every generator character $\chi \in \widehat{G}$,*

$$\chi_{|K} = (\chi^K)^{a/d}$$

where a is the coefficient of q in the divisor D_χ.

The way to calculate the coefficients in the divisor D_χ given by this corollary is similar to the rotation data defined in Fuchsian group theory (see [**Gu**], [**H1**]).

Let us now assume that the cyclic covering $\pi\colon \overline{C} \to C$ is not irreducible. Since G acts transitively on the fibres and C is connected, the connected components of $\overline{C}$ are isomorphic to a smooth irreducible curve $\widetilde{C}$. Let $\overline{r}$ be the number of the components of $\overline{C}$ and let H be the subgroup of G that acts on a connected component; since G is abelian, H acts on all components of $\overline{C}$. Moreover, H is the unique subgroup (cyclic) of G of order r. We have that $\widetilde{\pi}\colon \widetilde{C} \to C$ is a H-covering of degree $r = \#H = n/\overline{r}$.

Furthermore, by definition of H, the (G/H)-covering

$$\widehat{\pi}\colon \overline{C}/H = \widehat{C} \to C$$

is the trivial covering of degree $\overline{r} = \#(G/H)$.

LEMMA 3.4. *For every character $\overline{\chi} \in \widehat{G/H} \subset \widehat{G}$, one has that the datum $(D_{\overline{\chi}}, \mathcal{O}_{\overline{\chi}}^{-1}, \overline{\chi})$ of degree n associated with the covering $\pi\colon \overline{C} \to C$ is equal to the datum $(0, \mathcal{O}_C, \overline{\chi})$ associated with the trivial covering $\widehat{\pi}$ of degree $\overline{r}$.*

PROOF. We have the morphism of (G/H)-modules and G-modules:

$$\widehat{\pi}_* \mathcal{O}_{\widehat{C}} = \mathcal{O}_C \oplus \cdots \oplus \mathcal{O}_C \hookrightarrow \mathcal{O}_{\chi_1} \oplus \cdots \oplus \mathcal{O}_{\chi_n} = \pi_* \mathcal{O}_{\overline{C}}$$

Hence, for every submodule of $\widehat{\pi}_* \mathcal{O}_{\widehat{C}}$ we have that $\mathcal{O}_C \hookrightarrow \mathcal{O}_{\chi_i}$ for some i. However, since H acts on $\mathcal{O}_C$ by identity, $\chi_i|_H = \mathrm{Id}$; that is, $\chi_i = \overline{\chi}$ for some character $\overline{\chi} \in \widehat{G/H}$. Therefore, we have morphism:

$$\mathcal{O}_C \hookrightarrow \mathcal{O}_{\overline{\chi}} \Rightarrow (\mathcal{O}_{\overline{\chi}}^{-1})^{\otimes n} \hookrightarrow \mathcal{O}_C$$

Furthermore, since $(D_{\overline{\chi}}, \mathcal{O}_{\overline{\chi}}^{-1}, \overline{\chi})$ is a datum associated with a cyclic covering of degree n, $\mathcal{O}_{\overline{\chi}}^{\otimes n} \hookrightarrow \mathcal{O}_C$. Hence, $\mathcal{O}_{\overline{\chi}}^{\otimes n} \simeq \mathcal{O}_C$, that is, $D_{\overline{\chi}} = 0$. In particular, $\mathcal{O}_{\overline{\chi}}$ has degree zero and since $\mathcal{O}_C \hookrightarrow \mathcal{O}_{\overline{\chi}}$, we have that $\mathcal{O}_{\overline{\chi}} \simeq \mathcal{O}_C$.

Since there is the same number of summands on the decomposition of $\widehat{\pi}_* \mathcal{O}_{\widehat{C}}$ as characters of G/H, one deduces that the above equality holds for all characters of G/H. □

COROLLARY 3.5. *The datum $(D_\chi, \mathcal{O}_\chi^{-1}, \chi)$ of degree n associated with the covering $\pi \colon \overline{C} \to C$ fulfills the condition that the coefficients in the divisor D_χ are divisible by $\overline{r}$ and $(\mathcal{O}_\chi^{-1})^{\otimes n/\overline{r}} \simeq \mathcal{O}_C(1/\overline{r}\ D_\chi)$, where $\overline{r}$ is the number of the connected components of $\overline{C}$.*

PROOF. We have that $\widehat{G/H} \subset \widehat{G}$ is the unique subgroup of $\widehat{G}$ of order $\overline{r}$. So, for every character $\chi \in \widehat{G}$, $\chi^r \in \widehat{G/H}$. If $D = \sum_{i=1}^{s} a_i q_i$, by Theorem 2.5,

$$0 = -D_{\chi^r} = -rD_\chi + \sum_{i=1}^{s} n \left[\frac{ra_i}{n}\right] q_i$$

Therefore, $a_i = \overline{r}\left[\frac{ra_i}{n}\right]$. The relationship between the invertible sheaves $\mathcal{O}_\chi$ and $\mathcal{O}_{\chi^r} = \mathcal{O}_C$ gives us the result. □

We now study the data associated with the covering $\widetilde{\pi} \colon \widetilde{C} \to C$.

LEMMA 3.6. *Let $\chi \in \widehat{G}$ be a generator character, $(D_\chi, \mathcal{O}_\chi^{-1}, \chi)$ the datum associated with the covering π and let $\widetilde{\chi} = \chi_{|H}$, which is a generator character of $\widehat{H}$. One has that*

$$\overline{C}(n/\overline{r}, C; 1/\overline{r}\ D_\chi, \mathcal{O}_\chi^{-1}, \widetilde{\chi}) \simeq \widetilde{C}$$

as H-coverings of C. In particular, the datum $(D_{\widetilde{\chi}}, \mathcal{O}_{\widetilde{\chi}}^{-1}, \widetilde{\chi})$ associated with the H-covering $\widetilde{\pi} \colon \widetilde{C} \to C$ defined by the character $\widetilde{\chi} \in \widehat{H}$ is equal to the construction datum $(1/\overline{r}\ D_\chi, \mathcal{O}_\chi^{-1}, \widetilde{\chi})$ of degree r.

PROOF. By Corollary 3.5, we have that $(\mathcal{O}_\chi^{-1})^{\otimes n/\overline{r}} \simeq \mathcal{O}_C(1/\overline{r} D_\chi)$. Hence, $(1/\overline{r} D_\chi, \mathcal{O}_\chi^{-1}, \widetilde{\chi})$ is a construction datum of degree $r = n/\overline{r}$. We fix an isomorphism $\varphi \colon \mathcal{O}_\chi^{\otimes n/\overline{r}} \xrightarrow{\sim} \mathcal{O}_C(-1/\overline{r}\ D_\chi) \hookrightarrow \mathcal{O}_C$ and we consider the sheaf of rings of the singular curve $C'(n, C; D_\chi, \mathcal{O}_\chi^{-1}, \chi)$:

$$\mathcal{A} = \mathcal{O}_C \oplus \mathcal{O}_\chi \oplus \cdots \oplus \mathcal{O}_\chi^{\otimes(n-1)}$$

Let $\mathcal{I} \subset \mathcal{A}$ be the sheaf of ideals generated by the elements of the following form $(-\varphi(s), 0, \dots, 0, s, 0, \dots, 0)$ where s is a section of the sheaf $\mathcal{O}_\chi^{\otimes n/\overline{r}}$. Then $\mathcal{A}/\mathcal{I}$ is isomorphic to the sheaf of rings of the singular curve $C'(n/\overline{r}, C; 1/\overline{r}\ D_\chi, \mathcal{O}_\chi^{-1}, \widetilde{\chi})$. Moreover, since $\chi_{|H} = \widetilde{\chi}$, the quotient morphism $\mathcal{A} \to \mathcal{A}/\mathcal{I}$ is an H-morphism. We therefore have a closed H-immersion:

$$\overline{C}(n/\overline{r}, C; 1/\overline{r}\ D_\chi, \mathcal{O}_\chi^{-1}, \widetilde{\chi}) \hookrightarrow \overline{C}(n, C; D_\chi, \mathcal{O}_\chi^{-1}, \chi) = \overline{C} = \widetilde{C} \coprod \cdots \coprod \widetilde{C}$$

If we consider the projection morphism $\overline{C} \to \widetilde{C}$, we have a morphism

$$\overline{C}(n/\overline{r}, C; 1/\overline{r}\ D_\chi, \mathcal{O}_\chi^{-1}, \widetilde{\chi}) \longrightarrow \widetilde{C}$$

of H-coverings, which is an isomorphism by degrees over C. □

DEFINITION 3.7. The index of a construction datum $(D = \sum_{i=1}^{s} a_i \cdot q_i, L, \chi)$ of degree n on the curve C is the greatest natural number α such that α divides $\mathrm{g.c.d.}(a_1, \ldots, a_s, n)$ and satisfies

$$L^{\otimes n/\alpha} \simeq \mathcal{O}_C(1/\alpha\ D)$$

THEOREM 3.8. *Let (D, L, χ) be a construction datum of degree n on C whose index is α. Let H be the unique subgroup of G of index α. Then the G-covering $\pi\colon \overline{C}(n, C; D, L, \chi) \to C$ has α connected components and each one is isomorphic to $\overline{C}(n/\alpha, C; 1/\alpha\ D, L, \chi_{|H})$. Moreover, the points in the support of D are the branch points of π and the number of distinct points in the fibre of a point $q \in C$ is equal to $\mathrm{g.c.d.}(a, n)$, where a is the coefficient of q in the divisor D; in particular, the ramification index of every point in the fibre of q is $n/\mathrm{g.c.d.}(a, n)$.*

PROOF. Since $(L^{\otimes n/\alpha})^{\otimes \alpha} \simeq \mathcal{O}_C(D)$, we have that $(D, L^{\otimes n/\alpha}, \chi^{n/\alpha})$ is a construction datum of degree α on C. Furthermore, the coefficients in the divisor D are divisible by α. Hence, the construction datum $(D, L^{\otimes n/\alpha}, \chi^{n/\alpha})$ defines the same covering as the construction datum $(0, L^{\otimes n/\alpha}(-1/\alpha\ D), \chi^{n/\alpha})$ by Lemma 2.7. Since the index of datum (D, L, χ) is α, we have that $L^{\otimes n/\alpha}(-1/\alpha\ D) \simeq \mathcal{O}_C$; therefore, the construction datum $(0, L^{\otimes n/\alpha}(-1/\alpha\ D), \chi^{n/\alpha})$ defines the trivial covering of degree α:

$$\overline{C}(\alpha, C; D, L^{\otimes n/\alpha}, \chi^{n/\alpha}) = C \coprod \overset{(\alpha}{\cdots} \coprod C \longrightarrow C$$

By the natural inclusions, we have an injective morphism between the sheaves of rings of the singular curves $C'(\alpha, C; D, L^{\otimes n/\alpha}, \chi^{n/\alpha})$ and $C'(n, C; D, L, \chi)$:

$$\begin{array}{c}
\mathcal{O}_C \oplus (L^{-1})^{\otimes n/\alpha} \oplus (L^{-1})^{\otimes 2n/\alpha} \oplus \cdots \oplus (L^{-1})^{\otimes(\alpha-1)n/\alpha} \\
\big\downarrow \\
\mathcal{O}_C \oplus L^{-1} \oplus (L^{-1})^{\otimes 2} \oplus \cdots \oplus (L^{-1})^{\otimes(n-1)}
\end{array}$$

which is a morphism of algebras (the inclusions into $\mathcal{O}_C$ are given in both cases by the divisor D) and G-modules by election of the characters. Therefore, we have a surjective morphism

$$\begin{array}{ccc}
\overline{C}(n, C; D, L, \chi) & \xrightarrow{(n/\alpha):1} & \overline{C}(\alpha, C; D, L^{\otimes n/\alpha}, \chi^{n/\alpha}) \\
 & {\scriptstyle n:1} \searrow \quad \swarrow {\scriptstyle \alpha:1} & \\
 & C &
\end{array}$$

In consequence, the number $\overline{r}$ of the connected components of the covering $\overline{C}(n, C; D, L, \chi)$ is greater or equal to the number of components of the covering $\overline{C}(\alpha, C; D, L^{\otimes n/\alpha}, \chi^{n/\alpha})$; that is, $\overline{r} \geq \alpha$.

Conversely, the datum (D, L, χ) satisfies $L^{\otimes n/\overline{r}} \simeq \mathcal{O}_C(1/\overline{r}\ D)$ by Corollary 3.5. Therefore $\overline{r} \leq \alpha$. Thus, $\alpha = \overline{r}$.

Let us now calculate the number of points in the fibre of a point $q \in C$. If we denote by $\widetilde{\chi} = \chi_{|H}$, by Theorem 3.6, the irreducible component $\widetilde{C}$ of the covering $\overline{C}(n, C; D, L, \chi)$ is isomorphic to $\overline{C}(n/\alpha, C; 1/\alpha\ D, L, \widetilde{\chi})$. Since $\widetilde{\chi}$ is a generator character of $\widehat{H}$ and the covering associated with the datum $(1/\alpha\ D, L, \widetilde{\chi})$ is irreducible, by Theorem 3.2, the points of $\sup(1/\alpha\ D) = \sup D$ are the branch points of the covering $\widetilde{\pi}\colon \widetilde{C} = \overline{C}(n/\alpha, C; 1/\alpha\ D, L, \widetilde{\chi}) \to C$ and the number of distinct points in $\widetilde{\pi}^{-1}(q)$ is equal to $\text{g.c.d.}(a/\alpha, n/\alpha)$, where a is the coefficient of q in the divisor D. Bearing in mind that the covering $\overline{C}(n, C; D, L, \chi) \to \widetilde{C}$ is étale of degree α, one concludes the proof. □

COROLLARY 3.9. *Let (D, L, χ) be a construction datum of degree n on C. Then, the covering $\overline{C} = \overline{C}(n, C; D, L, \chi)$ is irreducible if and only if the index of the datum (D, L, χ) is one. In this case, the genus $\overline{g}$ of the complete irreducible smooth curve $\overline{C}$ is*

$$\overline{g} = n\left(g - 1 + \frac{1}{2}\sum_{i=1}^{s}(1 - \text{g.c.d.}(a_i, n)/n)\right) + 1$$

where g is the genus of C.

PROOF. One deduces the proof from above theorem and the Riemann-Hurwitz formula. □

REMARK 3.10. By Lemma 2.7, it is easy to check that the construction data of degree n such that the coefficients in the divisor are greater than or equal to n fulfill the same conditions concerning the irreducibility and the number of distinct points in the fibre as those given in Theorem 3.8.

From these results, we now deduce some properties of curves with non-trivial automorphisms.

COROLLARY 3.11. *Let $\overline{C}$ be a complete irreducible smooth curve with an automorphism σ of order n. Then, there cannot be only one orbit by σ with less than n distinct points. In particular, if n is a prime number, then σ has no fixed points or has at less two fixed points.*

PROOF. We assume that there exists only one orbit by σ with less than n distinct points; that is, the cyclic covering $\pi\colon \overline{C} \to \overline{C}/ < \sigma >$ of degree n has only one branch point. Therefore, the support of a divisor D associated with this covering has only one point q; so, $D = a \cdot q$ with $1 \leq a < n$ and $\deg D \equiv 0 \pmod{n}$, which is impossible. □

COROLLARY 3.12. *Given natural numbers $\{\overline{g}, n, s\}$, there exists a complete irreducible smooth curve of genus $\overline{g}$ that has an automorphism of order n with s fixed points if and only if $s \neq 1$ and $\overline{g} - 1 - \frac{s(n-1)}{2}$ is a integer number greater or equal to $-n$ and divisible by n*

PROOF. One deduces the necessary condition from the Riemann-Hurwitz formula.

Conversely, we have that $\overline{g} - 1 - \frac{s(n-1)}{2} = \lambda \cdot n$ with $\lambda \geq -1$. Let $g = \lambda + 1 \geq 0$ and let C be a complete irreducible smooth curve of genus g. By Corollary 3.9, we search for a cyclic covering of degree n over C with s branch points such that the ramification index of each one is n. Therefore, by Theorem 2.10, we must find natural numbers $\{a_i\}_{i=1}^{s}$ with $a_1 + \cdots + a_s$ divisible by n. Moreover, since the number of distinct points in the fibre of the branch points has to be one, these numbers must satisfy that $\text{g.c.d.}(a_i, n) = 1$ (Theorem 3.8). In particular, by this condition, the covering associated with these data is irreducible.

Since $s(n-1)$ has to be a even natural number, if n is a even number, then s is a even number. Bearing in mind this fact, it is easy to check by induction on s

that given natural numbers n and $s \geq 2$, there exist natural numbers $\{a_i\}_{i=1}^s$ prime with n such that their sum is divisible by n. □

4. Irreducible components of $\mathcal{M}_g^{\mathbb{Z}/n\mathbb{Z}}$

We shall now construct irreducible closed subschemes of the locus $\mathcal{M}_g^{\mathbb{Z}/n\mathbb{Z}}$ of complete irreducible smooth curves of genus g with an automorphism of order n. Moreover, among these subschemes are the irreducible components of $\mathcal{M}_g^{\mathbb{Z}/n\mathbb{Z}}$.

By Remark 2.12, the generator character of the dual group of $\mathbb{Z}/n\mathbb{Z}$ is not necessary for the construction of coverings that admit an action of $\mathbb{Z}/n\mathbb{Z}$. Therefore, the construction data of complete smooth curves with an automorphism of order n are the pairs (D, L) such that $L^{\otimes n} \simeq \mathcal{O}_C(D)$.

THEOREM 4.1. *Let n be a natural number, $\{a_1, \dots, a_s\}$ natural numbers with $a_i < n$ and $\sum_{i=1}^s a_i \equiv 0 \ (mod\ n)$, and $g' \leq g$ natural numbers with*

$$2g - 2 = n\left(2g' - 2 + \sum_{i=1}^{s}(1 - \text{g.c.d.}(a_i, n)/n)\right)$$

There exists a closed irreducible subscheme $S(n, g'; a_1, \dots, a_s) \subseteq \mathcal{M}_g^{\mathbb{Z}/n\mathbb{Z}}$ of dimension $3g' - 3 + s$ whose geometric points are the complete irreducible smooth curves of genus g which are coverings of the form $\overline{C}(n, X'; \sum_{i=1}^s a_i q_i, L)$ for some curve X' of genus g', $\{q_1, \dots, q_s\}$ distinct points of X' and L an invertible sheaf on X' such that $L^{\otimes n} \simeq \mathcal{O}_{X'}(\sum_{i=1}^s a_i q_i)$.

PROOF. This is similar to the proof of Theorem 2.1 in [**G**]: When n is a prime number, for $s > 0$, the author constructs, via the Hilbert scheme, a relative curve $f\colon \mathcal{C}_{g'} \to \mathcal{S}$, disjoint points $\{\mathfrak{q}_1, \dots, \mathfrak{q}_s, \mathfrak{q}_{s+1}\}$ of $\mathcal{C}_{g'}$ over $\mathcal{S}$ and an invertible sheaf $\mathcal{L}$ on $\mathcal{C}_{g'}$ such that $\mathcal{L}^{\otimes n} \simeq \mathcal{O}(\sum_{i=1}^s a_i \mathfrak{q}_i)$. Each geometric point of $\mathcal{S}$ represents $(X'; q_1, \dots, q_{s+1}; L)$, where X' is a complete irreducible smooth curve of genus g' with a tricanonical immersion, $\{q_1, \dots, q_{s+1}\}$ are distinct points of X' and L is an invertible sheaf on X' with $L^{\otimes n} \simeq \mathcal{O}_{X'}(\sum_{i=1}^s a_i q_i)$. Therefore, each geometric point of $\mathcal{S}$ corresponds to one construction datum of coverings of degree n and genus g.

When n is not a prime number, the problem is that the associated covering can be non-irreducible. Since an invertible sheaf on a complete scheme is trivial if and only if the invertible sheaf and its dual invertible sheaf have non-null global sections, by the semicontinuity theorem, if there exists a natural number r that divides to the natural numbers $\{a_1, \dots, a_s, n\}$, the set of points $z \in \mathcal{S}$ such that the invertible sheaf $\mathcal{L}^{\otimes n/r}(-1/r \sum_{i=1}^s a_i q_i)$ is trivial on the fibre of z by f is a closed subscheme of $\mathcal{S}$. There is a finite number of numbers r with the above condition and therefore there exists an open subscheme $\mathcal{S}' \hookrightarrow \mathcal{S}$ whose points fulfill the condition that for every r that divides to $\{a_1, \dots, a_s, n\}$, the invertible sheaf $\mathcal{L}^{\otimes n/r}(-1/r \sum_{i=1}^s a_i q_i)$ is non-trivial on the fibre of these points. Hence, by Corollary 3.9, the cyclic covering of degree n defined by a point of $\mathcal{S}'$ is irreducible.

If we restrict all the objects which appear in [**G**] to the open subscheme $\mathcal{S}'$, then the proof of our theorem is equal to the proof of Theorem 2.1 in [**G**].

Proof of the case $s = 0$ is reduced to the case when n is a prime number by a similar argument to the previous one. □

By construction,

$$\mathcal{M}_g^{\mathbb{Z}/n\mathbb{Z}} = \bigcup_{g',a_1,\ldots,a_s} S(n,g';a_1,\ldots,a_s)$$

with $\{g',a_1,\ldots,a_s\}$ fulfilling the conditions of Theorem 4.1. Since g is fixed, $\mathcal{M}_g^{\mathbb{Z}/n\mathbb{Z}}$ is a finite union (possibly redundant) of irreducible closed subschemes $S(n,g';a_1,\ldots,a_s)$. Therefore, among these subschemes are the irreducible components of $\mathcal{M}_g^{\mathbb{Z}/n\mathbb{Z}}$.

An interesting problem will be to calculate the inclusions and equalities existing among these irreducible closed subschemes (see [**R**]). By Corollary 2.11,

$$S(n,g';a_1,\cdots,a_s) = S(n,g';a_1',\cdots,a_s')$$

if there are $r \in \mathbb{N}$ with $1 \le r < n$ and $\text{g.c.d.}(r,n) = 1$, and a permutation σ such that $a'_{\sigma(i)}$ is congruent to $r \cdot a_i$ modulo n for every i. In [**C**], Cornalba finds the inclusions and equalities existing among these irreducible subschemes when n is a prime number.

References

[C] Cornalba, M., *On the locus of curves with automorphisms*, Ann. Mat. Pura Appl. **149** (1987), pp. 135–151

[D] Deuring, M., *Zur arithmetischen Theorie der algebraischen Funktionenkörper*, Math. Ann. **106** (1932), pp. 77–106

[G] Gómez González, E., *Irreducible components and isolated points in the branch locus of the moduli space of smooth curves*, Boletín de la Sociedad Matemática Mexicana (3) Vol. **2** (1996), pp. 115–128

[Gu] Guerrero, I., *Holomorphic families of compact Riemann surfaces with automophisms*, Illinois J. Math. **26** (1982), pp. 212–225

[H1] Harvey, W. J., *Cyclic groups of automorphisms of a compact Riemann surface*, Quart. J. Math. Oxford (2) **17** (1966), pp. 86–97

[H2] Harvey, W. J., *On branch loci in Teichmüler space*, Trans. Amer. Math. Soc. **153** (1971), pp. 387–399

[K] Kuribayashi, I., *On automorphism groups of a curve as linear groups*, J. Math. Soc. Japan **39**, n. 1 (1987), pp. 51–77

[L] Lønsted, K., *The singular points on the moduli spaces for smooth curves*, Math. Ann. **266** (1984), pp. 397–402

[P] Popp, H., *The singularities of the moduli scheme of curves*, J. Number Theory **1** (1969), pp. 90–107

[RR] Riera, G. and Rodríguez, R. E., *Riemann surfaces and abelian varieties with an automorphism of prime order*, Duke Mathematical Journal **69** (1993), pp. 199–217

[R] Ries, J. F. X., *Subvarieties of moduli space determined by finite groups acting on surfaces*, Trans. Am. Math. Soc. **335** (1993), pp. 385–406

Departamento de Matemáticas, Universidad de Salamanca. Plaza de la Merced 1-4. E-37008 Salamanca. Spain.

E-mail address: `esteban@gugu.usal.es`

Contemporary Mathematics
Volume **240**, 1999

SUBVARIETIES OF MODULI SPACE FOR RIEMANN SURFACES

G. GONZÁLEZ-DÍEZ AND W. J. HARVEY

ABSTRACT. This is a brief survey of current knowledge about the complex geometric character of the modular varieties $\mathcal{M}_g$ for compact Riemann surfaces with genus $g \geq 2$.

We describe results of various authors on algebraic curves inside these varieties and in their completions; by recasting the basic construction in terms of Teichmüller families, a more coherent view of their structure is achieved. In a final section we discuss general subvarieties of $\mathcal{M}_g$ and, in particular, address the question whether complex homogeneous structures exist on higher dimensional modular subvarieties.

SYNOPSIS.

The modular variety for a compact Riemann surface (or algebraic curve over the complex numbers) with positive genus may be constructed as a normal complex space from the Teichmüller theory of topologically marked surfaces: for a given surface of genus $g \geq 1$, a preliminary space T_g with an intrinsic structure of contractible complex manifold furnishes a (ramified) covering space blueprint for the modular variety $\mathcal{M}_g$, which is then given as the quotient with respect to the proper discrete action of the mapping-class group Mod_g on T_g by change of marking. A structure of (non-complete) projective algebraic variety is obtained by the provision of an ample line bundle. Each of the quasiprojective varieties $\mathcal{M}_g$ is a complex orbifold - a V-manifold in the sense of Satake - and has an orbifold singular set, the *ramification locus* Λ_g. This important algebraic subset corresponds to Riemann surfaces admitting non-trivial automorphisms and it exerts significant control over various aspects of the geometrical structure of the modular space itself. For instance, most (if not all) of the explicit examples of complete subvariety of $\mathcal{M}_g$ either lie in this subvariety or have a construction derived from it.

In this approach to the study of moduli spaces and families of algebraic curves, one makes systematic use of the fundamental group enshrined in Teichmüller theory, in both the standard and the orbifold sense. It is an essential ingredient of this part of complex geometry that geometric structure is largely dictated by the purely topological nature of the manifolds concerned. The importance of this perception was perhaps first emphasized by Mostow's rigidity theorem, and it underlies our view that the study of fundamental groups arising in this very special setting may yield important insight into the geometric structure of complex manifolds in general.

1980 *Mathematics Subject Classification* (1985 *Revision*). Primary 32G15, 14H15; Secondary 53C55.

Expanded version of a talk presented at the Congress on Geometry in Olmué, Chile; supported partially by grants from the Conference Fund and from the Royal Society.

The subspaces which interest us here are usually contained in a particular type of irreducible subvariety $\mathcal{M}_g(G)$ of the singular set Λ_g, characterised by a finite group G of mapping-classes whose action on each surface S is prescribed topologically; in some cases, the quotient surface is the complex projective line and the modular subvarieties concerned will then parametrise Galois coverings of the line with given group action.

After some preliminary discussion of the setting and elementary properties of Teichmüller theory, in section 2 we outline the original construction of complete curves in moduli space, due to Kodaira and (independently) Atiyah. In the third section, we summarise and expand on our earlier work ([G-H1,G-H2,G-H3]), where certain normal complex subvarieties of Λ_g are determined and a particular class of complete algebraic curve arising naturally as a subvariety of these is constructed by a geometrical procedure which extends that employed by Kodaira and Atiyah to all values $g \geq 4$ for the fiber genus.

In the final section, we discuss the general pattern of closed subvarieties which a moduli space has and pose a few open questions: following work of various authors, classical and modern, we examine briefly families of curves for which the base (a complex orbifold) is uniformised by a quotient of a Hermitian symmetric space. Such families include several important classes of complex two-dimensional surfaces with origins in arithmetic geometry, and in particular there is the classical example of the Picard family in genus 3, an algebraic surface parametrising the projective equivalence classes of curves which have an automorphism of order 3 with 5 fixed points.

1. BACKGROUND.

The study of complex subvarieties of modular varieties has many motivations, of which the most obvious is to achieve a better understanding of these classifying spaces themselves as interesting examples of complex algebraic varieties. However, the particular geometric properties uncovered by Teichmüller theory have led to several lines of approach within complex analytic geometry which make use of general analytic methods as well as more specific structural properties of the Teichmüller spaces. A good instance of this is the article by P.A. Griffiths [Gr] which provides a holomorphic representation of the universal covering of a Zariski-open subset of any projective variety as a bounded domain of Bergman type, obtained by an iterated fibration process, using the Ahlfors-Bers theory of Riemann surface families; there is also related work of G. Riera [Ri] which gives a reformulation and extension, again using the work of Ahlfors and Bers, of the Kodaira-Atiyah examples.

Two projective embeddings of the modular variety, viewed in this manner as a complex analytic space, are of special interest. The first one, constructed via the period mapping of T_g to the space of Jacobians $\mathcal{J}_g$, a closed submanifold of the Siegel space $\mathfrak{S}_g$ of Riemann period matrices, was given by W.L. Bailey [Bai] around 1960 using earlier work of Bailey and A. Borel on the Siegel modular variety $\mathcal{A}_g = \mathfrak{S}_g/Sp(2g,\mathbb{Z})$ of principally polarised Abelian varieties of dimension g. The completion of the variety $\mathcal{A}_g$ has a description in terms of the addition of a partial boundary (bordification) to the Siegel space $\mathfrak{S}_g$, following the work of I. Satake. Because of the very existence of this complex orbifold completion of $\mathcal{A}_g$, in which the boundary component is a closed subvariety of (complex) codimension at least 2 when $g > 1$, *ergo* negligible from the viewpoint of holomorphic functions by Riemann's singularity theorem, one may conclude (for instance) that the open modular varieties $\mathcal{A}_g$ and $\mathcal{M}_g$ behave like compact complex manifolds in the following sense: *all global holomorphic functions on $\mathcal{A}_g$ or $\mathcal{M}_g$ are constant.*

A second method of compactification, initiated by A. Mayer and D. Mumford in the complex analytic setting in the late 1960s and completed by P. Deligne and Mumford [D-Mum], embeds the nonsingular curves inside the space $\widehat{\mathcal{M}_g}$ of *stable curves* of genus g: such a curve may be singular and/or reducible, but can have at most nodal singularities and each nonsingular component must have negative Euler characteristic. By virtue of the existence of an ample line bundle on this variety (see for instance [Mum] or [Harr-Mo]), the variety $\mathcal{M}_g$ of nonsingular curves becomes the quasi-projective complement of a specific boundary divisor $\Delta_g \subset \widehat{\mathcal{M}_g}$. The divisor Δ_g decomposes into a union of $1+[\frac{g}{2}]$ irreducible components corresponding to the topologically distinct partitions of a genus g surface by a single closed loop.

With these projective embeddings of $\mathcal{M}_g$, techniques of classical algebraic geometry can be brought to bear on these varieties: for instance, by taking hyperplane sections one can obtain a wide range of closed subvarieties provided the dimension $3g-3$ is sufficiently large. For a wider discussion of this approach, see the text on moduli of curves [Harr-Mo].

2. Kodaira's irregular surfaces and generalisations.

We sketch the original elegant construction given in Kodaira's paper [Kod] of a sequence of compact complex surfaces $M_{m,n}$, for integers $m, n \geq 2$. Each one is obtained as a holomorphic family of Riemann surfaces with genus $g(m,n)$ over some compact surface S of genus $h(m,n)$, constructed by a kind of bootstrap process from the initial choice of a Riemann surface R_0 with genus n and a smooth double covering $R \to R_0$. By varying u on R_0 we also vary the pair of inverse image points $P_1(u)$, $P_2(u)$ in the covering R, and along with them the corresponding m-fold cyclic covering surface C_u of R with ramification over the points $P_1(u)$, $P_2(u)$; in this way a complex analytic one-parameter family with local fiber product structure is created over R_0. It is then possible to solve the monodromy problem for continuation of this structure by passing from base surface R to the mod m homology covering $f : S \to R$ and thus to produce a connected, holomorphically fibered family over the surface S defined by this covering: the fibers are surfaces of genus $g = m(2n-1)$, each isomorphic to some C_u. The base surface S has fundamental group the kernel of the composite epimorphism

$$\pi_1(R) \to H_1(R, \mathbb{Z}) \cong \mathbb{Z} \oplus \ldots \mathbb{Z} \to \mathbb{Z}/m \oplus \ldots \mathbb{Z}/m$$

and one finds from the Riemann-Hurwitz formula that the genus of S is given by $h(m,n) = m^{2n}(n-1)+1$.

An alternative way to view the total space of this fibration $M_{m,n}$ geometrically is as a ramified cyclic covering of the direct product complex surface $R \times S$, with branching over the disjoint pair of embedded Riemann surfaces which forms the graph in $R \times S$ of the smooth covering map $f : S \to R$. The composite mapping from $M \to R \times S \to S$ gives the fibration.

NOTE 1. It follows readily that this fibration is holomorphically non-trivial, i.e. that $M_{m,n}$ is not a direct product, since there are only finitely many places on S where the fibre is conformally a given genus g Riemann surface C_u.

NOTE 2. Kodaira calculates directly the deformation class in $H^1(C_u, \mathcal{O}(K^{-1})$ as a Cech cocycle representing $\frac{\partial}{\partial u}C_u$. He also computes the Chern numbers of $M_{m,n}$ explicitly in terms of m and n, showing thereby that the signature invariant is nonzero.

This construction may be extended in several ways: one can modify the ingredients, by changing to more general covering surfaces R and S, or restrict the type of structure used, for instance by exploiting the special nature of complex tori as we shall do in the next section.

Furthermore, it is clear that iteration of this process is possible. For instance – see the survey article [Har] for more details – one may take the tower of smoooth coverings R_ℓ of R with degree $2^{2\ell}$ for $\ell > 0$, defined by the kernels of the standard epimorphisms of $\pi_1(R) \to H_1(R, \mathbb{Z}) \to \mathbb{Z}/2^\ell \oplus \mathbb{Z}/2^\ell$. This tower has the form

$$\cdots \to R_{\ell+1} \to R_\ell \to \cdots \to R_1 \to R$$

and there is an inductive procedure for creating a family of curves at each successively higher level, which begins with the construction of a new fibration X^3 over the Kodaira-Atiyah surface $X^2 = M_{2,2}$ with fibers the Riemann surfaces Y_2 obtained as double coverings of the C_u already occurring in the family X^2. This is the second stage in the tower: successive coverings of the basic surface S, coordinated by the sequence of normal subgroups $K_\ell = \pi_1(R_\ell) \le \pi_1(R)$, produce progressively higher dimensional families X^ℓ of more symmetrical Riemann surfaces, for which a further fibration process is possible by virtue of the fact that this portion of the tower carries the extra topological data necessary for the formation of each next stage.

One important fact which distinguishes this construction of iterated families is the computation of the signatures of the compact manifolds X^n. This was used by E. Miller [Mi], and independently by S. Morita [Mor], to show that the Mumford *tautological characteristic classes* κ_n for surface bundles are (stably) nontrivial for all $n > 0$: these classes are obtained from the projection to $\mathcal{M}_g$ of tensor powers $\omega_g^{\otimes(n+1)}$ of ω_g, the 'cotangent bundle along the fiber' of the genus g universal curve family $\mathcal{C}_g$. See [Har],[Harr-Mo] or [Mum] for more details, including a discussion of the fascinating conjecture of Mumford that the tautological classes generate the stable cohomology ring $H^*(\mathcal{M})$, an algebra over the rationals obtained from taking the limit of the cohomology of $\mathcal{M}_g$ as $g \to \infty$.

3. Recent work on subvarieties in $\mathcal{M}_g$.

Using the Bers fiber space theorem and some elementary results concerning inclusions of Teichmüller spaces, together with analysis of *relative mapping class groups* due to Harvey and Maclachlan [H-Mac], it is possible to develop and expand on the results of Kodaira and Atiyah (section 2) along lines similar in some respects to Riera's earlier work.

By way of illustration, we explain briefly how to obtain the Kodaira construction by interweaving sundry inclusion relations between Teichmüller subspaces

$$T_{n,1} \subset T_{2n-1,2} \subset T_{m(2n-1)} = T(C)$$

representing Riemann surfaces C with an involutory automorphism σ such that $R = C/\langle\sigma\rangle$ is a smooth double cover of the genus n surface R_0. As background for the general facts we use below, the reader might consult the survey article of Bers [Be]; for a more detailed account, see the textbook [N]. For a specific choice of compact surface R_0, to achieve a suitable smooth base space S as outlined in $ 2 (for the family of fiber surfaces C given as branched covers of R with order m branching at the two points of R covering a single point of R_0) one must find an appropriate torsion-free normal subgroup of the Fuchsian group representing the orbifold surface R with two order m elliptic generators at designated points. This step can be done either by employing homology considerations as in [Kod] or by invoking more refined methods available for finding torsionfree subgroups of Fuchsian groups; in any case, the compact base surface S covering R_0 will occur as a quotient surface in a modular variety $\mathcal{M}_{g(m,n)}$ with the fiber genus $g(m,n)$ determined by the Riemann-Hurwitz branching formula.

To specify the surface S as a complex analytic modular subvariety, we employ a combination of the Bers fiber space theorem and the Bers-Greenberg isomorphism theorem to define a Teichmüller family $F_n \cong T_{n,1}$ and produce deformations of R_0 with a marked point. Let $F_n = F(\Gamma)$ be the Bers fiber space of the Fuchsian group representing the compact surface $R_0 = U/\Gamma$, with $\pi : F_n \to T_n$ the projection mapping. Interpreting F_n as the Teichmüller space $T_{n,1}$, we select a Fuchsian group Γ' representing the punctured surface R_0' with $T(\Gamma') \cong F_n$). In this fibration we consider the fiber hyperbolic disc $\Delta_{x_0} \cong \mathbb{D}^2$ over the marked Teichmüller point $x_0 = [R_0]$: this disc represents the universal covering of our chosen reference surface R_0 and is a complex analytic submanifold of F_n. For the sequel, it is convenient to work within the Bers fiber space over universal Teichmüller space, using instead co-compact Fuchsian groups Γ_0 with one conjugacy class of torsion generator designed to supply one marked (cone) point on R_0, an elliptic element of order $m \geq 2$ fixing that point; by the Bers-Greenberg isomorphism theorem, the Teichmüller space $T(\Gamma_0$ is biholomorphically equivalent to $T(\Gamma')$. With this switch of viewpoint, then, the fiber disc Δ_{x_0} has become a holomorphic submanifold of $T(\Gamma_0)$ and parametrises all marked Fuchsian groups with one elliptic generator (and cone point) of order m which uniformise the fixed surface R_0.

Next we pass from R_0 to a smooth 2-sheeted covering R: this corresponds, with the marked choice of elliptic points $P_1(u)$, $P_2(u)$ on R lying over $P(u) \in R_0$, to an inclusion of Fuchsian groups $\Gamma_1 \leq \Gamma_0$ with Γ_1 a normal subgroup of topological type $\{2n-1; m, m\}$ – genus $2n-1$ with two elliptic generators – and a holomorphic inclusion $T(\Gamma_0) \hookrightarrow T(\Gamma_1)$. Now we embed this Teichmüller space in the genus $g = g(m,n)$ space $T(K_m)$ by specifying a torsion-free normal subgroup $K_m \leq \Gamma_1$ with quotient cyclic of order m: via the resulting holomorphic inclusion of $T(\Gamma_1)$ in $T(K_m)$, the disc Δ_{x_0} determines a complex analytic family of (orbits of) marked elliptic fixed points in the holomorphic family of uniformisations of the two-pointed surface R. Using the models of these spaces within the Bers universal family $F(1) \cong T(1) \times \mathbb{H}^2$, each elliptic generating element γ in the group Γ_1 is acting as an element of the extended modular group mod (Γ_1); such an element preserves the submanifold $F(\Gamma_1) \subset F(K_m)$, fixes a single point of each fiber hyperbolic disc over $T(K_m)$ and has fixed point set a complex submanifold (of codimension 1) formed by the union of these points.

NOTE. The submanifold fixed by an elliptic element γ of a Fuchsian group Γ is a holomorphic section of the *Teichmüller curve* $\mathcal{V}(\Gamma) \cong F(\Gamma)/\Gamma$, and when γ represents a genus g Riemann surface automorphism as here, one obtains in this way a section of the restriction of the Teichmüller curve $\mathcal{V}_g$ to the submanifold of T_g fixed by that automorphism. This interpretation of the hyperbolic geometry in the fiber space over a Teichmüller space was employed first by Earle and Kra [Ea-Kr] to construct global holomorphic sections of pointed Teichmüller families, in analogy with the Hubbard sections of $\mathcal{V}_2$ which arise from the (continuous) selection of one Weierstrass point on each genus 2 curve.

On restricting to the fiber space over the submanifold $\Delta_{x_0} \subset T(K_m)$, we thus obtain as fixed point set of the element γ a complex analytic 1-disc D which is a holomorphic section of the base disc Δ_{x_0}.

The stabiliser of D in the group $\mathrm{Mod}_g = \mathrm{Mod} K_m$ is the monodromy group of the family under construction: this will be a finite index subgroup of the original group Γ_0, comprising those Möbius transformations which induce modular automorphisms of the Teichmüller space $T_g = T(K_m)$ preserving a certain subset $T_g(H)$. The latter is a *relative Teichmüller subspace*, a connected complex submanifold consisting of all (marked) surfaces topologically conjugate to $C_{x_0} = \mathbb{D}^2/K_m$ which carry the action of the cyclic group $H = \Gamma_1/K_m$. To make this construction precise on the level of the moduli space, one needs the analysis of relative modular groups made in [H-Mac] combined with passage to a suitable finite index subgroup of Mod_g, for instance the level m homology congruence subgroup. The crucial point is that in order to achieve a *connected* base surface S for the family, we must employ a suitable finite sheeted covering of the moduli space for which all the cover transformations (mapping classes on the marked fiber surfaces C_x, $x \in \mathbb{D}^2$) of the family are compatible with the selections made along the way, in the sense that they normalise each of the subgroups Γ_1, K_m.

In this way, we infer that the image of the disc Δ_{x_0} under projection to moduli space $\mathcal{M}_g$ is a compact complex analytic curve S lying inside an irreducible subvariety $\mathcal{M}_g(G)$ of the ramification locus Λ_g in $\mathcal{M}_g$, representing surfaces of genus g with a fixed automorphism group $G \geq H \cong \mathbb{Z}/m$, the finite subgroup of Mod_g we obtained as the quotient group Γ_1/K_m. The base surface S is therefore specified by the monodromy of the covering surface structure of C over R_0, and hence determined uniquely by the data of the various covering subgroups within the (purely combinatorial) structure of the Fuchsian group Γ_0.

For further details of this approach and analysis of the ramification locus, see for instance [H-Mac] and [G-H1]. By similar methods, one can prove the following more general result:-

Theorem 1. *Let a finite group G operate as automorphisms of a compact Riemann surface Y with quotient X of genus $h = \ell(k-1)+1$ and with ramification over a finite set $\Sigma_\ell \subset X$ with $\ell \geq 2$ points. In the Teichmüller space $T(X) = T_h$ there is a relative Teichmüller subspace T_k of marked surfaces representing smooth ℓ-sheeted coverings $f : X \to X_0$, for which each choice of a reference (genus k) surface $[X_0] \in T_k$ determines a holomorphic embedding of some finite sheeted Riemann surface S over X_0 into the modular variety $\mathcal{M}_g$; here $g = g(G, \ell, k)$ is the genus of the surface Y.*

Typically, the surface S will be a *characteristic covering surface*, a covering which corresponds to a characteristic normal subgroup of $\pi_1(X)$. A detailed proof of this result will appear in a future article.

We indicate next a similar type of construction given in [G-H2], which involves special properties of elliptic curves in place of the smooth covering of R_0 by R. The key step, as in Kodaira's method, is to produce a compact surface inside the symmetric product $X^{(r)}$ of distinct $r-$tuples of points on a specific curve X: one can then specify a ramified covering with this variable branching set of cardinality r by prescribing the branching data, thereby achieving a holomorphic family of surfaces with genus determined by r alone. Let E be a complex torus regarded as an elliptic curve, and $0 \in E$ the zero element for the group law $+$ on E. Choose X to be a genus 2 surface such that $\Pi : X \to E$ is a two sheeted covering ramified at two points of the torus, denoted 0, Q, and let $0 = Q_1, Q_2, \ldots, Q_r$ be distinct points of E such that $Q_j \neq \pm Q$ for all j. Define $S_r \subset X^r$ as follows:-

$$S_r = \{(P_1, \ldots, P_r \in X^r | \Pi(P_j) = \Pi(P_1) + Q_j, \, j = 2, \ldots, r\}.$$

Then by elementary complex geometry, S_r determines a compact complex analytic curve in $X^{(r)}$. A choice of ramification structure at the r points then determines the holomorphic family of covering surfaces. Thus, for instance, there is a family $\mathcal{Y}$ of genus 4 Riemann surfaces $\{ \, Y_u \mid u \in S_2 \, \}$ which are double covers of X with two ramification points. An inductive computation on the tower of coverings S_{2n}, $n \geq 1$ of X shows that the base curves S_{2n} have genus $2^{2n}n+1$.

As an easy consequence of the Riemann-Hurwitz formula for double coverings of X branched over $2n$ points, we see that the fiber surfaces Y_u obtained have genus $3+n$; from this it follows at once that complete families occur in every genus apart from $\mathcal{M}_2$ and $\mathcal{M}_3$. Thus our result in [G-H2] – which employs this line of reasoning on a specific genus 2 Riemann surface X with affine equation $y^2 = x^6 - 1$ – is strengthened to the following statement.

Theorem 2. *Every value of $g \geq 4$ arises as the fiber genus in a Riemann surface fibration of the above type, with base an arbitrary elliptic curve (i.e. complex torus with choice of origin).*

Using a modification of these ideas, which brings in the special properties of Prym varieties associated with ramified double coverings (the case $n = 2$), together with the intricate related *tetragonal curve* construction of S. Recillas, the production of complete curves in modular varieties can be carried through also in genus 3; see for instance [Z]. One suspects that it may be possible to express this type of construction too in Teichmüller -theoretic terms, using the results of [Ea] on the Teichmüller theory of the universal family of Jacobians.

NOTE. We can see directly that 3 is the lowest possible value of the genus, as no complete curve can exist in genus 2 moduli space – it is well known that $\mathcal{M}_2$ is an affine rational variety, hence has no complete subvariety (of positive dimension). In any case, it also follows immediately from a theorem of S. Diaz (see for instance [Harr-Mo]): *the dimension of any complete subvariety of $\mathcal{M}_g$ is at most $g-2$.*

It is not so easy to decide whether the embedded curves S obtained by these methods are nonsingular; this requires the precise calculation of the homotopic monodromy, which lies buried within the structure of the modular subgroup stabilising the relevant component of the branch locus, and so far it has defied analysis. However one can give a more topological prescription of curves in $\mathcal{M}_g$ for which the monodromy is torsion free [G-H3], which implies that the curves are smooth subvarieties; the price paid seems to be a lack of explicit analytic or algebraic description, and in particular the genus of these smooth base curves is not known in general.

4. Fundamental groups of some Kähler varieties.

This section, written by the second author, is intended for general motivation; we discuss some problems, results and ideas from complex geometry which may guide further investigation of Teichmüller families of Riemann surfaces.

A geometric problem of some interest is to discover (necessary and sufficient) conditions for a discrete group to be the fundamental group of a complex projective variety. This would shed light on the general classification problem for varieties, as well as delineating the place of such objects (as complex spaces) within the topological framework of manifold theory. For more details on this topic, one may consult the excellent text [Amorós] or [Arap]. One simple fact of interest in the present context of surface fundamental groups is that a free group of rank at least 2 cannot be the fundamental group of a complete Kähler variety, since it has finite index subgroups with odd Betti number. Current knowledge is fairly sparse, however, and not designed for applicability to our specific (modular subvariety) situation.

As an alternative to seeking insight from the general classification of varieties, one possible approach is to limit the class of the varieties/manifolds to be considered by some specific simplifying topological assumption: here, for instance, we shall restrict attention to the class of $K(\pi,1)$-manifolds, although it should be noted that the algebraic implications of this restriction may prove very subtle. Let $G=\pi_1(X)$ for some complex projective orbifold variety X: we ask when there is a 'classifying' projective variety Y with contractible universal cover having the same fundamental group as X. One would then hope further (see for instance the introductory section in Kollár's book [Kol]) to have a morphism $X \to Y$ inducing the fundamental group isomorphism. Particular interest is often directed at uniformisable complex varieties which have some local homogeneous structure, so that they are expressible as discrete quotients of some classical complex symmetric space. In the special case of lattices Γ in higher dimensional complex hyperbolic Lie groups, which determine complex hyperbolic varieties $\mathbb{D}^n/\Gamma$ as quotients of the complex ball of dimension

$n > 1$, it was proved by Toledo [Tol] that non-cocompact lattice subgroups Π do occur as fundamental group of a complete variety X; however, there cannot be a universal classifying (projective) variety Y for such an X which would direct all finite coverings as above since these lattice groups have odd cohomological dimension. Also there are reasons to suspect that such spaces seldom, if ever, occur as subvarieties of a moduli space for Riemann surfaces.

In complex algebraic geometry, one employs the notion of *algebraic fundamental group* due to Grothendieck, which amounts to taking the profinite completion – the inverse limit of all finite quotients – of the topologist's π_1. There is (see [Kol]) a notion of size for algebraic varieties based on this group: the (algebraic) fundamental group of a variety is called *large* if *for all morphisms of varieties* $f : Y \to X$ *the image* $f^*(\pi_1(Y))$ *in* $\pi_1(X)$ *is infinite.* One may compare this with Waldhausen's purely topological notion of sufficiently large for an irreducible three-manifold. Such 'large' varieties occur, for instance, as proper discrete quotients of a Stein manifold; the converse statement is a conjecture of Shafarevich. By a theorem of Bers and Ehrenpreis [Be-Ehr], the Teichmúller spaces T_g and their associated fiber spaces $T_{g,n}$ are Stein manifolds, so that the quotient modular varieties are a natural hunting ground for examples of large algebraic variety. Within the broad class of closed subvarieties of moduli spaces $\mathcal{M}_{g,n}$, it is natural to restrict consideration to the *incompressible* ones, complex orbifolds for which the inclusion into some moduli space induces an injection of the fundamental group into $\mathrm{Mod}_{g,n}$: this condition is analogous to the requirement of *minimality* for a variety in algebraic geometry.

It is a well known theorem of H. Royden that Teichmüller spaces of dimension greater than 1 are not homogeneous spaces in the sense of complex differential geometry. The following question, a refinement of one posed by the second author in [H1], concerns existence of homogeneous geometric structures on modular subvarieties:

Question 1. *Are there incompressible (minimal) complex subvarieties of* $\mathcal{M}_g$ *with* π_1 *a lattice in some (semisimple) Lie group?*

Let Γ be a subgroup of some mapping-class group Mod_g which is realised as the homotopical monodromy of an algebraic family of curves, perhaps with extra data such as a set of marked points. We note here that no procedure is known for recognising such monodromy groups among the general class of (finitely generated) subgroups of Mod_g – this is a problem of great relevance to contemporary developments in symplectic geometry. Then via the classical period mapping (see [Ea] for instance) there is a corresponding family of Jacobi varieties with base a subspace B of the Siegel space $\mathfrak{S}_g$ of marked Abelian varieties, for which the fundamental group is determined as the image of Γ under the morphism ϕ from the Teichmüller mapping class group to the integral symplectic group : the switch from a homotopy marking to homology induces a homomorphism from the mapping class group Γ_g into *Siegel's modular group*, the integral symplectic group $Sp(2g, \mathbb{Z})$ representing the action of surface mappings on integral homology of the surface. By Borel's Density Theorem, the (virtually isomorphic) image $\phi(\Gamma)$ inside the Siegel modular group has an *algebraic hull* $G(\phi(\Gamma))$, which is an algebraic subgroup of the corresponding algebraic group, the rational symplectic group $Sp(2g, \mathbb{Q})$, and which forms the stability group of the Zariski closure of the Jacobian subfamily B. Now the Mostow-Margulis Superrigidity theorems rule out such injective morphisms between distinct pairs comprising symmetric spaces with lattice subgroups, except when the hull is itself a symplectic group of smaller rank. Furthermore, the only case possible in this setting is when the base has rank 1. This latter case, however, produces a range of examples: for instance, there is a broad class of embeddings of arithmetic subgroups of $SL_2(\mathbb{R})$ in each Siegel modular group, occurring first in work of A. Weil and of G. Shimura, in which the base of the corresponding family is a complex 1-disc. A variety of methods for producing these is given in [Co-Wol]. In another approach, founded on pioneering work of Veech, families of Riemann surfaces are constructed whose base is a Teichmüller geodesic disc and which project to complete curves in $\widehat{\mathcal{M}_g}$ intersecting the boundary divisor Δ_g; see [H2] or [V] for instance. Earlier work of Kra [Kr2] on embedding Teichmüller discs in Siegel space is also relevant here. In the light of the above discussion, and taking into account a result of H. Masur that there are no Teichmüller discs with a co-compact Fuchsian group as their stabiliser in Mod_g, it seems likely that the answer to Question 1 is negative, i.e. that there are no complete homogeneous complex subvarieties of $\mathcal{M}_g$ with dimension greater than 0.

As we have already seen, modular varieties do provide a broad class of higher dimensional inhomogeneous compact (Kähler) submanifolds by iteration of the surface fibration construction of section 2. These are examples of varieties with 'large fundamental group,' having a special kind of non-abelian filtration with surface group subquotients. They tie in closely with the topological classification of polysurface groups and iterated surface bundles given by F.E.A Johnson [J1],[J2]. So far as we can say at present, the techniques described here (or some simple modifications of them) underlie all known explicit examples of complete holomorphic families of nonsingular higher genus curves, and one may ask the further question:-

Question 2. *Is every complete family of nonsingular curves finitely covered by an iterated surface bundle fibration?*

In the paper [G-H3] it is shown that the monodromy of the families constructed contains certain mapping-classes known as *spins*, which are a particular type of *reducible* surface mapping in the sense of Thurston's classification theorem – see for instance [Kr1] for a discussion of this. Because all of these fibered families turn out to have monodromy which contains reducible elements, it is pertinent to ask here a related question:

Question 3. *Do there exist complete families of compact surfaces with no reducible elements in the monodromy?*

As a final comment, it would be highly significant to know all representations of lattice subgroups of Lie groups in mapping class groups. Following the discussion in this section and recent results announced by B. Farb and H. Masur [preprint, Univ. of Chicago], one may now expect that any such representation of a lattice subgroup of a Lie group of rank ≥ 2 has finite image in Mod_g. Even in rank 1, no isomorphic examples seem to be known. The work of Deligne and Mostow [D-Mo] on monodromy of hypergeometric equations is relevant here. In particular, their methods encompass the classical example of the *Picard modular surface*, base space of the family of genus 3 curves admitting an automorphism of order 3 with 5 fixed points and quotient $\mathbb{C}P^1$. This modular variety has a much-studied representation as the quotient of the hyperbolic complex 2-disc by an arithmetic subgroup Γ of the Picard-Eisenstein modular group $U(2,1;\mathcal{O})$, where $\mathcal{O}$ is the integers of $\mathbb{Q}(\sqrt{-3})$ – see, for instance the lecture notes [Holz]. On the other hand, using Teichmüller theory it may be regarded as the quotient of $T_{0,5}$ by the Hurwitz braid/modular group $\mathrm{Mod}_{0,5}$; the epimorphism between these groups is induced by the period mapping for the corresponding curve family. We observe however that this variety is not incompressible: the homotopic monodromy group, which is the stabiliser in Mod_3 of the Picard family, is a noncentral extension of the lattice group Γ by the cyclic group of order 3.

References

[**Amoros**] J. Amorós et al., *Fundamental groups of compact Kähler manifolds*, AMS Surveys **44**, AMS, 1996.

[**Arap**] Arapura, D., *Fundamental groups of smooth projective varieties*, Current Topics in Complex Algebraic Geometry, MSRI Publ. **28**, 1995, pp. 1-16.

[**At**] Atiyah, M.F., *The signature of fiber bundles*, Global Analysis (Eds. D.C. Spencer & S. Iyanaga), Princeton Univ. Press (1969), pp. 73-84..

[**Bai**] Baily, W.L., *On the theory of θ-functions, the moduli of abelian varieties, and the moduli of curves*, Ann. of Math. **75** (1962), 342-381..

[Be] Bers, L., *Uniformisation, moduli and Kleinian groups*, Bull. London Math. Soc. **4** (1972), 257-300.

[Be-E] Bers, L. and Ehrenpreis, L., *Holomorphic convexity of Teichmüller spaces*, Bull. Amer. Math. Soc. **70** (1964), 761-764..

[Bir] Birman, J.S., *The algebraic structure of mapping class groups*, Discrete Groups & Automorphic Functions, (edited by W.J.Harvey), Academic Press, London & N.Y., 1977, pp. 163 - 198.

[Co-Wol] Cohen, P. B. & Wolfahrt J., *Modular embeddings for some non-arithmetic Fuchsian groups*, Acta Arithmetica **56** (1991), 93-110.

[D-Mo] Deligne, P. & Mostow, G. D., *Monodromy of hypergeometric functions and non-lattice integral monodromy*, Publ. IHES (1986), 5 - 89.

[D-Mum] Deligne, P. & Mumford D., *The irreducibility of the moduli space of curves of given genus*, Publ. IHES (1969), 75 - 110.

[Ea] Earle, C.J., *Families of Riemann surfaces and Jacobi varieties*, Ann. of Math.(2) (1978), 225-286.

[Ea-Kr] Earle, C.J. & Kra, I., *Holomorphic mappings between Teichmüller spaces*, Contributions to Analysis,, Papers dedicated to Lipman Bers (Eds. L.V. Ahlfors et al.), Academic Press, NY, 1974, pp. 107-124.

[G-H1] González-Diez, G. & Harvey, W.J., *Moduli of Riemann surfaces with symmetry*, Discrete Groups and Geometry, London Math. Soc. Lecture Notes, **173**, Cambridge U. Press, 1992, pp. 75-94..

[G-H2] González-Díez, G. & Harvey, W.J., *On complete curves in moduli space I,II*, Math. Proc. Camb. Phil. Soc. **110** (1991), 461-466, 467-472..

[G-H3] González-Díez, G.& Harvey, W.J., *Surface groups inside mapping class groups*, Topology **38** (1999), 57 - 69.

[Gr] Griffiths, P.A., *Complex analytic properties of certain Zariski open sets on algebraic varieties*, Ann. of Math. **94** (1971), 21-51..

[Har] Harer, J.L., *Cohomology of moduli space of curves*, Theory of Moduli, Springer LN in Math., vol. 1337, 1986, pp. 138-221.

[Harr-Mo] Harris, J. & Morrison, I., *Moduli of Curves*, Graduate Texts in Math. **187**, Springer Verlag, New York, 1998.

[H1] Harvey, W.J., *Modular groups and representation spaces*, Geometry of Group Representations, Contemp. Math. **74**, AMS, 1988, pp. 205-214..

[H2] Harvey, W.J., *On certain families of compact Riemann surfaces*, Mapping class groups and moduli spaces, Contemp. Math. **150**, AMS, 1993, pp. 137-148.

[H-Mac] Harvey, W.J. & Maclachlan C., *On mapping-class groups and Teichmüller spaces*, Proc. London Math. Soc. **30** (1975), 496-512..

[Holz] Holzapfel, R.-P., *The Ball and some Hilbert Problems*, Birkhäuser (Basel), 1995.

[J1] Johnson, F.E.A., *Surface fibrations and automorphisms of non-Abelian extensions*, Quart. J. Math. (2) **44** (1993), 199-214.

[J2] Johnson, F.E.A., *A group theoretic analogue of Parshin-Arakelov rigidity theorem*, Arch. Math. **63** (1994), 354-361..

[J-R] Johnson F.E.A. & Rees, E.G., *On the fundamental group of a complex algebraic manifold*, Bull. London Math. Soc. **19** (1987), 463-466..

[Kod] Kodaira, K., *A certain type of irregular algebraic surface*, J. Analyse Math. **19** (1967), 207-215.

[Kol] Kollár, J., *Shafarevic Maps and Automorphic Forms*, Princeton University Press, 1995..

[Kr1] Kra, I., *On the Nielsen-Thurston-Bers type of some self-maps of Riemann surfaces*, Acta Math. (1981), 231-270.

[Kr2] Kra, I., *The Caratheodory metric on Abelian Teichmüller discs*, J. Analyse Math. **40** (1981), 129- 143..

[Mi] Miller, E.Y., *The homology of the mapping class group*, J. Diff. Geom. **24** (1986), 1-16..

[Mor] Morita, S., *Characteristic classes of surface bundles*, Bull.A.M.S. **11** (1984), 386-389..

[Mum] Mumford, D., *Towards an enumerative geometry of the moduli space of curves*, Arithmetic & Geometry, (Shafarevich Festchrift) Vol. II, Birkhäuser, 1983, pp. 271-328.

[N] Nag, S., *The Complex Analytic Theory of Teichmüller Spaces*, John Wiley & Sons, 1988.

[Ri] Riera, G., *Semi-direct products of Fuchsian groups and uniformisation*, Duke Math. J. **44** (1977), 291-304.

[Tol] Toledo, D., *Examples of fundamental groups of compact Kähler manifolds*, Bull. London Math. Soc. **22** (1990), 339-343.

[V] Veech, W., *Teichmüller curves in moduli space*, Invent. Math. **97** (1989), 553 - 583.

[Z] Zaal, C.G., *Explicit complete curves in genus 3 moduli space*, Preprint 94-03, University of Amsterdam..

DEPARTAMENTO DE MATEMATICAS, UNIVERSIDAD AUTONOMA DE MADRID, CANTOBLANCO, E28049, MADRID, SPAIN.

DEPARTMENT OF MATHEMATICS, KING'S COLLEGE LONDON, STRAND, LONDON, WC2R 2LS, U.K.

E-mail: gabino@ccuam3.sdi.uam.es bill.harvey@kcl.ac.uk

Contemporary Mathematics
Volume **240**, 1999

Noded Function Groups

Rubén A. Hidalgo

Abstract. Given a Kleinian group G, we denote by $\Omega(G)$ and $P(G)$ its region of discontinuity and its set of double-cusped parabolic fixed points, respectively. The extended region of discontinuity of G is defined as $\Omega^{ext}(G) = \Omega(G) \cup P(G)$, together the cusp topology. In the present note, we consider noded function groups, that is, finitely generated kleinian groups having an invariant component of their extended region of discontinuity. We sketch an alternative proof of a theorem of Kra and Maskit [**9**] which says that a noded function group with two invariant components is in fact a noded fuchsian group and, in particular, geometrically finite (extending Maskit's result that a finitely generated Kleinian group with two invariant components is in fact a quasifuchsian group). The main purpose of this note is to show that the commutator subgroup of a torsion-free, non-elementary noded function group G determines it uniquely (obtaining in this way a generalization of the results in [**3**], [**4**], [**5**] and [**10**]). For a general non-elementary torsion-free Kleinian group with no invariant component the above rigidity property still yet unknown.

Preliminaries and Main Results

Let us consider a group G of Möbius transformations. We say that G acts discontinuously at some point $x \in \widehat{\mathbb{C}}$, if

(i) $G_x = \{g \in G : g(x) = x\}$ (the stabilizer of x in G) is finite; and

1991 *Mathematics Subject Classification.* Primary 30F40.
Key words and phrases. Kleinian groups, Function groups.
This work is partially supported by projects UTFSM 971223, Fondecyt 8970007 and Presidential Science Chair on Geometry.

(ii) there is an open set U containing x such that $g(U) \cap U = \emptyset$, for all $g \in G - G_x$.

The set of all points of the Riemann sphere on which G acts discontinuously is called the region of discontinuity of G and denoted by $\Omega(G)$. Its complement $\Lambda(G)$ is the limit set of G. We have that $\Omega(G)$ (respectively, $\Lambda(G)$) is an open (respectively, closed) invariant set under the action of G. The group G is said to be a Kleinian group if $\Omega(G) \neq \emptyset$.

Given a subgroup H of G, we say that a set $B \subset \widehat{\mathbb{C}}$ is precisely invariant under H in G if:

(i) $h(B) = B$, for all $h \in H$; and
(ii) $g(B) \cap B = \emptyset$, for all $g \in G - H$ (see [**11**]).

Let p be a fixed point of a parabolic transformation in G. We say that p is a double-cusped parabolic fixed point in G if there are disjoint round open discs D_1 and D_2 contained in $\Omega(G)$, both tangent at the parabolic fixed point p, so that $D_1 \cup D_2$ is precisely invariant under G_p in G. We denote by $P(G)$ the set of double-cusped parabolic fixed points of G. This set is invariant under the action of G.

The extended region of discontinuity of G is defined as $\Omega^{ext}(G) = \Omega(G) \cup P(G)$, which is an invariant set for G. On this extended region we consider the cusp topology, generated by the usual open sets in $\Omega(G)$ and sets of the form $D_1 \cup D_2 \cup \{p\}$, where $D_i \subset \Omega(G)$ is an open round disc, $D_1 \cap D_2 = \emptyset$, $p \in P(G)$ and D_1 and D_2 are tangent at p. In this way, $\Omega(G)$ (with the usual topology) is a dense open subspace of $\Omega^{ext}(G)$.

The group G acts as a group of homeomorphisms on $\Omega^{ext}(G)$, and its restriction to $\Omega(G)$ is conformal.

The quotient space $\Omega^{ext}(G)/G$ turns out to be a complex analytic space of complex dimension one. The points where it fails to be a complex manifold are nodes (have a neighborhood homeomorphic to the algebraic set $\{(z, w) \in \mathbb{C}^2 : z \cdot w = 0\}$). The nodes of $\Omega^{ext}(G)/G$ are in one-to-one correspondence with the G-classes of double-cusped parabolic fixed points with cyclic parabolic stabilizer.

We must mention at this point that there are parabolic fixed points which are double-cusped but with stabilizer a $\mathbb{Z}_2$-extension of a cyclic parabolic subgroup of G. The projection of such points on $\Omega^{ext}(G)/G$ are not nodes as defined above. They are called phantom nodes.

The natural map $\pi : \Omega^{ext}(G) \to \Omega^{ext}(G)/G$ is G-invariant, $\pi(x) = \pi(y)$ if and only if x and y are G-equivalent, and its restriction $\pi|_{\Omega(G)} : \Omega(G) \to \Omega(G)/G$ is a holomorphic (branched) covering. The set of nodes and phantom nodes of $\Omega^{ext}(G)/G$ is exactly $\pi(P(G))$.

A finitely generated Kleinian group for which there is an invariant component of the extended region of discontinuity (in the cusp topology) is called a *noded function group*. If the group has no double-cusped parabolic fixed points, then this is the definition of a function group.

As it happens for function groups, noded function groups may be geometrically finite or geometrically infinite. Examples of geometrically finite noded function groups are given by the noded fuchsian groups (geometrically finite discrete groups algebraically isomorphic to a fuchsian group) [**6**] and [**7**].

Noded function groups appear in the boundary of the deformation space of a fixed function group (a dense subset in the boundary). These boundary groups will uniformize degenerations of the Riemann surface uniformized by the original function group. To be more precise, let us assume we have a function group G with invariant component Δ. The quotient space $S = \Delta/G$ is a Riemann surface, marked with special points (the branched values of the natural projection $\pi : \Delta \to S$ induced by the holomorphic action of G on Δ). Consider a simple loop

γ in S so that it does not lift to a loop on Δ and such that the stabilizer of any component γ_i of $\pi^{-1}(\gamma)$ contains some loxodromic element $L_i \in G$. In the moduli space of S we can approach the boundary by the process of pinching the loop γ. The limit will represent a noded Riemann surface. We can think of such a process as making L_i (and its conjugates) to approach to a parabolic transformation when moving in the deformation space of G. Nice papers dealing originally with this idea are for instance [**8**] and [**14**].

The following result is an easy consequence of Maskit's arguments in [**13**] which use the Marden-Maskit cusp construction in [**14**]. We recall these arguments in the next section.

Theorem 1. *Let G be a noded function group with invariant component $\Delta^* \subset \Omega^{ext}(G)$. Then there is a function group K with an invariant component $\Delta \subset \Omega(K)$, a surjective map $w : \Delta \to \Delta^*$ and an isomorphism $\theta : K \to G$ satisfying:*

(i) *$w \circ k = \theta(k) \circ w$, for all $k \in K$.*

(ii) *There is a K-invariant collection $\mathcal{F} \subset \Delta$ of pairwise disjoint simple arcs, necessarily with end points in the limit set $\Lambda(K)$, each one invariant by some loxodromic element of K, so that*

(ii.1) *two different arcs in $\mathcal{F}$ have disjoint closures in $\widehat{\mathbb{C}}$;*

(ii.2) *$w : \Delta - \mathcal{F} \to \Delta^* \cap \Omega(G)$ is an homeomorphism;*

(ii.3) *w restricted to each $\gamma \in \mathcal{F}$ is a constant $p \in P(G)$; and*

(ii.4) *for different arcs γ_1 and γ_2, the values $w(\gamma_1)$ and $w(\gamma_2)$ are also different.*

(iii) *If $\gamma \in \mathcal{F}$ and $L \in K$ is a loxodromic element such that keeps γ invariant, then $\theta(L)$ is a parabolic element in G with $w(\gamma)$ as fixed point.*

A well known result due to B. Maskit [**12**] asserts that a finitely generated Kleinian group with two invariant components in its region of discontinuity is in fact a quasifuchsian group. The following result is a first extension of Maskit's result.

Theorem 2 [**7**]. *Let G be a geometrically finite discrete group with $\Omega^{ext}(G)$ containing two invariant components. Then G is a noded fuchsian group.*

In Maskit's result it is not assumed that the group is geometrically finite. The following theorem proved by Kra and Maskit in [**9**] asserts that geometric finiteness can be dropped from the hypothesis of theorem 2. We sketch an alternative proof of this fact.

Theorem 3. *Let G be a finitely generated Kleinian group with two invariant components in $\Omega^{ext}(G)$ (in the cusp topology). Then G is necessarily a noded fuchsian group.*

A rigidity property, related to Torelli's theorem, asserts that torsion-free non-elementary function groups are uniquely determined by their commutator subgroup [**3**], [**4**], [**5**] and [**10**]. In the case that the group has no invariant component, the above rigidity property is still unknown. The following, the main result of this paper, asserts that for noded function groups this property holds.

Theorem 4. *Let G be a finitely generated non-elementary torsion-free noded function group. Then its commutator subgroup $[G, G]$ determines it uniquely.*

The above result, for torsion-free noded fuchsian groups, has been obtained in [**7**].

The homological cover $\widetilde{M}$ of a manifold M is the one determined by the commutator subgroup of $\pi_1(M)$. For hyperbolic three-manifolds the above says that the hyperbolic structure of M is determined by the hyperbolic structure of $\widetilde{M}$. Theorem 4 then can be written in the language of hyperbolic three-manifolds as follows.

Theorem 5. *Let M be a hyperbolic-three manifold with fundamental group $\pi_1(M)$ containing no finite index abelian subgroup. Let M^* be the manifold with boundary obtained after the procedure of adding pairing tubes to the geometrically finite ends of M. Assume that M^* has a connected component in its boundary, say S, so that the natural inclusion $i_S : S \hookrightarrow M^*$ induces a surjective homomorphism $i_* : \pi_1(S) \rightarrow \pi_1(M^*)$. Then the hyperbolic structure of M is uniquely determined, up to isometries, by the hyperbolic structure of its homological cover.*

The Marden-Maskit Cusp Construction

In this section, we recall the Marden-Maskit cusp construction as appeared in [**13**].

Let G be an analytically finite Kleinian group with extended region of discontinuity $\Omega^{ext}(G) = \Omega(G) \cup P(G)$, where $\Omega(G)$ is its region of discontinuity and $P(G)$ is the set of its double-cusped parabolic fixed points.

The analytically finite condition asserts that we can choose a maximal collection $x_1, \ldots, x_r$ of non-G-equivalent double-cusped parabolic fixed points.

Set $G_{x_i} = \{g \in G : g(x_i) = x_i\}$. We have that either G_{x_i} is a cyclic parabolic subgroup, say generated by a primitive parabolic element $P_i \in G$, or it is generated by a primitive parabolic element P_i and an involution E_i so that $E_i \circ P_i = P_i^{-1} \circ E_i$.

For each $i = 1, \ldots, r$, choose a Möbius transformation B_i so that $B_i \circ P_i \circ B_i^{-1}(z) = z+1$. We may also assume that $B_i \circ E_i \circ B_i^{-1}(z) = -z$.

The double-cusped condition on x_i asserts that there is some $\alpha_i > 1$ so that

$\{z \in \mathbb{C} : |Im(z)| > \alpha_i\}$ is precisely invariant under $B_i G_{x_i} B_i^{-1}$ in $B_i G B_i^{-1}$.

Set $\alpha = max\{\alpha_1, ..., \alpha_r\}$ and $\mathcal{S}_\alpha = \{(z,t) : t > \alpha\} \cup \{(z,t) : |Im(z)| > \alpha\}$.

We have that the boundary of $\mathcal{S}_\alpha$ can be decomposed into $\partial\mathcal{S}_\alpha = \partial_1\mathcal{S}_\alpha \cup \partial_2\mathcal{S}_\alpha$, where

$$\partial_1\mathcal{S}_\alpha = \{(z,t) : |Im(z)| = \alpha, 0 \leq t \leq \alpha\} \cup \{(z,t) : |Im(z)| \leq \alpha, t = \alpha\}$$

and

$$\partial_2\mathcal{S}_\alpha = \{(z,t) : |Im(z)| \geq \alpha, t = 0\} \cup \{\infty\}$$

The set $\partial_1\mathcal{S}_\alpha$, called the infinity shoe-box, contains a center line

$$L_\alpha = \{(z,t) : Im(z) = 0, t = \alpha\}$$

There is a map $\phi : \partial_1\mathcal{S}_\alpha \to \partial_2\mathcal{S}_\alpha$, defined as follows. If $(z,t) = (x+yi, t) \in \partial_1\mathcal{S}_\alpha$, then $\phi(z,t)$ is the end point of the hyperbolic geodesic in $\mathbb{H}^3$ passing through the point (z,t) with the other end being $(x,0)$.

We have $\phi(L_\alpha) = \infty$ and $\phi : \partial_1\mathcal{S}_\alpha - L_\alpha \to \partial_2\mathcal{S}_\alpha - \{\infty\}$ a homeomorphism satisfying $\phi \circ k_i = k_i \circ \phi$ for all $k_i \in B_i G_{x_i} B_i^{-1}$.

Each one of the sets $\mathcal{S}_\alpha$, $\partial_1\mathcal{S}_\alpha$ and $\partial_2\mathcal{S}_\alpha$ is precisely invariant under $B_i G_{x_i} B_i^{-1}$ in $B_i G B_i^{-1}$.

For each $g \in G$, the set $g(B_i^{-1}(\mathcal{S}_\alpha))$ is precisely invariant under $gG_{x_i}g^{-1}$ in G. Two such sets are either equal or disjoint. Set

$$\mathcal{S}_\alpha^* = \mathbb{H}^3 - \mathcal{S}_\alpha = \{(z,t) : |Im(z)| \leq \alpha, 0 < t \leq \alpha\}$$

$$\mathbb{H}_\alpha = \bigcap_{g \in G} \cap_{i=1}^{r} g(B_i^{-1}(\mathcal{S}_\alpha^*)).$$

We have that $\mathbb{H}_\alpha$, $\partial\mathbb{H}_\alpha$ and $\partial\mathbb{H}_\alpha - P(G)$ are invariant under G. The set $\partial\mathbb{H}_\alpha - P(G)$ is planar surface. There is a map $\phi_\alpha : \partial\mathbb{H}_\alpha \to \widehat{\mathbb{C}}$, defined as follows.

Set $\phi_\alpha = I$ on $\partial\mathbb{H}_\alpha \cap \widehat{\mathbb{C}}$, where I denotes the identity map. If $(z,t) \in \partial\mathbb{H}_\alpha \cap \mathbb{H}^3$, then there is some i and some $g \in G$ so that $(z,t) \in g(B_i^{-1}(\partial_1\mathcal{S}_\alpha))$; we set $\phi_\alpha(z,t) = g \circ B_i^{-1} \circ \phi \circ B_i \circ g^{-1}(z,t)$.

In this way, the map ϕ_α commutes with all elements of G. Moreover, ϕ_α is a homeomorphism from the complements of the G-translates of the center lines onto $\widehat{\mathbb{C}} - P(G)$.

The group G acts on $\partial\mathbb{H}_\alpha$ discontinuously on a region Ω_α which results to be exactly $\phi_\alpha^{-1}(\Omega^{ext}(G)) - P(G)$. The planarity of $\partial\mathbb{H}_\alpha - P(G)$ ensures the planarity of Ω_α. Since lines are removable singularities for conformal structures, the action of G on Ω_α is by conformal automorphisms. It follows that Ω_α/G is a finite collection of Riemann surfaces. We have a natural bijection between this collection of Riemann surfaces and the collection of noded Riemann surfaces $\Omega^{ext}(G)/G$. The bijection is essentially given by fattening the nodes and replacing a small neighborhood disc of the phantom nodes (the projections of those double-cusped parabolic fixed points whose stabilizer is not cyclic) with a disc with two points of order two. The center lines go to loops (if the double-parabolic fixed point has cyclic stabilizer) and to simple arcs joining two branched values of order two (if the double-parabolic fixed point has non-cyclic stabilizer) that after pinching them we get the nodes and phantom nodes back, respectively. Since we have that $\Omega(G)/G$ is a finite number of analytically finite Riemann surfaces by the Ahlfor finiteness theorem, we obtain that Ω_α/G is a finite number of analytically finite Riemann surfaces.

The boundary of each connected component of Ω_α is then contained in the limit set for the action of G on $\partial\mathbb{H}_\alpha$ which is the same limit set for G in its action on the Riemann sphere.

The Proof of the Theorems

In what follows, we keep the notations of the last section.

Proof of Theorem 1. Let G be a noded function group with Δ^* an invariant component of $\Omega^{ext}(G)$ (in the cusp topology).

The component Δ^* determines a unique planar region $\Delta_1 \subset \partial\mathbb{H}_\alpha$ which is also invariant under the action of G. The Riemann surface Δ_1/G is analytically finite.

In [**12**] is shown the existence of a conformal embedding $Q_1 : \Delta_1 \hookrightarrow \widehat{\mathbb{C}}$, so that $K_1 = Q_1 G Q_1^{-1}$ is a group of Möbius transformations.

The region $Q_1(\Delta_1)$ is invariant under K_1 and contained in its region of discontinuity. It follows that there is an invariant component Δ for K_1 containing $Q_1(\Delta_1)$. Since both $Q_1(\Delta_1)/K_1 \subset \Delta/K_1$ and Δ_1/G are analytically finite and conformally equivalent to $Q_1(\Delta_1)/K_1$, we must have that $Q_1(\Delta_1) = \Delta$.

If we consider the map $w = \phi_\alpha \circ Q_1^{-1} : \Delta \to \Delta^*$, then we get the desired result taking $K = K_1$.

Sketch of Proof of Theorem 3. We may assume, after passing to a finite index normal subgroup if necessary, that G is torsion-free. If G has no double-cusped parabolics, then G is a quasifuchsian group from Maskit's result in [**12**]. From now on, we assume that G has double-cusped parabolic transformations. In particular, double-cusped parabolic fixed points are dense in the limit set of G.

Let Δ_* and Δ_{**} be different invariant components in $\Omega(G)^{ext}$ in the cusp topology. As in the proof of theorem 1, we have that Δ_* and Δ_{**} both determine disjoint invariant components Δ_1 and Δ_2 inside Ω_α.

The planarity property of $\Omega_\alpha - P(G)$ together the invariance of each Δ_i assert that Δ_1 and Δ_2 are simply-connected components. It follows that Δ_1/G and Δ_2/G are analytically finite Riemann surfaces of same type (γ, k). Hyperbolicity ensures that $2\gamma - 2 + k > 0$. It follows that G is isomorphic to either a free group of rank $2\gamma + k - 1$ if $k \geq 1$ or isomorphic to the fundamental group of a closed surface of genus $\gamma \geq 2$ if $k = 0$.

Using area arguments, we can obtain that $\Omega_\alpha = \Delta_1 \cup \Delta_2$. In fact, assume we have another component U in Ω_α. Then we have that

$$(*) \qquad \begin{aligned} \text{Area}(\Delta_1/G) + \text{Area}(\Delta_2/G) + \text{Area}(U/G_U) &\leq \text{Area}(\Omega_\alpha/G) \\ &= \text{Area}(\Omega(G)/G), \end{aligned}$$

where G_U is the subgroup of G that keeps U invariant.

The first area inequality [**2**] asserts that

$$(**) \quad \text{Area}(\Omega_\alpha/G) = \text{Area}(\Omega(G)/G) \leq 4\pi(l-1),$$

where l is the number of generators of G, that is, $l = 2\gamma + k - 1$ for $k \geq 1$ and $l = 2\gamma$ for $k = 0$.

For $k \geq 1$, both inequalities $(*)$ and $(**)$ assert that $\text{Area}(U/G_U) = 0$ a contradiction.

For $k = 0$, we have that G is isomorphic to the fundamental group of some closed surface of genus γ. It follows that G_U is isomorphic to the fundamental group of some closed surface of genus $g_0 \geq 2$. Then inequalities $(*)$ and $(**)$ assert that $g_0 = 2$ and, in particular, we obtain an equality in $(*)$. But equality of $(*)$ implies that G is a free group, a contradiction.

We obtain in this way that $\Omega^{ext}(G) = \Delta_* \cup \Delta_{**}$, and each component of $\Omega(G)$ is a simply connected Jordan domain. If G has an invariant component in $\Omega(G)$, then G is actually a regular B-group, in which case we are done. Let us assume that G has no invariant component in $\Omega(G)$.

We must remark at this point that the case $k \geq 1$ is not possible. In fact, the number k is the number of maximal parabolic cyclic non-conjugate subgroups of G which are not double-cusped. These cyclic groups produce punctures in both Δ_1/G and Δ_2/G. This obligates them to be double-cusped. In what follows, we may still assume that $k \geq 0$.

For each double-cusped parabolic fixed point $x \in P(G)$, say $x \in \Delta_*$, we have the existence of a simple arc $A_x \subset \Delta_{**}$ precisely invariant under $G_x = \langle P \rangle$, a cyclic double-cuspidal parabolic subgroup. We call A_P an axis of P. Using this, one is able to obtain the same result in [**1**] (theorem 2), that is, if $\lambda \in \Lambda(G)$, then either: (i) λ belongs to the boundary of some component in $\Omega(G) \cap \Delta_*$ or (ii) there is a double-cuspidal parabolic transformation $P \in G$ and a sequence $(P_n) \subset G$ such that if A_P is the (closed) axis of P, then $P_n(A_P)$ is a spherical nest about λ and $\operatorname{diam}(P_n(A_P)) \to 0$. From this, one can follows the arguments also in [**1**] to obtain that the limit set of G is locally connected.

For each $i = 1, 2$, we can choose $Q_i : \Delta_i \to \widehat{\mathbb{C}}$ conformal embedding so that $Q_i G Q_i^{-1} = K_i$ is a torsion-free fuchsian group of the first kind acting on the unit disc $\Delta = Q_i(\Delta_i)$. In particular, the groups K_1 and K_2 are both isomorphic to G.

The map $Q_i^{-1} : \Delta \to \Delta_i$ defines an isomorphism $\theta_i : K_i \to G$ so that $Q_i^{-1} \circ k_i = \theta_i(k_i) \circ Q_i^{-1}$, for all $k_i \in K_i$. Moreover, if $\theta_i(k_i) \in G$ is parabolic, then k_i is parabolic if and only if $\theta_i(k_i)$ is not double-cuspidal.

We have an isomorphism $\theta = \theta_1 \circ \theta_2^{-1} : K_1 \to K_2$, which is a type-preserving g-isomorphism (see definition in [**12**]). Nielsen's isomorphism theorem asserts the existence of a homeomorphism $W : \Delta \to \Delta$ so that $W \circ k_1 = \theta(k_1) \circ W$, for all $k_1 \in K_1$.

Let j be the reflection on the unit circle, and let us consider the map $R : \Delta \cup j(\Delta) \to \partial\mathbb{H}_\alpha$, defined by Q_1^{-1} on Δ and by $Q_2^{-1} \circ W \circ j$ on $j(\Delta)$.

Since the limit set of G is locally connected and it is the common boundary of Δ_1 and Δ_2, we have that R restricted to each Δ and $j(\Delta)$ extends continuously to the boundary circle. Since R, restricted to each of the discs, defines the same isomorphism $\theta_1 : K_1 \to G$, we have that both extensions coincide in the boundary unit circle.

The image of the unit circle under R is all the limit set of G, and the preimage of any double-cusped parabolic fixed point x consists of exactly two points, the end points of the hyperbolic element $\theta^{-1}(P)$, where $P \in G$ is some parabolic transformation fixing x.

The set $R(\widehat{\mathbb{C}})$ is exactly $\partial\mathbb{H}_\alpha$. Composing this map with $\phi_\alpha : \partial\mathbb{H}_\alpha \to \widehat{\mathbb{C}}$, we obtain a surjective map $T : \widehat{\mathbb{C}} \to \widehat{\mathbb{C}}$, defining an isomorphism $\eta : K_1 \to G$ so that $T \circ k_1 = \eta(k_1) \circ T$, for $k_1 \in K_1$, so that the preimages of double-cusped parabolic fixed points are geodesics (together their end points) which have pairwise disjoint closures.

Each parabolic fixed point in G either comes from a parabolic fixed point of K_1 or it is a double-cusped parabolic fixed point. But the parabolic fixed points of a Fuchsian group are double-cusped. It follows that all parabolic fixed points of G are double-cusped. Since the action of G on $\partial\mathbb{H}_\alpha - P(G)$ is topologically quasifuchsian, the action is geometrically finite. Now we apply theorem 2 to our group G to obtain the desired result.

Proof of Theorem 4. Let us assume we have two torsion-free non-elementary noded function groups G_1 and G_2 so that they have the same commutator subgroup K. The normality of K in both G_1 and G_2 and the fact that K is non-elementary assert that the region of discontinuity of the three groups G_1, G_2 and K are the same, say Ω.

Since K is non-elementary, the normalizer N_K of K in the group of Möbius transformations is a Kleinian group with region of discontinuity

Ω. In particular, the group J generated by G_1 and G_2 is a Kleinian group with region of discontinuity Ω.

The fact that $\Lambda = \widehat{\mathbb{C}} - \Omega$ is infinite asserts that each connected component of Ω is hyperbolic, that is, covered by the hyperbolic plane. It follows that each connected component of Ω/K, Ω/G_1, Ω/G_2 and Ω/J is hyperbolic.

The groups G_1, G_2 and J are finitely generated, so the Ahlfor finiteness theorem asserts that Area(Ω/J), Area(Ω/G_1) and Area(Ω/G_2) are finite. It follows that G_1 and G_2 have finite index in J. From this we obtain that $H = G_1 \cap G_2$ has finite index in G_1, G_2 and J.

The finite index property of H asserts that $P(G_1) = P(H) = P(G_2)$ and, in particular, $\Omega^{ext}(G_1) = \Omega^{ext}(G_2) = \Omega^{ext}(H) = \Omega^{ext}$.

Let us denote by $\Delta_i \subset \Omega^{ext}$ an invariant component of G_i, for $i = 1, 2$. If we have that $\Delta_1 \neq \Delta_2$, then H will have two invariant components. It follows from theorem 3 that $\Omega^{ext} = \Delta_1 \cup \Delta_2$ (we only need the first part of the proof of theorem 3). Since Δ_1 is invariant under G_1, we must have that Δ_2 is also invariant under G_1. As a consequence, we may assume that $\Delta_1 = \Delta_2 = \Delta^*$.

In the Marden-Maskit cusp construction we have that Δ^* defines an planar surface Δ_α in $\partial\mathbb{H}_\alpha$, invariant under J.

We can choose a conformal embedding $g : \Delta_\alpha \to \widehat{\mathbb{C}}$ so that $F = g \circ J \circ g^{-1}$ is a group of Möbius transformations. Set $\Delta = g(\Delta_\alpha)$. Since Δ is F invariant, there is an invariant component X of $\Omega(F)$ containing Δ. We have then that Δ/F is conformally embedded in X/F. Since F is finitely generated, the Ahlfor finiteness theorem asserts that X/F is an analytically finite Riemann surface. On the other hand, Δ/F is conformally equivalent to Δ_α/J which is analytically finite. It follows that $X = \Delta$.

Now, we are dealing with two torsion-free non-elementary function groups $N_1 = gG_1g^{-1}$ and $N_2 = gG_2g^1$ with the same commutator subgroup gKg^{-1}. From the results in [**5**] we have that $N_1 = N_2$ and, in particular, $G_1 = G_2$ as desired.

References

[1] W. Abikoff. On boundaries of Teichmüller spaces and on kleinian groups, III. *Acta math.* **134** (1975), 211-237.

[2] L. Bers. Inequalities for finitely generated Kleinian groups. *J. Ann. Math.* **18** (1967), 23-41.

[3] R. Hidalgo. Homology coverings of Riemann surfaces, *Tôhoku Math. J.* **45** (1993), 499-503.

[4] ______ Kleinian groups with common commutator subgroup, *Complex variables* **28** (1995) 121-133.

[5] ______ A Commutator Rigidity for Function Groups. Preprint.

[6] ______ Noded Schottky Space, *Proc. London Math. Soc.* **73** (1996), 385-403.

[7] ______ Noded Fuchsian Groups I, To appear in *Complex Variables.*

[8] L. Keen, B. Maskit and C. Series. Geometric finiteness and uniqueness for Kleinian groups with circle packing limit sets, *J. reine angew. Math.* **436** (1993), 209-219.

[9] I. Kra and B. Maskit. Pinched 2−component kleinian groups. To appear in Analysis and Topology.

[10] B. Maskit. The homology covering of a Riemann surface, *Tôhoku Math. J.* **38** (1986), 561-562.

[11] ______ *Kleinian Groups* Grundlehren der Mathematischen Wissenschaften, Vol. 287, Springer-Verlag, 1988.

[12] ______ On boundaries of Teichmüller spaces and on kleinian groups II, *Ann. of Math.* **91** (1970), 607-639.

[13] ______ On Free Kleinian Groups, *Duke Math. J.* **48** (1981), 755-765.

[14] ______ Parabolic elements in Kleinian groups, *Ann. of Math.* **117** (1983), 659-668.

Departamento de Matemática, UTFSM, Casilla 110-V Valparaiso, Chile

E-mail address: rhidalgo@trantor.mat.utfsm.cl

Contemporary Mathematics
Volume **240**, 1999

THE HYPERELLIPTIC FIBRE OF THE PRYM MAP IN GENUS FOUR

L. Hidalgo-Solís and S. Recillas-Pishmish

ABSTRACT. Let $\bar{\mathfrak{P}}_3 : \bar{\mathfrak{R}}_4 \to \mathfrak{A}_3$ be the Prym map in genus 4. We give a biregular description of the fibre $\bar{\mathfrak{P}}_3^{-1}(JX)$, where JX is a hyperelliptic Jacobian of genus 3.

1. Introduction

If $\pi : \tilde{C} \to C$ is an étale double cover of a smooth curve of genus g. The Prym variety $P = \mathfrak{p}(\tilde{C}, C, \pi)$ is defined to be the connected component of the origin in kerNm, where $Nm : J\tilde{C} \to JC$ is the norm map.

By Mumford [**M**, pg. 342] P is a $(g-1)-$dimensional abelian variety, with a natural principal polarization $\Xi = \frac{1}{2}(P.\tilde{\Theta})$, where $\tilde{\Theta}$ is a theta divisor on $J(\tilde{C})$.

Let denote by $\mathfrak{R}_g$ and $\mathfrak{A}_g$ respectively the coarse moduli space of curves C of genus g, together with an unramified double cover $\tilde{C}$ and of principally polarized abelian varieties of dimension g.

The above construction yields a morphism, the Prym map

$$\mathfrak{p}_{g-1} : \mathfrak{R}_g \to \mathfrak{A}_{g-1}.$$

The problem of extending $\mathfrak{p}_{g-1}$ to possibly singular and ramified covers was attempted by Fay [**F**], Mumford [**M**] and Masiewicki [**Ma**]. Beauville in [**Be.1**], gave the precise definiton of allowable double cover and in §6 he obtained a proper map $\bar{\mathfrak{P}}_{g-1}$ which on a dense open set factors through the previously described $\mathfrak{p}_{g-1}$.

The map $\bar{\mathfrak{P}}_{g-1} : \bar{\mathfrak{R}}_g \to \mathfrak{A}_{g-1}$ is generically injective for $g \geq 7$ [**FS**], generically finite for $g = 6$ [**DS**] and dominant for $g \leq 6$ [**DS**].

For $g = 5$, Donagi and Smith show that the fibre of $\bar{\mathfrak{P}}_5$ has the structure of the 27 lines on a cubic surface [**DS**]. Donagi [**Do.**] studied the fibre of $\bar{\mathfrak{P}}_4$ at a general (A, Θ), and he shows that it is isomorphic to a double cover of the Fano surface of the lines of a cubic threefold, and for $g = 3$ Verra [**V**] proves that the fibre of $\bar{\mathfrak{P}}_2$ is biregular to a Siegel modular quartic threefold V. In [**HR**] we studied the fibre of $\bar{\mathfrak{P}}_3$ at a general (A, Θ), and showed that it is isomorphic to a sequence of blow-ups of the Kummer variety of A.

In this paper we give a description of $\bar{\mathfrak{P}}_3^{-1}(A)$ in the case in which A is a hyperelliptic Jacobian.

1991 *Mathematics Subject Classification.* Primary:14H40 , secondary 14H45.

First author was supported by PAPIIT-IN101296, second author was supported by PAPIIT-IN108996 and both by proyect CONACyT-CONICyT "variedades abelianas con automorfismos".

We will work over the complex field.

One should recall [**Be.1**, Prop. 5.2 pg. 174] that if $\tilde{C}$ is a stable curve and ι an involution on it, then the pair $(\tilde{C}, \iota)$ is allowable if the following equivalent conditions hold:

1. $\bar{\mathfrak{P}}(\tilde{C}, \iota)$ is a $(g-1)-$dimensional abelian variety;
2. The only fixed points of ι are nodes where the two branches are not exchanged. The number of nodes exchanged by ι equals the number of irreducible components exchanged by ι.

The authors wish to thank to G. Kempf and J. Muñoz-Porras for their several helpful conversations.

2. The fibre of the Prym map on the hyperelliptic Jacobian case.

In this section we study the case in which the principally polarized abelian variety (p.p.a.v) is the Jacobian variety of a hyperelliptic curve X. This case is similar to the general case given in [**HR**], in particular we prove the following:

THEOREM 1. *Let X be a hyperelliptic curve, let (JX, Θ) denote its Jacobian variety and assume $Aut(JX) = \mathbb{Z}/2\mathbb{Z}$, then the inverse image of (JX, Θ) by the Prm map $\bar{\mathfrak{P}}_3$ is obtained from the Kummer variety KX, by a sequence of two blow-up's σ_1 and σ_2, where $\sigma_1 : KX(0) \to KX$ blow up KX at the "origin" point and $\sigma_2 : \widetilde{KX(0)} \to KX(0)$ is centered along a curve $\tilde{\mathfrak{X}}$.*

The proof of the theorem is divided in 3 parts, in the first, we give a set theoretical description of the fibre, and we study the subset of the fibre consisting of the hyperelliptic covers. In the second, we study the trigonal construction for a general linear series of type g^1_4 on the curve X, and we give a description of the non-general linear series of type g^1_4.

In the third we study the trigonal construction for the nongeneral linear series g^1_4, and then we give a biregular description of the fibre.

Let X be a smooth, reduced, irreducible, projective hyperelliptic curve of genus 3, let $j : X \to X$ denote the hyperelliptic involution and $h : X \overset{2:1}{\to} \mathbb{P}^1 = X/ < j >$ the map induced by the hyperelliptic involution. Let $JX = Pic^0 X$ be the Jacobian variety of X with $\Theta \subset JX$ a symmetric theta divisor such that $Sing\Theta = \{0\}$, $Pic^n X$ the divisors on X of degree n modulo linear equivalence, $W_n \subset Pic^n X$ the image of the $n^{th}-$Abel-Jacobi map, if $n = 2m$ for some $m \in \mathbb{Z}$, then $u_n(x_1 + \ldots + x_n) = \mathcal{O}(x_1 + \ldots + x_n - m g^1_2)$, $X - X \subset JX$ the image of the map $X^2 \to JX$, $(p, q) \mapsto \mathcal{O}(p - q)$, let R be the image of the map $X^{(2)} \to JX$, $p + q \mapsto [\omega_X(-2p - 2q)]$, and let $JX(0)$ denote the blow-up of JX at the origin. Let $G^1_4(X)$ denote the variety which parametrizes the linear series of degree 4 and dimension 1 on X, this variety is cannonicaly isomorphic to $JX(0)$.

In this part our aim is to describe set-theoretically the fibre of the Prym map over a hyperelliptic Jacobian JX of genus 3.

As a consequence of Mumford [**M**] and Beauville [**Be.1**] work, it is easy to see that $\bar{\mathfrak{P}}_3^{-1}(JX)$ consists of the following types of allowable covers:

- The general point of $\bar{\mathfrak{P}}_3^{-1}(JX)$ is a smooth double cover $\pi : Z \to Y$, where Y is a genus 4 curve with an even theta characteristic L such that $\Gamma(Y, L) \neq \{0\}$ and $L \otimes \eta$ also an even theta characteristic, where $\eta \in J_2 Y$ is the order two point which defines the cover π.

- The covers of hyperelliptic type, which consists of smooth double covers $\pi : Z \to Y \in \mathcal{R}_4$, with Y a hyperelliptic curve of genus 4 and $\eta = \mathcal{O}_Y(p-q)$, where $p \neq q$ are Weierstrass points of Y.
- Irreducible allowable double covers of lower geometric genus.
- Allowable double covers of Wirtinger type [**Be.1**, pg. 175].
- The elliptic tails [**Be.1**, pg. 74].

In the next part of this section we study *the hyperelliptic component.*

Let $X \in \mathfrak{M}_3$ be a hyperelliptic curve of genus 3.

Let $\bar{\mathfrak{R}}_g^h \subset \bar{\mathfrak{R}}_g$ be the closure of the étale covers $\pi : Z \to Y$, with Y a hyperelliptic curve. Then we have [**DS**, pg. 62].

LEMMA 1.

$$\bar{\mathfrak{P}}_3 \mid_{\bar{\mathfrak{R}}_4^h}{}^{-1}(JX) \simeq \mathbb{P}^1 \times \mathbb{P}^1$$

for a hyperelliptic Jacobian JX of genus 3. □

LEMMA 2. *Let $\bar{\mathfrak{R}}_E$ denote the hypersurface in $\bar{\mathfrak{R}}_4$ parametrizing the elliptic tails, then the fibre of the restricted Prym map $\bar{\mathfrak{R}}_E \to \mathfrak{A}_3$ over a generic hyperelliptic Jacobian JX is isomorphic to $\mathbb{P}^1 \times \mathbb{P}^1$.* [**DS**, Lemma 1.1 pg. 62]

□

Let us assume that JX is a general hyperelliptic Jacobian, in the sense that $Aut(JX) = \pm 1$, and let us recall the construction of the following family [**HR**, Prop. 2] .

Consider the divisor

$$\mathcal{D} = \{\theta(z-c) = 0 \mid (c,z) \in JX \times \Theta\} \subset JX \times \Theta,$$

and the morphism $\pi : JX(0) \times \Theta \to JX \times \Theta$ defined as the blow up of the origin in the first factor and the identity on the second.

Then $\pi^*\mathcal{D} = S + 2(E \times \Theta)$, where S is the strict transform of $\mathcal{D}$ and $E \subset JX(0)$ is the exceptional divisor. The first projection gives us a flat family $S \to JX(0)$ of effective algebraic 1−cycles with fundamental class $[Z] = \Lambda^2[\Theta]$ in $H^4(JX, \mathbb{Z})$, and its members are $Z_c = \Theta.(\Theta_c)$, if $c \in JX - \{0\}$, and $Z_c = \mathcal{G}^{-1}(\mid K + c \mid)$, where (by abuse of language) $\mid K + c \mid \subset \mid K \mid$ denotes the line in $\mid K \mid = E^\vee$ corresponding to the point $c \in E$ and $\mathcal{G} : \widehat{\Theta} \to \mid K \mid$ is the Gauss map. Moreover $Z_c = \Theta.\Theta_c$ has an involution $\iota_c(x) = c - x$, and since the involution $-1 : JX \to JX$ commutes with the Gauss map, it induces an involution on $Z_c = \mathcal{G}^{-1}(\mid K + c \mid)$.

The family S exhausts all possible $1-$ *cycles* of the above type.

PROPOSITION 1. *(*[**Be.2**]*, for a proof see* [**HR**]*) Let X be a curve of genus* 3*, (JX, Θ) its Jacobian variety. Let Z be an effective algebraic $1-$cycle in JX with fundamental class $[Z] = \bigwedge^2[\Theta]$ in $H^4(JX, \mathbb{Z})$, then Z is one of the following:*

(i) $Z = \Theta \cap \Theta_a$ *for some* $a \in JX - \{0\}$, *or*

(ii) $Z = \mathcal{G}^{-1}(\ell)$ *for some line* $\ell \subset \mid K \mid$.

Let us describe the g_4^1's by considering the Abel map $u_4 : X^{(4)} \to JX$, $D \mapsto [\omega_X(-D)]$, then if $c \in JX - \{0\}$, $u_4^{-1}(c) = \mid K + c \mid$ is a complete g_4^1, and the non complete ones are the lines $\mid K + c \mid \subset \mid K \mid$.

Let us consider now the sum map: $s : X^{(2)} \times X^{(2)} \to X^{(4)}$, then $Z_c' := s^{-1}(\mid K + c \mid) \to \mid K + c \mid = \mathbb{P}^1$ is a curve with a g_6^1, this map commutes with the exchange of coordinates, so we get an involution $\iota_c' : Z_c' \to Z_c'$ and a trigonal curve $Y_c' \to \mid K + c \mid$ where $Y_c' = Z_c'/\iota_c'$.

So to prove our theorem we have to describe the fundamental locus of the rational map $\phi : KX(0) \to \mathfrak{P}^{-1}(JX)$.

Now we want to describe the closed subset of $KX(0)$ where the rational map $\phi : KX(0) \to \bar{\mathfrak{P}}^{-1}(JX)$ cannot be extended, that is, we want to describe:

$$\{ \ c \in JX(0) \ \mid \ (Z_c, \iota_c) \text{ is not allowable } \}.$$

Observe now that if $v : X^{(2)} \times X^{(2)} \overset{pr_1}{\to} X^{(2)} \overset{u_2}{\to} JX$ then $v(Z_c') = Z_c$ and ι_c' descends to ι_c.

This is the trigonal construction, and one has a bijection [**Rec.1**]:

$$\{f : X \to \mathbb{P}^1, \ \deg f = 4, \ f \text{ has no ramification of type } 2p + 2q\}$$
$$\updownarrow$$
$$\{Z \overset{\pi}{\to} Y \overset{g}{\to} \mathbb{P}^1, \deg \pi = 2, \ \deg g = 3, \ \pi \text{ étale}\}$$

PROPOSITION 2. *There exists a birational map*

$$\phi : KX(0) \to \bar{\mathfrak{P}}_3^{-1}(JX).$$

PROOF. The condition that $\mid K + c \mid$ be free of base points and of divisors of type $2p + 2q$ is clearly open, and since the family $S \to JX(0)$ is flat, we have a map $KX(0) \to \bar{\mathfrak{R}}_4$, but the involution $-1 : JX \to JX$ induces an isomorphism $Z_c \simeq Z_{-c}$, so in fact we have a rational map $KX(0) \to \bar{\mathfrak{R}}_4$.

By the Matsusaka's criterion for Prym's $\mathfrak{P}(Z_c, \iota_c) \simeq JX$ as p.p.a.v., hence we have a rational map $\phi : KX(0) \to \bar{\mathfrak{P}}_3^{-1}(JX)$.

The fact that the map is dominant follows from the set theoretical description given above, by counting dimensions, but will prove a stronger statement.

We describe now the inverse of ϕ, to do this we consider level structures in order to have fine moduli (see [**DS**] and [**HR**]).

Let $\bar{\mathfrak{M}}_g^{(n)}$ denote the moduli space of stable curves of genus g with a level $n-$structure ($n \geq 3$). There exist a $\bar{\mathfrak{M}}_{2g-1}^{(n)}-$scheme $\bar{\mathfrak{R}}_g^{(n)}$ a universal curve $q : \tilde{\mathfrak{C}} \to \bar{\mathfrak{R}}_g^{(n)}$ and a $\bar{\mathfrak{R}}_g^{(n)}-$involution $\iota : \tilde{\mathfrak{C}} \to \tilde{\mathfrak{C}}$ such that:

(a) For each $s \in \bar{\mathfrak{R}}_g^{(n)}$, the induced involution $\iota_s : \tilde{\mathfrak{C}}_s \to \tilde{\mathfrak{C}}_s$ is different from the identity on each component of $\tilde{\mathfrak{C}}_s$.
(b) $\tilde{\mathfrak{C}}_s$ has arithmetic genus $2g - 1$, and the quotient curve $\tilde{\mathfrak{C}}_s/\iota_s$ has arithmetic genus g.
(c) Any pair $(\tilde{C}, \iota)$ where $\tilde{C} \in \mathfrak{M}_{2g-1}$, ι a fixed-point free involution is isomorphic to $(\tilde{\mathfrak{C}}_s, \iota_s)$ for some $s \in \mathfrak{R}_g^{(n)} \subset \bar{\mathfrak{R}}_g^{(n)}$.

By [**Be.1**] there is a proper morphism

$$\mathfrak{p} : \bar{\mathfrak{R}}_g^{(n)} \to \mathfrak{A}_{g-1}$$

which descends to the Prym map

$$\bar{\mathfrak{P}} : \bar{\mathfrak{R}}_g \to \mathfrak{A}_{g-1}.$$

Let $\mathcal{P}$ denote the subscheme of $\bar{\mathfrak{R}}_4^{(n)}$ parametrizing the inverse image of $[JX] \in \mathfrak{A}_3$ by the morphism $\mathfrak{p}$, and let $\bar{\Gamma} \to \mathcal{P}$ be the restriction of the universal curve $\tilde{\mathfrak{C}}$ to $\mathcal{P}$ and denote also by ι the restriction of the $\bar{\mathfrak{R}}_4^{(n)}-$ involution to $\mathcal{P}$.

Then one considers the Jacobian map

$$\begin{array}{ccc} \varphi : \bar{\Gamma} & \longrightarrow & J(\bar{\Gamma}/\mathcal{P}) =: \mathfrak{J} \\ & \searrow \quad \swarrow & \\ & \mathcal{P} & \end{array}$$

The $\mathfrak{R}_4^{(n)}$–involution ι induces (by descent theory) a $\mathcal{P}$– involution $\iota^* : \mathfrak{J} \to \mathfrak{J}$. Denote by α the $\mathcal{P}$–morphism $(1_{\mathfrak{J}} - \iota^*) \cdot \varphi : \bar{\Gamma} \to \mathfrak{J}$ and let $P := Im(1_{\mathfrak{J}} - \iota^*)$.

So on the diagram

$$\begin{array}{ccc} \bar{\Gamma} & \xrightarrow{\alpha} & P \\ & \searrow \quad \swarrow & \\ & \mathcal{P} & \end{array}$$

over closed points of $\mathcal{P}$, we get on P the Prym varieties and α is the corresponding Abel Prym map.

Moreover if $b \in \mathcal{P}$ is a closed point, $\mathcal{P}_b$ is the Prym of $(\bar{\Gamma}_b, \iota_b)$ so $\alpha_\star(\bar{\Gamma}_b) \subset \mathcal{P}_b$ has class $[\Theta.\Theta]$, and by definition we know that $\mathcal{P}_b \simeq JX$ (as p.p.a.v), so $\alpha_\star(\bar{\Gamma}_b) = \Theta.\Theta_{a(b)}$ for some $\pm a(b) \in KX(0)$. By proposition 2, $a(b)$ is unique (here we use $Aut(JX) = \pm 1$), so this defines a morphism $\mathcal{P} \to KX(0)$ which descends to a morphism $\psi : \mathfrak{P}^{-1}(JX) \to KX(0)$, which is clearly the inverse of ϕ. $\square$

Let $\widehat{R}$ and $\widehat{X - X}$ be the proper transforms of R and $X - X$ respectively under the blow up $JX(0) \to JX$. Then we observe that

1. Given $c \in JX - \{0\}$, $\mid K + c \mid$ has a fixed point iff, $c \in (X - X) - \{0\}$.
   If $c \in \mid K \mid$, then $\varphi_K(X)$ has a fixed point iff $c \in \varphi_K(X) \subset E = \mid K \mid^\vee$ (here $\varphi_K : X \to \mid K \mid^\vee$ is the canonical map and $\varphi_K(X)$ is a conic).
   So $\{c \in JX(0) \mid \mid K + c \mid$ has a fixed point $\} = \widehat{X - X}$ and $\widehat{X - X} \cap E = \varphi_K(X) \simeq \mathbb{P}^1$.
2. By definition, a complete $\mid K + c \mid$ has a divisor of type $2p + 2q$ iff $c \in R - \{0\}$. Observe that since $\varphi_K(X) \subset \mid K \mid^\vee$ is a conic, every $\mid K + c \mid \subset \mid K \mid$ has a divisor of type $2p + 2q$.

PROPOSITION 3. *For $c \in JX(0)$, let $Z_c = \Theta.\Theta_c$ and $i_c(x) = c - x$ be as before. Then*

1. *If $c \in \widehat{X - X} - \widehat{R}$ then Z_c has two components, each isomorphic to X. The components are exchanged under the involution i_c, they intersect transversally in two points, which in turn are exchanged by i_c. Wirtinger type.*
2. *If $c \in \widehat{X - X} \cap \widehat{R}$, then either Z_c has two components, each isomorphic to X, those components are exchanged by i_c and they intersect non transversally in a unique point, which is a fixed point of i_c, or $Z_c = 2X$. Non allowable type.*
3. *If $c \in \widehat{R} - (\widehat{X - X})$, then Z_c is irreducible but not smooth, has ordinary double points and at most 4 of them, such points are fixed under the involution, which does not interchange the branches at the point. Lower genus type.*
4. *If $c \notin \widehat{R} \cup (\widehat{X - X})$, then Z_c is irreducible and non singular. Generic type.*
5. *The above are the only possible fixed points of the involutions i_c.*

PROOF. For $c \notin (\widehat{X - X}) \cap \widehat{R}$ the proof is similar to the non-hyperelliptic case, (see [**Rec.2** , Proposition 2.14]).

Let $c \in ((X - X) \cap R) - \{0\}$, then $\mid K + c \mid = g_2^1 + 2p$ with $p \neq jp$. Then $Z = X_p \cup X_p^v$, where

$$\begin{array}{ccc} X & \rightarrow & X_p \\ x & \mapsto & u_2(x+p) \end{array} \qquad \begin{array}{ccc} X & \rightarrow & X_p^v \\ x & \mapsto & u_2(p+jx) \end{array}$$

In particular the hyperelliptic involution on X implies that $X_p = X_p^v$, and the involution ι fixes the points of the form $u_2(p_i + p)$ with $p_i \in X$ a Weierstrass point. So $Z = 2X$, in particular (Z, ι) is not allowable.

We note that this curve Z is in the image of the Abel-Prym map of the hyperelliptic curve $C = normalizaton\ of\ (X \times_{\mathbb{P}^1} \tilde{\mathbb{P}}^1)$, where $\tilde{\mathbb{P}}^1 \rightarrow \mathbb{P}^1$ is a ramified double cover, with exactly two branch points (see [**M**]).

Let $c \in (\widehat{X - X}) \cap \widehat{R} \cap E$, once again one has $\mid K + c \mid = g_2^1 + p + jp$.

- Assume $p \neq jp$, then $Z = X_p \cup X_p^v$, where

$$\begin{array}{ccc} X & \rightarrow & X_p \\ x & \mapsto & u_2(x+p) \end{array} \qquad \begin{array}{ccc} X & \rightarrow & X_p^v \\ x & \mapsto & u_2(jp+jx) \end{array}$$

  with $X_p \cap X_p^v = \{u_2(p + jp)\} = \{0\}$. The involution ι exchanges the curves X_p, X_p^v, and fixes the point $\{0\}$, so this cover is not allowable (see [**Be.1**]).
  This curve Z is the image of the Abel-Prym map of an elliptic tail $X \cup_{p \sim f} F \cup_{\iota f \sim p'} X'$
- Let assume that $p = jp$, i.e., p is a Weierstrass point. Once again one has that $Z = X_p \cup X_p^v$ where $X_p = X_p^v$. The involution ι fixes 8 points, so this cover is not allowable (see [**Be.1**]).

□

As a consequence of the above we have that

COROLLARY 1. *The pair (Z_c, ι_c) is allowable if and only if $c \notin \widehat{X - X} \cap \widehat{R}$.*

PROOF OF THE THEOREM: Let $\mathfrak{X} \subset KX(0)$ denote the image of the closed subscheme with the reduced structure $\widehat{X - X} \cap \widehat{R} \subset JX(0)$ by the induced morphism $JX(0) \rightarrow KX(0)$.

In particular, the morphism $\phi : KX(0) \rightarrow \bar{\mathfrak{P}}^{-1}(JX)$ has the curve $\mathfrak{X}$ as its fundamental locus. Let $R" := KX(0) - \mathfrak{X}$. The morphism $\psi_{|\psi^{-1}(R")} : \psi^{-1}(KX(0) - \mathfrak{X}) \rightarrow KX(0) - \mathfrak{X}$ is an isomorphism, and $\psi^{-1}(R'')$ is the maximal open set $U \subset \bar{\mathfrak{P}}^{-1}(JX)$ such that the restriction $\psi_{|U}$ is an isomorphism. Let $\mathfrak{J}$ denote a sheaf of ideals on $KX(0)$ with support $\mathfrak{X}$. Then $\bar{\mathfrak{P}}^{-1}(JX)$ is isomorphic to the blown up $p : \widetilde{KX(0)} \rightarrow KX(0)$ with respect to $\mathfrak{J}$ and the following diagram commutes:

$$\begin{array}{ccc} \bar{\mathfrak{P}}^{-1}(JX) & \xrightarrow{\ \simeq\ } & \widetilde{KX(0)} \\ & {\scriptstyle\psi} \searrow \quad \swarrow {\scriptstyle p} & \\ & KX(0) & \end{array}$$

By Lemma 1 and 2, $\mathfrak{J}$ is the ideal of the reduced scheme $\mathfrak{X}$, with this we finish the proof of the theorem. □

References

[Be.1] Beauville, A., *Prym varietes and the Schottky problem.* Invent. Math. vol **41**, (1977), 149-196.

[Be.2] Beauville, A., *Variétés de Prym et Jacobiennes intermédiares.* Ann. Sci. École Norm. Sup. vol **10**, (1977), 309-391.

[Do.] Donagi, R., *The fibers of the Prym map.* Cont. Math., Amer. Math. Soc. **No. 136**, (1992).

[DS] Donagi, R., Smith, R. C., *The structure of the Prym map.* Acta Math. vol **146**, 1982, 25–102.

[F] Fay, J. *Theta functions on Riemann Surfaces.* Lecture Notes vol **352**, (1973), Springer-Verlag.

[FS] Friedman, R., Smith, R., *The generic Torelli theorem for the Prym map.* Invent. Math. vol **74**, (1982), 473-490.

[HR] Hidalgo-Solís, L., Recillas, S., *The fibre of the Prym map in genus four.* To appear at Bolletino U.M.I.

[Ma] Masiewicki, L., *Universal properties of Prym varieties with an aplication to algebraic curves of genus five.* Trans. Amer. Math. Soc. vol **222**, (1976), 221-240.

[M] Mumford, D., *Prym varietes I.* Contributions to Analysis. Academic Press, New York. (1974).

[Rec.1] Recillas, S. *Jacobians of curves with g^1_4's are Prym's varietes of trigonal curves.* Bol. Soc. Mat. Mexicana. vol **19**, (1974), 9-13.

[Rec.2] Recillas, S. *Symmetric Cubic Surfaces and Curves of Genus 3 and 4.* Bolletino U.M.I. (7), vol **7-B** , (1993), 787-819.

[V] Verra, A. *The Fibre of the Prym Map in Genus Three.* Math. Ann. vol **276**, (1987), 433-448.

Laura Hidalgo-Solís, UAM, Unidad Iztapalapa, Departamento de Matemáticas, Av. La Purisima s/n, México D.F., México, C.P. 09340

E-mail address: `lhso@xanum.uam.mx`

Sevin Recillas-Pishmish, Instituto de Matemáticas, UNAM, Nicolás Romero No. 150, Col. Centro, Morelia, Mich., México, C.P. 5800

CIMAT, Callejón Jalisco s/n, Valenciana, Gto., México

E-mail address: `sevinrec@servidor.unam.mx`

Contemporary Mathematics
Volume **240**, 1999

From the Shift Loci to the Connectedness Loci of Complex Polynomials

Jan Kiwi

ABSTRACT. When complex polynomials, in one complex variable, are regarded as dynamical systems the set of all polynomials is subdivided according to how many critical points escape to infinity. One extreme of this subdivision is the Shift locus $\mathcal{S}_d$ while the other extreme is the Connectedness locus $\mathcal{C}_d$. This paper is an expository account of some work done by the author in **[K]** which describes how polynomials with all cycles repelling are organized in $\partial\mathcal{S}_d \cap \mathcal{C}_d$ according to their "combinatorics".

1. Introduction

In the study of complex polynomial as dynamical systems from the complex plane onto itself the orbit of the critical points play a key role. Thus, parameter space is naturally subdived according to how many critical points escape to infinity. Following Branner and Hubbard **[BH]** we work in the space of monic centered polynomials $\mathcal{P}_d \cong \mathbb{C}^{d-1}$ of degree d. That is, polynomials of the form:

$$z^d + a_{d-2}z^{d-2} + \cdots + a_0.$$

One extreme of this subdivision is the connectedness locus $\mathcal{C}_d$ formed by the polynomials $f \in \mathcal{P}_d$ with all its critical points having bounded orbits under f. Equivalently, a polynomial f lies in $\mathcal{C}_d$ if and only if its Julia set $J(f)$ is connected. The other extreme is the set of polynomials with all critical points having unbounded orbit called the shift locus $\mathcal{S}_d$.

Since quadratic polynomials have only one critical point this two sets cover the whole parameter space $\mathcal{P}_2$. The quadratic connectedness locus is better known as the Mandelbrot set $\mathcal{M}$. In the early 80's Douady and Hubbard **[DH0, DH1]**, after constructing a conformal isomorphism between the punctured disk and the quadratic shift locus proceeded to study the boundary behavior of this conformal map. In the midst of a wealth of results they showed that polynomials are carefully organized in $\partial\mathcal{M}$ according to their "combinatorics". In this paper we discuss a partial generalization of this phenomenon to parameter spaces of higher degree

1991 *Mathematics Subject Classification.* 58F23, 30D05, 39B12.

The author's research was partially supported by a "Beca Presidente de la República". The author was supported by "Fundación Andes, Chile" and "Cátedra Presidencial en Geometría" while writing this paper.

polynomials. The emphasis will be to describe where in $\partial\mathcal{S}_d \cap \mathcal{C}_d$ a polynomial with all cycles repelling and a given "combinatorics" lies.

For us, the combinatorics of a polynomial $f \in \mathcal{C}_d$ with all cycles repelling will be described by its rational lamination $\lambda_{\mathbb{Q}}(f)$ (compare with [**McM**]). The rational lamination is an equivalence relation in $\mathbb{Q}/\mathbb{Z}$ that captures the pattern in which rational external rays of f land in its Julia set $J(f)$. After discussing some preliminaries in Section 2 we proceed to introduce rational laminations and discuss some of their basic properties in Section 3.1 .

To establish a connection between parameter space and rational laminations we will use "critical portraits". Critical portraits were introduced by Fisher in [**F**]. Loosely, a critical portrait Θ (see Section 3.2) is a finite collection of finite subsets of $\mathbb{R}/\mathbb{Z}$ which abstractly captures the location of the critical points of a polynomial. Inspired by the work of Bielefield, Fisher and Hubbard [**BFH**] and Poirier [**P**], each critical portrait generates and equivalence relation $\Lambda_{\mathbb{Q}}(\Theta)$ in $\mathbb{Q}/\mathbb{Z}$ (see Section 3.2). This equivalence relation is a natural candidate to be the rational lamination of a polynomial. Critical portraits which generate candidates to be the rational lamination of a polynomial with all cycles repelling have "aperiodic kneading". Thus, in Section 3.2, a basic dichotomy is established between critical portraits having "periodic kneading" and those having "aperiodic kneading".

On the other hand in order to prescribe a location in $\partial\mathcal{S}_d \cap \mathcal{C}_d$ we cover this set by "critical portrait impressions" (Section 4). That is, motivated by Goldberg [**G**], we will show that each critical portrait defines a direction to go from the shift locus $\mathcal{S}_d$ towards the connectedness locus $\mathcal{C}_d$. The set of points in $\mathcal{C}_d$ accumulated by the sequences escaping of $\mathcal{S}_d$ in the direction of a critical portrait Θ is the impression $I_{\mathcal{C}_d}(\Theta)$ of the critical portrait Θ (see Section 4). For quadratic polynomials, critical portrait impressions coincide with the usual notion of prime end impressions of the complement of the Mandelbrot set on $\partial\mathcal{M}$.

Our main result shows that impressions of critical portraits with aperiodic kneading are formed by polynomials with all cycles repelling and a common "combinatorics" which only depends on the critical portrait. Vis-versa, a polynomial f with all cycles repelling in $\partial\mathcal{S}_d \cap \mathcal{C}_d$ can only lie in an impression of a critical portrait that generates the rational lamination of f

THEOREM 1. *Consider a map f in the impression $I_{\mathcal{C}_d}(\Theta) \subset \mathcal{C}_d$ of a critical portrait Θ.*

If Θ has aperiodic kneading then $\lambda_{\mathbb{Q}}(f) = \Lambda_{\mathbb{Q}}(\Theta)$ and all the cycles of f are repelling.

If Θ has periodic kneading then at least one cycle of f is non-repelling.

In Douady and Hubbard's results about the Mandelbrot set a crucial fact is to establish that "rational parameter rays" land. Today, simplified proofs of these landing results are available in the literature (see [**GM, Sch, M2**]). When restricted to quadratic parameter space, the first part of the Theorem above implies the Douady and Hubbard landing Theorem for "quadratic parameter rays" with strictly pre-periodic arguments. From a different viewpoint, Petersen and Ryd [**PR**] have also established generalizations to higher dimensional parameter spaces and new proofs of Douady and Hubbard's results for quadratic polynomials.

For degrees higher than two, we do not expect impressions of critical portraits with aperiodic to be formed by a single element. In contrast, for quadratic polynomials, the Mandelbrot Local Connectivity Conjecture states that every quadratic impression is a single polynomial. Recent work of Willumsen [**W**] shows that there are cubic impressions with periodic kneading which are non-trivial.

A construction due to Blanchard, Devaney and Keen [**BDK**] shows that for each critical portrait Θ its impression $I_{\mathcal{C}_d}(\Theta)$ is non-empty. It is worth mentioning that this together with the Theorem 1 gives a new proof of the Bielefield-Fisher-Hubbard [**BFH**] Theorem which shows that polynomials with all critical points pre-repelling and a given combinatorics do exist. The novelty of our proof is that using parameter space techniches we are able to bypass the use of Thurston's characterization of rational maps.

The aim of this paper is to summarize the main definitions and results leading to the proof of Theorem 1. Also, we provide a sketch of the proof in Section 5. The paper is mostly expository and the interested reader may refer to [**K**] for details.

Acknowledgments: The results summarized here form part of the author's Thesis [**K**]. The author is grateful to John Milnor for his enlightening comments and suggestions during his graduate studies.

2. Preliminaries

Here we recall some facts about polynomial dynamics. For more background material we refer the reader to [**M1**] and references therein.

Consider a monic polynomial $f : \mathbb{C} \to \mathbb{C}$ of degree d. Basic tools to understand the dynamics of f are the Green function g_f and the Böttcher map ϕ_f.

The Green function g_f measures the escape rate of points to ∞:

$$\begin{array}{rccl} g_f : & \mathbb{C} & \to & \mathbb{R}_{\geq 0} \\ & z & \mapsto & \lim \frac{\log_+ |f^{\circ n}(z)|}{d^n} \end{array}$$

It is a well defined continuous function which vanishes on the filled Julia set $K(f)$ and satisfies the functional relation:

$$g_f(f(z)) = dg_f(z).$$

Moreover, g_f is positive and harmonic in the basin of infinity

$$\Omega(f) = \{z : f^{\circ n}(z) \to \infty\}.$$

In $\Omega(f)$, the derivative of g_f vanishes at z if and only if z is a pre-critical point of f. In order to avoid confusions, we say that z is a **singularity** of g_f.

The Böttcher map ϕ_f conjugates f with $z \mapsto z^d$ in a neighborhood of ∞. The germ of ϕ_f at ∞ is unique up to conjugacy by $z \mapsto \zeta z$ where ζ is a $(d-1)st$ root of unity. Since f is monic we can normalize ϕ_f to be asymptotic to the identity:

$$\frac{\phi_f(z)}{z} \to 1$$

as $z \to \infty$. Observe that near infinity, $g_f(z) = \log |\phi_f(z)|$.

For the purpose of simplicity, we make a distinction according to whether the Julia set $J(f)$ is connected or disconnected.

2.1. Connected Julia sets. The Julia set $J(f)$ is connected if and only if all the critical points of f are non-escaping. That is, the forward orbit of the critical points remains bounded. Thus, g_f has no singularities in the basin of infinity $\Omega(f)$. Moreover, the Böttcher map extends to the basin of infinity $\Omega(f)$,

$$\phi_f : \Omega(f) \to \mathbb{C} \setminus \overline{\mathbb{D}}$$

and $\phi_f(f(z)) = (\phi_f(z))^d$ for $z \in \Omega(f)$. Furthermore,

$$g_f(z) = \log_+ |\phi_f(z)|$$

for $z \in \Omega(f)$.

An **external ray** R_f^t is the pre-image of radial line $(1, \infty)e^{2\pi i t}$ under ϕ_f, i.e.

$$R_f^t = \phi_f^{-1}((1, \infty)e^{2\pi i t}).$$

Thus, external rays are curves that run from infinity towards the Julia set $J(f)$. If R_f^t has a well defined limit $z \in J(f)$ as it approaches the Julia set $J(f)$ we say that R_f^t **lands** at z.

External rays are parameterized by the circle $\mathbb{R}/\mathbb{Z}$ and f acts on external rays as multiplication by d. (i.e. $f(R_f^t) = R_f^{dt}$). A ray R_f^t is said to be rational if $t \in \mathbb{Q}/\mathbb{Z}$. Rational rays can be either periodic or pre-periodic according to whether t is periodic or pre-periodic under $m_d : t \mapsto dt$. A periodic ray always lands at a repelling or parabolic periodic point. A pre-periodic ray lands at a pre-repelling or pre-parabolic point [**DH1**]. Conversely, putting together results of Douady, Hubbard, Sullivan and Yoccoz [**H, M1**], we have the following:

THEOREM 2.1. *Let z be a parabolic or repelling periodic point in a connected Julia set $J(f)$. Then there exists at least one periodic ray landing at z. Moreover, all the rays that land at z are periodic of the same period.*

2.2. Disconnected Julia sets. A polynomial f has disconnected Julia set $J(f)$ if and only if some critical point of f lies in the basin of infinity $\Omega(f)$. In this case, the Böttcher map does not extend to all of $\Omega(f)$. It extends, along flow lines of $-\text{grad} g_f$ to the basin of infinity under the gradient flow $\text{grad} g_f$. Following Levin and Sodin [**LS**], the **reduced basin of infinity** $\Omega^*(f)$ is the basin of infinity under the gradient flow $\text{grad} g_f$. Now

$$\phi_f : \Omega^*(f) \to U_f \subset \mathbb{C} \setminus \overline{\mathbb{D}}$$

is a conformal isomorphism from $\Omega^*(f)$ onto a star-like (around ∞) domain U_f. A maximal flow line of $\text{grad} g_f$ in $\Omega^*(f)$ is an **external radius**. An external radius maps into an external radius by f. Thus, $f(\Omega^*(f)) \subset \Omega^*(f)$.

External radii are parameterized by $\mathbb{R}/\mathbb{Z}$. More precisely, for $t \in \mathbb{R}/\mathbb{Z}$ let $(r, \infty)e^{2\pi i t}$ be the maximal portion of $(1, \infty)e^{2\pi i t}$ contained in U_f. The external radius R_f^{*t} with argument t is

$$R_f^{*t} = \phi_f^{-1}((r, \infty)e^{2\pi i t}).$$

As one follows the external radius R_f^{*t} from ∞ towards the Julia set $J(f)$ one might hit a singularity z of g_f or not. In the first case, $r > 1$ and we say that R_f^{*t} **terminates** at z. In the second case, $r = 1$ and R_f^{*t} is in fact the smooth external ray R_f^t with argument t. Notice that, from the point of view of the gradient flow, an external radius which terminates at a singularity z is an unstable manifold of z.

Under iterations of f, each point z in the basin of infinity $\Omega(f)$ eventually maps to a point in the reduced basin of infinity $\Omega^*(f)$. Say that $f^{\circ n}(z) \in \Omega^*(f)$ and that the local degree of $f^{\circ n}$ at z is k. After a conformal change of coordinates, the gradient flow lines nearby z are the pre-image under $w \mapsto w^k$ of the horizontal flow lines near the origin. Thus, at a singularity z of g_f, there are exactly k local unstable and k local stable manifolds which alternate as one goes around z. A local unstable manifold is contained in $\Omega^*(f)$ if and only if it is part of an external radius that terminates at z.

Now let $\theta_1, \ldots, \theta_l$ be the arguments of the external radii that terminate at critical points of f. Recall that every pre-critical point of f is a singularity of g_f, the external radii with arguments in

$$\Sigma = \bigcup_{n \geq 0} m_d^{-n}(\{\theta_1, \ldots, \theta_l\})$$

also terminate at a singularities. Since every singularity is a pre-critical point we have smooth external rays defined for arguments in $\mathbb{R}/\mathbb{Z} \setminus \Sigma$. Following Goldberg and Milnor, for $t \in \mathbb{R}/\mathbb{Z}$ let

$$R_f^{t^{\pm}} = \lim_{s \to t^{\pm}} R_f^s.$$

If $t \notin \Sigma$ then $R_f^{t^{\pm}}$ coincide, and we say that R_f^t is a smooth external ray. If $t \in \Sigma$ then $R_f^{t^{\pm}}$ do not agree, and we say that they are **non-smooth or bouncing rays** with argument t.

3. Rational Laminations

3.1. Rational laminations. The pattern in which rational external rays land in a connected Julia set is captured by the rational lamination:

DEFINITION 3.1. *Consider a polynomial f with connected Julia set $J(f)$. The equivalence relation $\lambda_{\mathbb{Q}}(f)$ in $\mathbb{Q}/\mathbb{Z}$ that identifies $t, t' \in \mathbb{Q}/\mathbb{Z}$ if the external rays R_f^t and $R_f^{t'}$ land at a common point is called the* **rational lamination** *of f.*

REMARK 3.2. We work with the definition of rational lamination which appears in [**McM**]. The word "lamination" corresponds to the usual representation of this equivalence relation in the unit disk $\mathbb{D}$. That is, each equivalence class A is represented as the convex hull (with respect to the Poincaré metric) of $A \subset \mathbb{R}/\mathbb{Z} \cong \partial\mathbb{D}$. The use of "laminations" to represent the pattern in which external rays of a polynomial land or can land was introduce by Thurston in [**Th**] (also see [**D**]).

For example, the external rays with arguments

$$11/216, 17/216, 83/216, 89/216, 155/216, 161/216$$

land at a pre-periodic critical point $z = 0$ of the polynomial

$$f : z \mapsto z^3 + 0.2203 + 1.1863I.$$

Hence,

$$\{11/216, 17/216, 83/216, 89/216, 155/216, 161/216\}$$

is an equivalence class of $\lambda_{\mathbb{Q}}(f)$ (see Figure 1). Also, the external rays with arguments $11/72, 51/72$ land at the critical value $f(0)$. Thus, $\{11/72, 51/72\}$ is another equivalence class of $\lambda_{\mathbb{Q}}(f)$.

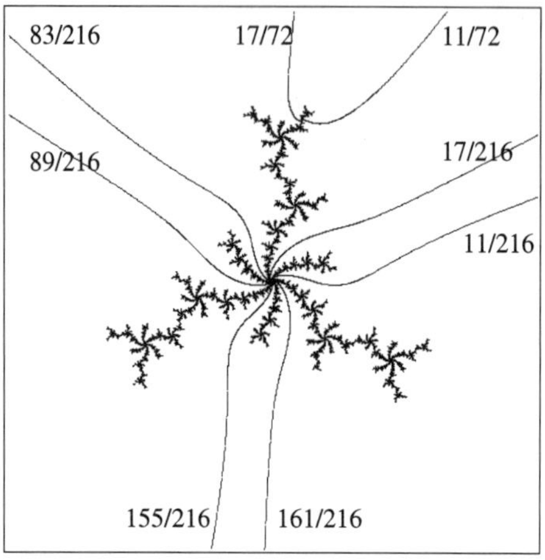

FIGURE 1

We will refer to an equivalence class of $\lambda_{\mathbb{Q}}(f)$ as a **rational type**. We denote the arguments of the rational external rays landing (if any) at a point z by $A(z)$ and say that $A(z)$ is the **type** of z. Let us summarize the basic properties of rational types $A(z)$.

By the landing Theorem 2.1:

• $A(z)$ is a finite set.

Continuity of f implies that:

• (invariance) $d \cdot A(z) = A(f(z))$

Since f is orientation preserving it can be shown that

• (consecutive preserving) If t_1, t_2 are consecutive elements of $A(z)$ then dt_1, dt_2 are consecutive elements of $A(f(z))$.

Now, since external rays are disjoint the Jordan curve Theorem implies that

• (unlinked) If $z \neq w$ then $A(z)$ is contained in a connected component of $\mathbb{R}/\mathbb{Z} \setminus A(w)$.

Rational types inherit this four basic properties but also have some extra important features contained in the statement of the following:

PROPOSITION 3.3. *Let $\lambda_{\mathbb{Q}}(f)$ be the rational lamination of a polynomial f with all cycles repelling and connected Julia set $J(f)$. Then:*
(R1) $\lambda_{\mathbb{Q}}(f)$ is a closed equivalence relation in $\mathbb{Q}/\mathbb{Z}$
(R2) Every $\lambda_{\mathbb{Q}}(f)$-equivalence class A is a finite set.
(R3) If A_1 and A_2 are distinct equivalence classes then A_1 is contained in a connected component of $\mathbb{R}/\mathbb{Z} \setminus A_2$.
(R4) If A is an equivalence class then $d \cdot A$ is an equivalence class.
(R5) If (t_1, t_2) is a connected component of $\mathbb{R}/\mathbb{Z} \setminus A$ where A is an equivalence class then (dt_1, dt_2) is a connected component of $\mathbb{R}/\mathbb{Z} \setminus d \cdot A$.

Moreover, $\lambda_{\mathbb{Q}}(f)$ is maximal with respect to properties (R2) and (R3).

The fact that $\lambda_{\mathbb{Q}}(f)$ is closed and maximal comes from the abundance of non-trivial equivalence classes of $\lambda_{\mathbb{Q}}(f)$. That is, polynomials with all cycles repelling have a wealth of non-trivial rational types. More precisely, the following Lemma shows that each rational type is well *approximated* by other rational types:

LEMMA 3.4. *Consider a polynomial f with all cycles repelling and connected Julia set $J(f)$. Let $A(z) = \{t_0, \ldots, t_{p-1}\} \subset \mathbb{Q}/\mathbb{Z}$ be a rational type (subscripts respecting cyclic order and mod p). For all sufficiently small $\epsilon > 0$, there exists rational types $A(w_0), \ldots, A(w_{p-1})$ such that $A(w_i)$ has elements both in $(t_i, t_i + \epsilon)$ and $(t_{i+1} - \epsilon, t_{i+1})$. Moreover, $w_0, \ldots, w_{p-1}$ can be chosen so that they do not belong to the grand orbit of a critical point.*

For a proof of these results see Chapter 3 of [**K**].

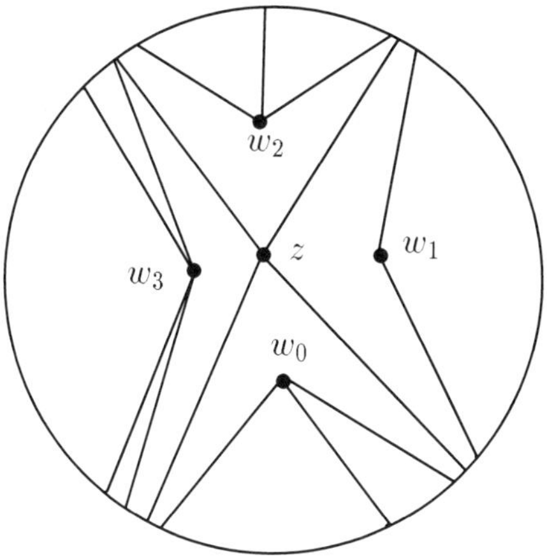

FIGURE 2. Illustration of Lemma 3.4

3.2. Critical portraits and Rational laminations. Critical points play a key role in the understanding of the dynamics of a polynomial. In order to capture the location of the critical points we will use critical portraits. Critical portraits were introduced by Fisher [**F**] to study critically pre-repelling maps and, since then, widely used in the literature to capture the location of the critical points (e.g. [**BFH, P, GM, G**]).

Here, we start with an abstract discussion about critical portraits. Afterwards, given a critical portrait we will construct an equivalence relation in $\mathbb{Q}/\mathbb{Z}$ which is a natural candidate to be the rational lamination of a polynomial.

Two subsets A and A' of $\mathbb{R}/\mathbb{Z}$ are **unlinked** if A is contained in a connected component of $\mathbb{R}/\mathbb{Z} \setminus A'$.

DEFINITION 3.5 (Critical Portraits). *A collection $\Theta = \{\Theta_1, \ldots, \Theta_m\}$ of finite subsets of $\mathbb{R}/\mathbb{Z}$ is called a* **critical portrait** *of degree d the following hold.*
(CP1) For every j, $|\Theta_j| \geq 2$ and $|m_d(\Theta_j)| = 1$,
(CP2) $\Theta_1, \ldots, \Theta_m$ are pairwise unlinked,
(CP3) $\sum(|\Theta_j| - 1) = d - 1$.

Critical portraits chop the circle into unlinked classes:

DEFINITION 3.6. *We say that $t, t' \in \mathbb{R}/\mathbb{Z}$ are $\Theta = \{\Theta_1, \ldots, \Theta_m\}$-***unlinked equivalent** *if $\{t, t'\}, \Theta_1, \ldots, \Theta_m$ are pairwise unlinked.*

EXAMPLE 1. Consider the cubic critical portrait

$$\Theta = \{\{1/3, 2/3\}, \{1/9, 7/9\}\}.$$

The Θ-unlinked classes are $L_1 = (7/9, 1/9)$, $L_2 = (1/9, 1/3) \cup (2/3, 7/9)$ and $L_3 = (1/3, 2/3)$.

EXAMPLE 2. Consider the cubic critical portrait

$$\Theta = \{\{11/216, 83/216, 161/216\}\}.$$

The Θ-unlinked classes are $L_1 = (11/216, 83/216)$, $L_2 = (83/216, 161/216)$ and $L_3 = (161/216, 11/216)$.

In general, given a critical portrait Θ, there are exactly d Θ-unlinked classes $L_1, \dots, L_d$. Moreover, each unlinked class L_j is the union of open intervals with total length $1/d$. Now symbolic dynamics of $m_d : t \to dt \bmod 1$ with respect to this partition give us the right and left itineraries $\text{itin}^{\pm}_{\Theta}$. That is, we let:

$$\begin{array}{rccc} \text{itin}^{\pm}_{\Theta}: & \mathbb{R}/\mathbb{Z} & \to & \{1, \dots, d\}^{\mathbb{N}\cup\{0\}} \\ & t & \mapsto & (j_0, j_1, \dots) \end{array}$$

if, for each $n \geq 0$, there exists $\epsilon > 0$ such that $(d^n t, d^n t \pm \epsilon)) \subset L_{j_n}$.

This allow us to generate from each critical portrait an equivalence relation in $\mathbb{Q}/\mathbb{Z}$:

DEFINITION 3.7. *Given a critical portrait Θ we say that two arguments t, t' in $\mathbb{Q}/\mathbb{Z}$ are $\Lambda_{\mathbb{Q}}(\Theta)$-equivalent if and only if there exists $t = t_1, \dots, t_n = t'$ such that one of the two itineraries $\text{itin}^{\pm}_{\Theta}(t_i)$ coincides with one of the two itineraries $\text{itin}^{\pm}_{\Theta}(t_{i+1})$ for $i = 1, \dots, n-1$.*

EXAMPLE 1. (continued) For $\Theta = \{\{1/3, 2/3\}, \{1/9, 7/9\}\}$:

$$\{1/9, 2/9, 7/9, 8/9\}$$

is a $\Lambda_{\mathbb{Q}}(\Theta)$-equivalence class. In fact,

$$\text{itin}^{+}_{\Theta}(7/9) = \text{itin}^{-}_{\Theta}(8/9) = 13111...$$

$$\text{itin}^{+}_{\Theta}(2/9) = \text{itin}^{-}_{\Theta}(7/9) = 22111...$$

$$\text{itin}^{+}_{\Theta}(8/9) = \text{itin}^{-}_{\Theta}(1/9) = 12111...$$

The abstract equivalence relation $\Lambda_{\mathbb{Q}}(\Theta)$ generated by a critical portrait is a candidate to be the rational lamination of a polynomial. In fact, it is not difficult to verify that $\Lambda_{\mathbb{Q}}(\Theta)$ has satisfies the conditions (R1) through (R5) of Proposition 3.3 (see Chapter 4 in [**K**]).

We are mainly interest on polynomials with all cycles repelling. A polynomial f with all cycles repelling has exactly d^p periodic points of period dividing p. Hence, $\lambda_{\mathbb{Q}}(f)$ must have d^p classes of period dividing p. Now a simple counting argument shows that $\Lambda_{\mathbb{Q}}(\Theta)$ has exactly d^p classes of period dividing p if and only if the elements of Θ do not have periodic itinerary. This motivates the following distinction.

DEFINITION 3.8. *We say that a critical portrait $\Theta = \{\Theta_1, \dots, \Theta_m\}$ has* **periodic kneading** *if there exists an element $\theta \in \Theta_1 \cup \dots \cup \Theta_m$ such that $\text{itin}^{+}_{\Theta}(\theta)$ or $\text{itin}^{-}_{\Theta}(\theta)$ is periodic under the one sided shift. Otherwise, we say that Θ has* **aperiodic kneading**.

The following Lemma characterizes which critical portraits generate the rational lamination of a given polynomial with all cycles repelling.

LEMMA 3.9. *Let $\Theta = \{\Theta_1, \ldots, \Theta_m\}$ be a critical portrait and f be a polynomial with all cycles repelling. The identity $\Lambda_{\mathbb{Q}} = \lambda_{\mathbb{Q}}(f)$ holds if and only if each Θ_i is either contained in a $\lambda_{\mathbb{Q}}(f)$-equivalence class or Θ_i is unlinked with all $\lambda_{\mathbb{Q}}(f)$-equivalence classes. In this case, Θ has aperiodic kneading.*

Given a polynomial f it is not difficult to show that there exist critical portraits Θ which satisfy the conditions of the Lemma above. When the Julia set $J(f)$ is locally connected these critical portraits are formed by the arguments of the external rays landing at the critical points. For example the critical portraits which generate the rational lamination of the polynomial shown in Figure 1 are:

$$\{11/216, 83/216, 155/216\}$$
$$\{17/216, 89/216, 161/216\}$$
$$\{\{17/216, 89/216\}, \{11/216, 155/216\}\}$$
$$\{\{89/216, 161/216\}, \{11/216, 83/216\}\}$$
$$\{\{17/216, 161/216\}, \{83/216, 155/216\}\}$$

Other example is given by the polynomial $g : z \mapsto z^3 - 9/4z + \sqrt{3}/4$ whose Julia set and rays landing at its critical points are shown in Figure 3. In this case, the critical portraits which generate $\lambda_{\mathbb{Q}}(g)$ are $\{\{1/3, 2/3\}, \{1/9, 7/9\}\}$ and $\{\{1/3, 2/3\}, \{2/9, 8/9\}\}$.

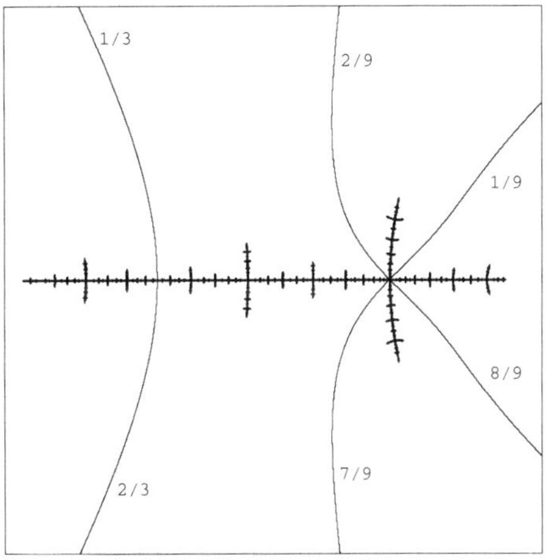

FIGURE 3

4. Impressions

In parameter space, following Branner and Hubbard [**BH**], we work in the set $\mathcal{P}_d \cong \mathbb{C}^{d-1}$ of monic centered polynomials of degree d. Namely, polynomials of the form:

$$z^d + a_{d-2}z^{d-2} + \cdots + a_0.$$

Parameter space $\mathcal{P}_d$ is stratified according to how many critical points escape to ∞. One extreme is the connectedness locus $\mathcal{C}_d \subset \mathcal{P}_d$, which is the set of polynomials f that have connected Julia set $J(f)$. Equivalently, all the critical points of f are non-escaping. The other extreme is the shift locus $\mathcal{S}_d \subset \mathcal{P}_d$, formed by the polynomials f that have all their critical points escaping. Here, we prepare ourselves to explore the set $\partial\mathcal{S}_d \cap \mathcal{C}_d$ where these two extremes meet.

The connectedness locus $\mathcal{C}_d$ is compact, connected and cellular (see [**DH0, BH, La**]). For $d \geq 3$, $\mathcal{C}_d$ is known not to be locally connected [**La**]. In contrast, the quadratic connectedness locus, better known as the Mandelbrot set $\mathcal{M}$, is conjectured to be locally connected (see [**DH1**]).

The dynamics of a polynomial f in the shift locus $\mathcal{S}_d$ is completely understood. In fact, f has a Cantor set as Julia set $J(f)$ and, f acts on $J(f)$ as a hyperbolic dynamical system which is topologically conjugate to the one sided shift in d symbols. The shift locus $\mathcal{S}_d$ is open, connected and unbounded. For $d \geq 3$, $\mathcal{S}_d$ has a highly non-trivial topology. More precisely, its fundamental group is infinitely generated [**BDK**]. In contrast, after Douady and Hubbard [**DH0**], the quadratic shift locus $\mathcal{S}_2 = \mathbb{C} \setminus \mathcal{M}$ is conformally isomorphic to the complement $\mathbb{C} \setminus \overline{\mathbb{D}}$ of the unit disk.

Here, we introduce objects that will allow us to explore the portion of $\partial\mathcal{S}_d$ contained in $\mathcal{C}_d$. More precisely, we define what it means to go from the shift locus $\mathcal{S}_d$ towards the connectedness locus $\mathcal{C}_d$ in a given direction. Each direction will be specified by a "critical portrait" and will determine an "impression" in the connectedness locus $\mathcal{C}_d$.

In the next Section, we are going to show that the "combinatorics" of a polynomial f in $\partial\mathcal{S}_d \cap \mathcal{C}_d$ is completely determined by the "impression(s)" to which f belongs, provided that f has all its cycles repelling.

For quadratic polynomials, we have a dynamically defined conformal isomorphism from $\mathcal{S}_2 = \mathbb{C} \setminus \mathcal{M}$ onto $\mathbb{C} \setminus \overline{\mathbb{D}}$ (see [**DH0, DH1**]). This map provides us with parameter rays and a dynamical parameterization of the prime end impressions of $\mathcal{S}_2$ in $\partial\mathcal{M}$. For higher degrees, we need to overcome the difficulties that stem from the non-trivial topology of the shift locus. Motivated by Goldberg [**G**], it is better to work with a dense subset of $\mathcal{S}_d$ where the critical points are easily located by the Böttcher coordinates. That is, the polynomials f such that each critical point of f is "visible" from ∞, in the sense defined below. Recall that an external radius is a gradient $\mathrm{grad} g_f$ flow line that reaches ∞ (see Section 2).

DEFINITION 4.1 (Visible Shift Locus). *Consider a polynomial f in the shift locus $\mathcal{S}_d$. We say that f belongs to the* **visible shift locus** $\mathcal{S}_d^{vis}$ *if for each critical point c of f:*

(a) there are exactly k external radii terminating at c, where k is the local degree of f at c;

(b) the critical value $f(c)$ belongs to an external radius.

Our definition has a slight difference with Goldberg's definition of the "generic shift locus". Thus, although we use a different name, there is a strong overlap with the ideas found in [**G**].

We keep track of the external radii that terminate at the critical points:

DEFINITION 4.2. *Let f be a polynomial in the visible shift locus $\mathcal{S}_d^{vis}$ with critical points $c_1, \ldots, c_m$ and $\Theta_i \subset \mathbb{R}/\mathbb{Z}$ be the set formed by the arguments of the external radii that terminate at c_i. We say that $\Theta(f) = \{\Theta_1, \ldots, \Theta_m\}$ is the* **critical portrait of** *f.*

It is not difficult to verify that $\Theta(f)$ is in fact a critical portrait in the sense of Definition 3.5.

For $f \in \mathcal{S}_d^{vis}$, the external radii which terminate at the critical points cut the plane into d components. This situation is captured in the circle at infinity, by the Θ-unlinked classes. That is while the external radii terminating at the critical points of f chop the plane into d components, the critical portrait $\Theta(f)$ chops the circle into d unlinked components $L_1, \ldots, L_d$. Intuitively, for polynomials in $\mathcal{S}_d^{vis}$ close to f, this partition does not change too much. Formally, we introduce a topology on the set of all critical portraits:

DEFINITION 4.3. *Let $\mathcal{A}_d$ be the set formed by all critical portraits endowed with the* **compact-unlinked** *topology which is generated by the sub-basis formed by*

$$V_X = \{\Theta \in \mathcal{A}_d : X \subset L_\Theta\}$$

where X is a closed subset of $\mathbb{R}/\mathbb{Z}$ and L_Θ is a Θ-unlinked class.

With this topology the map $f \in \mathcal{S}_d^{vis} \mapsto \Theta(f) \in \mathcal{A}_d$ is continuous. Moreover, a construction of Blanchard, Devaney and Keen [**BDK**] shows that this map is onto. The exact version of this construction needed to establish epijectivity can be found in Goldberg's paper [**G**].

The set of quadratic critical portraits $\mathcal{A}_2$ is homeomorphic to $\mathbb{R}/\mathbb{Z}$. The homeomorphism is given by $\{\theta, \theta + 1/2\} \mapsto 2\theta$. The set of cubic critical portraits $\mathcal{A}_3$ can be obtained from a Möebius band M as follows. Parameterize the boundary of M by $\mathbb{R}/\mathbb{Z}$ and identify $\beta \in \mathbb{R}/\mathbb{Z}$ with $\beta + 1/3$ and $\beta + 2/3$. The resulting topological space is homeomorphic to $\mathcal{A}_3$. In general, for $d \geq 3$, $\mathcal{A}_d$ is compact and connected but it is not a manifold. It can be shown that the set of critical portraits $\mathcal{A}_d$ is homeomorphic to the subset $\mathcal{E} \subset \mathcal{S}_d$ of polynomials f such that all the critical points c of f escape to ∞ at a fixed rate $\rho = g_f(c)$. The topology of $\mathcal{E}$, for cubic polynomials, has been previously described by Branner and Hubbard in [**BH**].

Now, the topology in the set of critical portraits allows us to introduce the impression of a critical portrait Θ in the connectedness locus $\mathcal{C}_d$:

DEFINITION 4.4. *Let Θ be a critical portrait. We say that f belongs to the* **impression** *$I_{\mathcal{C}_d}(\Theta)$ of the critical portrait Θ if there exists a sequence of maps $f_n \in \mathcal{S}_d^{vis}$ converging to f such that the corresponding critical portraits $\Theta(f_n)$ converge to Θ.*

REMARK 4.5. For quadratic polynomials, $I_{\mathcal{M}}(\{\theta, \theta/2\})$ is a prime end impression. More precisely, it is the prime end impression corresponding to 2θ under the Douady-Hubbard map $\Phi : \mathbb{C} \setminus \mathcal{M} \to \mathbb{C} \setminus \overline{\mathbb{D}}$.

From the compactness of $\mathcal{A}_d$ and the density of $\mathcal{S}_d^{vis}$ in $\mathcal{S}_d$ it follows that

$$\bigcup_{\Theta \in \mathcal{A}_d} I_{\mathcal{C}_d}(\Theta) = \partial \mathcal{S}_d \cap \mathcal{C}_d.$$

That is, the critical portrait impressions cover all of $\partial \mathcal{S}_d \cap \mathcal{C}_d$. Moreover, it can be shown that each critical portrait impression $I_{\mathcal{C}_d}(\Theta)$ is a non-empty, compact and connected subset of $\partial \mathcal{S}_d \cap \mathcal{C}_d$.

5. Combinatorial Continuity

In this Section we sketch the proof of Theorem 1. Roughly, the idea is that for polynomials in the visible shift locus external rays land in a pattern very closely related to that of $\Lambda_{\mathbb{Q}}(\Theta)$. Then we show that, as we approach the connectedness locus, this landing pattern is preserved in the limit.

The next well known Lemma (e.g. see [**G**]) describes how external rays land in the visible shift locus.

LEMMA 5.1. *Consider f in the visible shift locus $\mathcal{S}_d^{vis}$ with critical portrait $\Theta = \Theta(f)$. Two external rays $R_f^{t^\epsilon}$ and $R_f^{s^\delta}$ land at a common point if and only if* $\operatorname{itin}_\Theta^\epsilon(t) = \operatorname{itin}_\Theta^\delta(t)$ *where $\epsilon, \delta = \pm$.*

A key step is to prove that impressions of critical portraits with aperiodic kneading are formed by polynomials with all cycles repelling. All the cycles of a polynomial in the shift locus are repelling. As we approach the connectedness locus in an aperiodic direction we need an obstruction for two distinct periodic orbits to coalesce and become a parabolic periodic orbit. This obstruction will eventually be a consequence of the following Separation Lemma which shows that two periodic points of a polynomial $f \in \mathcal{S}_d^{vis}$ with aperiodic kneading are separated by pre-periodic rays.

LEMMA 5.2 (Separation). *Consider a critical portrait Θ with aperiodic kneading and let A_1 and A_2 be two distinct periodic $\Lambda_{\mathbb{Q}}(\Theta)$-equivalence classes. Then there exists a strictly pre-periodic $\Lambda_{\mathbb{Q}}(\Theta)$-equivalence class C such that A_1 and A_2 lie in different connected components of $\mathbb{R}/\mathbb{Z} \setminus C$. Moreover, C can be chosen such that, for all $n \geq 0$, $m_d^{\circ n}(C)$ is contained in a Θ-unlinked class.*

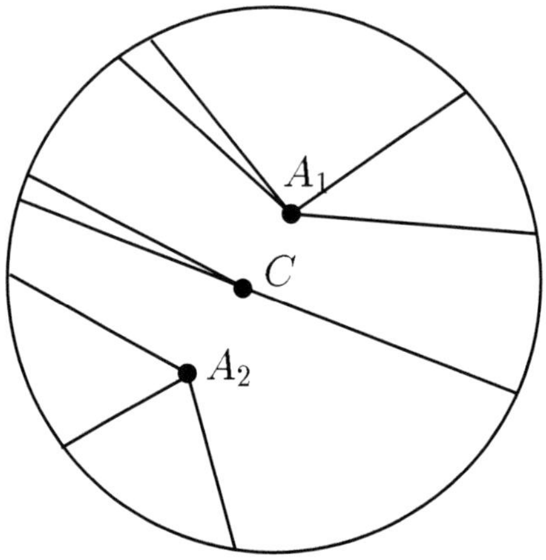

FIGURE 4. Illustration of Lemma 5.2

To show that the separation given by the above lemma persists in the limit we consider the lim sup of external rays in the sense of the Hausdorff topology on compacts subsets of $\hat{\mathbb{C}}$. More precisely, we say that $z \in \limsup \overline{R}^t_{f_n}$ if every neighborhood of z intersects non-trivially infinitely many $\overline{R}^t_{f_n}$.

LEMMA 5.3. *Let Θ be a critical portrait with aperiodic kneading. Consider a sequence $f_n \in \mathcal{S}_d^{vis}$ such that $\Theta(f_n) \to \Theta$ and $f_n \to f \in \mathcal{C}_d$.*

If $t \in \mathbb{Q}/\mathbb{Z}$ is periodic and $z \in (\limsup \overline{R}^t_{f_n}) \cap J(f)$ then z is periodic under f.

If $t \in \mathbb{Q}/\mathbb{Z}$ is such that t is not periodic and $z \in (\limsup \overline{R}^t_{f_n}) \cap J(f)$ then z is not periodic under f.

The last ingredient in the proof of Theorem 1 is a Lemma due to Goldberg and Milnor [**GM**]:

LEMMA 5.4. *Consider $f_0 \in \mathcal{P}_d$ such that $J(f_0)$ is connected. Assume that $R^t_{f_0}$ is a smooth external ray which lands at a pre-repelling or repelling periodic point z_{f_0}. Also, assume that z_{f_0} is not a pre-critical point. Then, for any f sufficiently close to f_0, the external ray R^t_f is smooth and lands at the analytic continuation z_f of z_{f_0}.*

PROOF OF THEOREM 1: Consider a critical portrait Θ with aperiodic kneading and a polynomial f in its impression $I_{\mathcal{C}_d}(\Theta)$. So let f_n be a sequence in the visible shift locus such that:

$$f_n \to f$$
$$\Theta(f_n) \to \Theta.$$

The next two claims show that f must have all cycles repelling:

Claim 1: f does not have a parabolic cycle.

Proof of Claim 1: By contradiction, suppose that z is a parabolic periodic point of f. It follows that there exists distinct repelling periodic points $z_1(n)$ and $z_2(n)$ of f_n that converge to z. Also, the periods of $z_1(n)$ and $z_2(n)$ divide a fixed number p. Since Θ has aperiodic kneading the itinerary of periodic elements of $\mathbb{Q}/\mathbb{Z}$ vary continuously. In particular, for Θ' in a sufficiently small neighborhood of Θ the periodic classes of $\Lambda_{\mathbb{Q}}(\Theta)$ with period dividing p coincide with those of $\Lambda_{\mathbb{Q}}(\Theta')$. After passing to a subsequence, we may assume that the type of $z_1(n)$ is A_1 and the type of $z_2(n)$ is A_2, where A_1 and A_2 are distinct equivalence classes of $\Lambda_{\mathbb{Q}}(\Theta)$. By Lemma 5.2 we know that there exists a strictly pre-periodic class C of $\Lambda_{\mathbb{Q}}(\Theta)$ such that it separates A_1 and A_2 (i.e. A_1 and A_2 lie in different connected components of $\mathbb{R}/\mathbb{Z} \setminus C$). Recall that none of the elements of C and its forward orbit participate of Θ. Hence, for Θ' in a sufficiently small neighborhood of Θ, the elements of C are also identified by $\Lambda_{\mathbb{Q}}(\Theta')$. Thus, for n large the external rays of f_n with arguments in C together with their landing points form a connected set Γ_n. Now the periodic points $z_1(n)$ and $z_2(n)$ lie in different connected components of $\mathbb{C} \setminus \Gamma_n$. Passing to the limit, $z \in \limsup \Gamma_n$ and therefore $z \in \limsup R^t_{f_n}$, for some $t \in C$. This contradicts Lemma 5.3 and shows that f cannot have parabolic cycles.

Claim 2: f does not have irrationally neutral cycles.

Proof of Claim 2: Again by contradiction we suppose that z is an irrationally neutral periodic point of f with period p. There exists a sequence $z(n)$ of periodic points of f_n that converge to z. As above, after passing to a subsequence, we may assume that $z(n)$ has type $A \subset \mathbb{Q}/\mathbb{Z}$, for all n. Pick an element of $t \in A$ and notice that R^t_f must land at a repelling periodic point w. Hence, for n large enough, $R^t_{f_n}$ has to land in the analytic continuation of w which is not $z(n)$.

Claim 3: If f is a polynomial with all cycles repelling and $f \in I_{\mathcal{C}_d}(\Theta)$ for some critical portrait $\Theta \in \mathcal{A}_d$ then $\Lambda_{\mathbb{Q}}(\Theta) = \lambda_{\mathbb{Q}}(f)$.

It follows from Lemma 3.9 that Θ has aperiodic kneading.

Proof of Claim 3: Consider f_n in the visible shift locus such that

$$f_n \to f$$

and

$$\Theta(f_n) \to \Theta = \{\Theta_1, \dots, \Theta_m\}$$

Now we show that each Θ_i is either contained in a $\lambda_{\mathbb{Q}}(f)$-class or it is unlinked with all $\lambda_{\mathbb{Q}}(f)$-classes. If Θ_i has elements in two distinct equivalence classes of $\lambda_{\mathbb{Q}}(f)$ then Θ_i is linked with the rational type $A(w)$ of a point w which is not in the grand orbit of a critical point. For Θ' close to Θ, we also have that $C = A(w)$ has points in two different Θ'-unlinked classes. In view of Lemma 5.4, for n large enough, the external rays of f_n with arguments in C are smooth and land at a common point. Therefore, they intersect the external radii of f_n terminating at some critical point. This is impossible. A similar situation occurs when $\Theta_i \subset \mathbb{R}/\mathbb{Z} \setminus \mathbb{Q}/\mathbb{Z}$ is linked with a $\lambda_{\mathbb{Q}}(f)$-class. By Lemma 3.9, $\Lambda_{\mathbb{Q}}(\Theta) = \lambda_{\mathbb{Q}}(f)$. □

References

[BFH] B. Bielifield, Y. Fisher. J. H Hubbard The classification of critically preperiodic polynomials as dynamical systems. *Journal AMS*, 5:721–762, 1992.

[BDK] P. Blanchard, R. Devaney, and L. Keen. The dynamics of complex polynomials and automorphism of the shift. *Invent. Math*, 104:545–580, 1991.

[BH] B. Branner and J. H. Hubbard. The iteration of cubic polynomials. part i: The global topology of parameter space. *Acta Math.*, 160:143 –206, 1988.

[DH0] A. Douady and J. H. Hubbard. Itération des polynomes quadratiques complexes. *C. R. Acad. Sci. Paris Ser. I Math.*, 294:123–126, 1982.

[DH1] A. Douady and J. H. Hubbard. Étude dynamique des polynomes complexes I, II. *Publ. Math. Orsay*, 1984 - 1985.

[D] A. Douady. Description of compact sets in C. In *Topological Methods in Modern Mathematics.* Publish or Perish, 1993.

[F] Y. Fisher. -. PhD thesis, Cornell, 1989.

[GM] L. R. Goldberg and J. Milnor. Fixed points of polynomial maps II: Fixed point portraits. *Ann. Scient. École Norm. Sup.*, 4e *série*, 26:51–98, 1993.

[G] L. R. Goldberg. On the multiplier of a repelling fixed point. *Invent. Math*, pages 85–108, 1994.

[H] J. H. Hubbard. Local connectivity of Julia sets and bifurcation loci: three theorems of J.-C. Yoccoz. In *Topological Methods in Modern Mathematics.* Publish or Perish, 1993.

[K] J. Kiwi. *Rational rays and Critical portraits of complex polynomials.* PhD thesis, SUNY at Stony Brook, 1997 (available as SUNY-IMS Prepint 1997/15).

[La] P. Lavaurs. *Systemes dynamiques holomorphes: explosion de points périodiques paraboliques.* PhD thesis, Université Paris-Sud, 1989.

[LS] G. Levin and M. Sodin. Polynomialsm with disconnected julia sets and green maps. *Preprint No 23, The Hebrew University of Jerusalem*, 1991.

[McM] C. T. McMullen. *Complex Dynamics and Renormalization.* Princeton University Press, 1994.

[M1] J. Milnor. Dynamics in one complex variable: Introductory lectures. Stony Brook IMS Preprint 1990/5.

[M2] J. Milnor. Periodic orbits, external rays and the Mandelbrot set. Manuscript, 1995.

[PR] C.L. Petersen and G. Ryd. Convergence of Rational Rays in Parameter Spaces. Preprint, 1998.

[P] A. Poirier. On postcritically finite polynomials, parts I, II. Stony Brook IMS Preprint 1993/5 and 7.

[Sch] D. Schleicher. *Internal addresses in the Mandelbrot set and irreducibility of polynomials.* PhD thesis, Cornell, 1994.

[Th] W. P. Thurston. On the combinatorics of iterated rational maps. Manuscript, 1985.

[W] P. Willumsen. *On accumulation of Stretching rays.* PhD thesis, Denmark Tech. Univ., 1997.

FACULTAD DE MATEMÁTICAS, PONTIFICIA UNIVERSIDAD CATÓLICA, CASILLA 306, CORREO 22, SANTIAGO, CHILE.

E-mail address: jkiwi@riemann.mat.puc.cl

Contemporary Mathematics
Volume **240**, 1999

Some Properties of the Operator of Exterior Derivation On Surfaces of Revolution and L_p-cohomology

Yaroslav A. Kopylov

Abstract. In section 1, the operator d of exterior derivation is studied on a smooth surface of revolution M in $\mathbb{R}^{n+2}$, $n \geq 2$. The operator d is considered as an unbounded one defined on a subspace of the Banach space of measurable differential forms on M that have their modulus integrable to the power $p \geq 1$. Some conditions of normal solvability of d are presented (a linear operator is called *normally solvable* if its range is closed). In section 2, L_p-cohomology of warped flat fiber bundles is investigated.

Introduction

Let M be a smooth real Riemannian manifold. For every differential j-form ω on M the modulus of ω is defined by the formula

$$\begin{aligned} |\omega(x)| = |\omega(x)|_M &= \sup\{|\omega(x, e_1 \wedge \dots \wedge e_j)| \mid e_i \in T_xM, |e_i| \leq 1\} \\ &= \sup_{e_i \in T_xM,\, e_i \neq 0} \frac{|\omega(x, e_1 \wedge \dots \wedge e_j)|}{\|e_1\| \dots \|e_j\|} \end{aligned} \tag{1}$$

(here $\omega(x, e_1 \wedge \dots \wedge e_j)$ is the value of the multilinear form $\omega(x)$ at a multivector $e_1 \wedge \dots \wedge e_j$).

Given an extended real number $p \in [1, \infty]$ and a positive continuous function $\tau : M \to \mathbb{R}$, denote by $L_p^j(M, \tau)$ the Banach space of measurable j-forms on M with the finite norm

$$\|\omega\|_{L_p^j(M,\tau)} = \begin{cases} \left\{ \int\limits_M |\omega(x)|_M^p \tau^p(x)\, dx \right\}^{1/p} & \text{if } 1 \leq p < \infty, \\ \operatorname{ess\,sup}_{x \in M} |\omega(x)|_M \tau(x) & \text{if } p = \infty, \end{cases} \tag{2}$$

1991 *Mathematics Subject Classification.* Primary 58A10, 58A12, 58A14, 58G05; Secondary 46E35, 55N35, 55R20.

Key words and phrases. Normal solvability, Riemannian manifolds, L_p-cohomology, warped flat fiber bundles.

The research was partially supported by the International Science Foundation and the Russian Government (Grants #NQ9000, #NQ9300).

where dx stands for the volume form of M.

Let $D^j(M)$ be the vector space of smooth differential j-forms on M having compact support in Int M. A form $\psi \in L^{j+1}_{1,\mathrm{loc}}(M)$ is called the (*weak*) *differential* $d\omega$ of $\omega \in L^j_{1,\mathrm{loc}}(M)$ if

$$\int\limits_U \omega \wedge du = (-1)^{j+1} \int\limits_U \psi \wedge u$$

for every orientable domain $U \subset \operatorname{Int} M$ and every form $u \in D^{n-j-1}(M)$ having support in U. We put

$$W^j_p(M,\tau) = \{\omega \in L^j_p(M,\tau) \mid d\omega \in L^{j+1}_p(M,\tau)\}.$$

The norm in $W^j_p(M,\tau)$ is given by the formula

$$\|\omega\|_{W^j_p(M,\tau)} = \|\omega\|_{L^j_p(M,\tau)} + \|d\omega\|_{L^{j+1}_p(M,\tau)}.$$

We denote the closure of $D^j(M)$ in $W^j_p(M,\tau)$ by $V^j_p(M,\tau)$.

The paper consists of two sections. In Section 1, we deal with the following problem.

Suppose that Γ is a closed subspace of $W^j_p(M,\tau)$ including $V^j_p(M,\tau)$. Denote by d_Γ the closed linear operator from $L^j_p(M,\tau)$ into $L^{j+1}_p(M,\tau)$ that coincides with the operator of exterior derivation d on Γ. Given $\varphi \in d\Gamma$, all solutions to the equation $dx = \varphi$ belonging to Γ are in the same equivalence class $d_\Gamma^{-1}\varphi \in \Gamma/\operatorname{Ker} d_\Gamma$. The operator $d_\Gamma^{-1} : d\Gamma \to L^j_p(M,\tau)/\operatorname{Ker} d_\Gamma$, defined on the subspace $d\Gamma$ of $L^{j+1}_p(M,\tau)$, is continuous if and only if $d\Gamma$ is closed in $L^{j+1}_p(M,\tau)$. In this case, the operator $d_\Gamma : L^j_p(M,\tau) \to L^{j+1}_p(M,\tau)$ is called *normally solvable.*

The topological linear space

$$H^j_p(M,\tau) = \operatorname{Ker} d_{W^j_p(M,\tau)}/\operatorname{Im} d_{W^{j-1}_p(M,\tau)}$$

is called *the jth L_p-cohomology space of the manifold M (with weight τ).*

For $\tau \equiv 1$, we just write $L^j_p(M)$, $W^j_p(M)$, $V^j_p(M)$, and $H^j_p(M)$ instead of $L^j_p(M,\tau)$, $W^j_p(M,\tau)$, $V^j_p(M,\tau)$, and $H^j_p(M,\tau)$ respectively.

It is well known that d_Γ is normally solvable on a compact manifold [1].

The aim of Section 1 is to find conditions of normal solvability of the operator $d_\Gamma : L^j_p(M) \to L^{j+1}_p(M)$ $(1 \le p < \infty)$ when M is a *surface of revolution*, i.e. a surface in $\mathbb{R}^{n+2}$, $n \ge 1$, defined by the equation

$$f^2(x_1) = x_2^2 + \dots + x_{n+2}^2, \quad (x_1,\dots,x_{n+2}) \in \mathbb{R}^{n+2}, \quad 0 \le x_1 < b, \tag{3}$$

where $0 < b \le \infty$ and $f : [0,b[\to \mathbb{R}$ is a positive smooth function. We prove that in the case of an unbounded function f, $d_\Gamma : L^j_p(M) \to L^{j+1}_p(M)$ is normally solvable for no j, $0 \le j \le n$. We also prove that when $p > 1$, $b = \infty$, f is bounded, $j \in \{0,n\}$, and $\Gamma \in \{V^j_p(M), W^j_p(M)\}$, for d_Γ to be normally solvable it is necessary that $\lim_{x\to\infty} f(x) = 0$. These results are the contents of Theorems 1 and 2.

The study is based on the results of [1] about normal solvability of the operator of exterior derivation on warped products. Alongside [1], the problem of normal solvability of the operator of exterior derivation on Riemannian manifolds was considered in [2–5]. Some properties of the operator on surfaces of revolution which relate to normal solvability were studied in [6, 7].

In Section 2, we introduce warped flat fiber bundles and study their L_p-cohomology. The warped product $X \times_f Y$ of two Riemannian manifolds (X, g_X) and (Y, g_Y) (f is a positive smooth function on X) is the product $X \times Y$ endowed with the metric $g_X + f^2 g_Y$. It turns out that an analogous metric can be considered on flat fiber bundles. Some results of algebraic topology exposed in [8] make it possible to construct a Leray type spectral sequence that converges to the L_p-cohomology of the bundle when the base is closed and the function $1/f$ is bounded. Moreover, we prove a generalization of the Künneth formulas [9,10] to warped flat fiber bundles. These two results constitute Theorem 3.

1. The operator of exterior derivation on surfaces of revolution

1.1. Normal solvability of the operator and quasi-isometries of Riemannian manifolds. Let $\mathfrak{X}$ and $\mathfrak{Y}$ be Banach spaces and let $T : \mathfrak{X} \to \mathfrak{Y}$ be a closed linear operator with domain $\mathfrak{A} = \operatorname{Dom} T$, kernel $\operatorname{Ker} T$ and range $\operatorname{Im} T$. We can turn $\mathfrak{A}$ into a Banach space by endowing it with the norm

$$\|x\|_{\mathfrak{A}} = \|x\|_{\mathfrak{X}} + \|Tx\|_{\mathfrak{Y}}.$$

The subspace $\operatorname{Ker} T$ is closed in $\mathfrak{X}$ as well as in $\mathfrak{A}$. Thus, $\mathfrak{X}/\operatorname{Ker} T$ is a Banach space. The induced operator $T : \mathfrak{X}/\operatorname{Ker} T \to \mathfrak{Y}$ has a densely defined inverse $T^{-1} : \overline{\operatorname{Im} T} \to \mathfrak{X}/\operatorname{Ker} T$ (the bar standing for closure in $\mathfrak{Y}$). T is called *normally solvable* if T^{-1} is bounded, i.e., the quantity

$$K = K(T) = \sup\{K_x(T) \mid x \in \operatorname{Dom} T,\ \|Tx\|_{\mathfrak{Y}} = 1\}$$

is finite, where

$$K_x(T) = \inf\{\|x'\|_{\mathfrak{X}} \mid x' \in \operatorname{Dom} T,\ Tx' = Tx\}.$$

Banach's Closed Graph Theorem implies that T is normally solvable if and only if $\operatorname{Im} T$ is closed in $\mathfrak{Y}$ [4].

For a Riemannian manifold M and a closed subspace $\Gamma \subset W_p^j(M)$, the operator $d_\Gamma : L_p^j(M) \to L_p^{j+1}(M)$ is closed. In the case of $T = d_\Gamma$, we write $K(T)$ as $K_\Gamma^j(M)$ and use the notation $K_{\omega,\Gamma}$ (simply K_ω when $\Gamma = W_p^j(M)$) instead of $K_\omega(T)$ for $\omega \in W_p^j(M)$.

LEMMA 1. *Let M be a Riemannian manifold and $\omega \in W_p^j(M)$, a form with $d\omega \neq 0$. Then $K_\omega > 0$.*

PROOF. Suppose the contrary. We can find a sequence $\varphi_\nu \in L_p^j(M)$, $\nu = 1, 2, \dots$, such that $d\varphi_\nu = d\omega$ and $\|\varphi_\nu\|_{L_p^j(M)} \to 0$ as $\nu \to \infty$. Since $d_{W_p^j(M)}$ is closed, it follows that $d\omega = 0$. This contradiction proves the lemma.

We also consider the constant

$$\overline{K}^j(M) = \sup\bigl\{K_\omega \bigm| \omega \in V_p^j(M),\ \|d\omega\|_{L_p^{j+1}(M)} = 1\bigr\}.$$

COROLLARY. $\overline{K}^j(M) > 0$ *whenever* $0 \leq j \leq \dim M - 1$.

LEMMA 2. *Let (M, g_M) and (S, g_S) be n-dimensional Riemannian manifolds and let $\varphi : M \to S$ be a quasi-isometry, i.e., a diffeomorphism with the following property:*

$$(\exists C_1, C_2 > 0)(\forall x \in M)(\forall \xi \in T_x M)$$
$$\left(C_1^2 g_M(\xi, \xi) \le g_S((d_x\varphi)\xi, (d_x\varphi)\xi) \le C_2^2 g_M(\xi, \xi)\right) \quad (4)$$

($d_x\varphi$ is the differential of φ at a point x). Then for all constants C_1 and C_2 meeting the inequalities of (4) the relation

$$C_1^{n/p+j} C_2^{-n/p-j-1} \overline{K}^j(S) \le \overline{K}^j(M) \le C_1^{-n/p-j-1} C_2^{n/p+j} \overline{K}^j(S)$$

holds.

PROOF. First, observe that every form $\omega \in L_p^j(S)$ meets the inequalities

$$C_1^j \left|\omega(\varphi(x))\right|_S \le \left|(\varphi^*\omega)(x)\right|_M \le C_2^j \left|\omega(\varphi(x))\right|_S. \quad (5)$$

Indeed, by (1) and the definition of φ^*, we have

$$\left|(\varphi^*\omega)(x)\right|_M = \sup_{e_i \in T_x M,\, e_i \ne 0} \frac{|\omega(\varphi(x), (d_x\varphi)e_1 \wedge \dots \wedge (d_x\varphi)e_j)|}{\|e_1\| \dots \|e_j\|}$$
$$\le C_2^j \sup_{e_i \in T_x M,\, e_i \ne 0} \frac{|\omega(\varphi(x), (d_x\varphi)e_1 \wedge \dots \wedge (d_x\varphi)e_j)|}{\|(d_x\varphi)e_1\| \dots \|(d_x\varphi)e_j\|} = C_2^j |\omega(\varphi(x))|_S.$$

The second inequality of (5) is proved likewise.

Furthermore, the inequalities

$$C_1^n \int_M |\gamma(\varphi(x))|\, dx \le \int_S |\gamma(y)|\, dy \le C_2^n \int_M |\gamma(\varphi(x))|\, dx \quad (6)$$

hold for every function $\gamma \in L_1^0(S)$.

From (5) and (6) it follows that

$$C_1^j C_2^{-n/p} \|\omega\|_{L_p^j(S)} \le \|\varphi^*\omega\|_{L_p^j(M)} \le C_2^j C_1^{-n/p} \|\omega\|_{L_p^j(S)}. \quad (7)$$

We now estimate $K_{\varphi^*\omega}$ in terms of K_ω. First, we note that φ^* maps $W_p^j(S)$ onto $W_p^j(M)$ and $V_p^j(S)$ onto $V_p^j(M)$ bijectively. Moreover, φ^* commutes with the differential d. Using these facts and (7), we infer

$$K_{\varphi^*\omega} = \inf\left\{\|\theta\|_{L_p^j(M)} \mid \theta \in W_p^j(M),\, d\theta = d(\varphi^*\omega)\right\}$$
$$\le C_2^j C_1^{-n/p} \inf\left\{\|\mu\|_{L_p^j(S)} \mid \mu \in W_p^j(S),\, d\mu = d\omega\right\} = C_2^j C_1^{-n/p} K_\omega.$$

The inequality $K_{\varphi^*\omega} \ge C_1^j C_2^{-n/p} K_\omega$ is proved in a similar way. Consequently,

$$C_1^j C_2^{-n/p} K_\omega \le K_{\varphi^*\omega} \le C_2^j C_1^{-n/p} K_\omega. \quad (8)$$

Relations (8) enable us to calculate

$$\overline{K}^j(M) = \sup\left\{K_{\varphi^*\omega} \mid \omega \in V_p^j(S),\, \|d(\varphi^*\omega)\|_{L_p^{j+1}(M)} = 1\right\}$$
$$\le C_2^j C_1^{-n/p} \sup\left\{K_\omega \mid \omega \in V_p^j(S),\, \|d\omega\|_{L_p^{j+1}(S)} \le C_1^{-j-1} C_2^{n/p}\right\}$$
$$= C_1^{-n/p-j-1} C_2^{n/p+j} \overline{K}^j(S).$$

The inequality

$$\overline{K}^j(M) \geq C_1^{n/p+j} C_2^{-n/p-j-1} \overline{K}^j(S)$$

can be proved in an analogous manner.

Lemma 2 is proved.

1.2. On normal solvability of the operator d_Γ on a surface of revolution. Let (M_1, g_1) and (M_2, g_2) be Riemannian manifolds and let $h : M_1 \to \mathbb{R}$ be a positive smooth function. The product manifold $M_1 \times M_2$ endowed with the Riemannian metric $g_1 + h^2 g_2$ is referred to as the *warped product* $M_1 \times_h M_2$.

In this section, we assume that M is a surface of revolution in $\mathbb{R}^{n+2}$, $n \geq 1$, defined by (3). Obviously, M can be thought of as the product $[0, b[\times \mathbb{S}^n$ with the metric

$$g_M = (1 + f'^2) g_{[0,b[} + f^2 g_n,$$

$g_{[0,b[}$ and g_n being the conventional metrics on the interval $[0, b[$ and the sphere $\mathbb{S}^n$ respectively. In other words, M may be considered as the warped product $[0, B[\times_F \mathbb{S}^n$ of the manifolds $[0, B[$ and $\mathbb{S}^n$ endowed with their conventional metrics. Here $B = \lim_{x \to b} G(x)$ and $F = f \circ G^{-1}$, where $G(x) = \int_0^x \sqrt{1 + f'^2(t)}\, dt$.

THEOREM 1. *Suppose that f is unbounded, $p \in [1, \infty[$, $0 \leq j \leq n$, and Γ is a closed subspace of $W_p^j(M)$ including $V_p^j(M)$. Then the operator $d = d_\Gamma : L_p^j(M) \to L_p^{j+1}(M)$ is not normally solvable.*

PROOF. Since f is unbounded, it follows that for every number $N > \inf_{x \in [0,b[} f(x)$ there exist numbers $x_1 = x_1(N)$ and $x_2 = x_2(N)$ in the interval $[0, b[$ such that $x_1 < x_2$, $f(x_1) = N$, $f(x_2) = 2N$, and $f([x_1, x_2]) = [N, 2N]$. It suffices to put $x_2 = \sup\{x \mid f(x) < 2N\}$ and $x_1 = \inf\{x \mid x < x_2,\ f(x) > N\}$. Consequently, $F(G(x_1)) = N$, $F(G(x_2)) = 2N$, and $F\big([G(x_1), G(x_2)]\big) = [N, 2N]$. Since $G(x_2) - G(x_1)$ is the length of the curve $z \mapsto f(z)$, $x_1 \leq z \leq x_2$, it follows that

$$G(x_2) - G(x_1) \geq f(x_2) - f(x_1) = N.$$

Let t_1 and t_2 be arbitrary points in $[G(x_1), G(x_2)]$ such that $t_2 - t_1 = N$. The identity mapping of the cylinder $Z_N = [t_1, t_2[\times \mathbb{S}^n$ is a quasi-isometry, with the constants C_1 and C_2 independent of N, of the warped products $Z'_N = \big(Z_N, g'_N\big) = [t_1, t_2[\times_F \mathbb{S}^n$ and $Z''_N = \big(Z_N, g''_N\big) = [t_1, t_2[\times_H \mathbb{S}^n$, where $H(t) \equiv N$ is a constant function.

Indeed, for every $(t, y) \in Z_N$ and every $\zeta = (\xi, \eta) \in T_{(t,y)} Z_N$, we have

$$\begin{aligned} g'_N(\zeta, \zeta) &= g_{[t_1,t_2[}(\xi, \xi) + F^2(t) g_n(\eta, \eta) \\ &\leq g_{[t_1,t_2[}(\xi, \xi) + 4N^2 g_n(\eta, \eta) \leq 4 g''_N(\zeta, \zeta). \end{aligned}$$

On the other hand,

$$g''_N(\zeta, \zeta) = g_{[t_1,t_2[}(\xi, \xi) + N^2 g_n(\eta, \eta) \leq g'_N(\zeta, \zeta).$$

Thus,

$$C_1^2 g'_N(\zeta, \zeta) \leq g''_N(\zeta, \zeta) \leq C_2^2 g'_N(\zeta, \zeta),$$

where $C_1 = 1/2$ and $C_2 = 1$.

Let Z be the cylinder $[0, 1[\times \mathbb{S}^n$ with the conventional metric and let ι be the inclusion mapping $Z_N \to M$.

Now, note that every $\gamma \in \Gamma$ satisfies the inequality $K_{\gamma,\Gamma} \geq K_\gamma$. Moreover, every differential form $\omega \in V_p^j(Z'_N)$ may be extended by zero to some $\tilde{\omega} \in V_p^j(M)$. If $\theta \in W_p^j(M)$ is such that $d\theta = d\tilde{\omega}$ then $\iota^*\theta \in W_p^j(Z'_N)$ and $d(\iota^*\theta) = \iota^* d\theta = \iota^*\tilde{\omega} = \omega$. Therefore,

$$\begin{aligned} K_\Gamma^j(M) &= \sup\{K_{\gamma,\Gamma} \mid \gamma \in \Gamma,\ \|d\gamma\|_{L_p^{j+1}(M)} = 1\} \\ &\geq \sup\{K_\gamma \mid \gamma \in \Gamma,\ \|d\gamma\|_{L_p^{j+1}(M)} = 1\} \\ &\geq \sup\bigl\{K_{\tilde{\omega}} \mid \omega \in V_p^j(Z'_N),\ \|d\omega\|_{L_p^{j+1}(Z'_N)} = 1\bigr\} \\ &\geq \sup\bigl\{K_\omega \mid \omega \in V_p^j(Z'_N),\ \|d\omega\|_{L_p^{j+1}(Z'_N)} = 1\bigr\} = \overline{K}^j(Z'_N). \end{aligned}$$

Furthermore, by the above-proven quasi-isometry, Lemma 2 implies that

$$\overline{K}^j(Z'_N) \geq C\overline{K}^j(Z''_N) = CN\overline{K}^j(Z)$$

for an appropriate positive constant C. Moreover, by the Corollary to Lemma 1, $\overline{K}^j(Z) > 0$. Consequently, $K_\Gamma^j(M) = \infty$ and d_Γ is not normally solvable.

The proof of Theorem 1 is over.

Now, we introduce the notation $m = m(j) = n/p - j$.

Following [1], for a number $p \in]1, \infty[$ and a positive function $\sigma : [0, B[\to \mathbb{R}$ we set by definition

$$\chi_0(\sigma) = \sup_{0<x<B} \left\{ \left(\int_x^B \sigma^p(x)\,dx \right)^{1/p} \left(\int_0^x \sigma^{-p'}(x)\,dx \right)^{1/p'} \right\},$$

$$\chi_B(\sigma) = \sup_{0<x<B} \left\{ \left(\int_0^x \sigma^p(x)\,dx \right)^{1/p} \left(\int_x^B \sigma^{-p'}(x)\,dx \right)^{1/p'} \right\},$$

where $p' = p/(p-1)$.

Thus,

$$\chi_0(F^m) = \sup_{0<x<b} \left\{ \left(\int_x^b f^{mp}(t)\sqrt{1+f'^2(t)}\,dt \right)^{1/p} \left(\int_0^x f^{-mp'}(t)\sqrt{1+f'^2(t)}\,dt \right)^{1/p'} \right\},$$

$$\chi_B(F^m) = \sup_{0<x<b} \left\{ \left(\int_0^x f^{mp}(t)\sqrt{1+f'^2(t)}\,dt \right)^{1/p} \left(\int_x^b f^{-mp'}(t)\sqrt{1+f'^2(t)}\,dt \right)^{1/p'} \right\}.$$

Suppose now that f is bounded. It follows from [1] that for $\Gamma \in \{V_p^j(M), W_p^j(M)\}$ and $0 < j < n$ the operator $d_\Gamma : L_p^j(M) \to L_p^{j+1}(M)$ is normally solvable, and, in the case $j \in \{0, n\}$, the normal solvability of d_Γ is equivalent to that of $\partial = d_\Gamma^0 : L_p^0([0, B[, F^m) \to L_p^1([0, B[, F^m)$. The operator ∂ is in turn normally solvable if and only if $\chi_0(F^m) < \infty$ or $\chi_B(F^m) < \infty$ [1, Lemma 11].

Assume that $b = \infty$ (and so $B = \infty$). For simplicity, we write χ_0 and χ_∞ instead of $\chi_0(F^m)$ and $\chi_\infty(F^m)$ respectively.

LEMMA 3. *The following assertions hold:*
1. *If* $\chi_0 < \infty$ *or* $\chi_\infty < \infty$ *then* $\lim_{x\to\infty} f(x) = 0$.
2. *If* $m \leq 0$ *then* $\chi_0 = \infty$.
3. *If* $m \geq 0$ *then* $\chi_\infty = \infty$.

PROOF. Suppose that $\chi_0 < \infty$. We have

$$\int_0^\infty f^{mp}(t)\sqrt{1+f'^2(t)}\,dt < \infty. \tag{9}$$

Furthermore,

$$f^{mp}(t)\sqrt{1+f'^2(t)} \geq f^{mp}(t)|f'(t)|.$$

Consequently, the integral

$$\int_0^\infty f^{mp}(t)f'(t)\,dt = \begin{cases} \frac{1}{mp+1}\lim_{x\to\infty}(f^{mp+1}(x) - f^{mp+1}(0)) & \text{if } mp \neq -1, \\ \lim_{x\to\infty}\log\frac{f(x)}{f(0)} & \text{if } mp = -1 \end{cases}$$

is finite.

There appear several possibilities:

(a) $m > 0$. The above implies that there exists a finite limit $\lim_{x\to\infty} f(x)$, which is zero by (9).

(b) $m = 0$. This is impossible because of (9).

(c) $-1/p < m < 0$. hen $mp+1 > 0$ and $f(x)$ has a finite limit as $x \to \infty$, which contradicts (9).

(d) $m = -1/p$. A contradiction to (9).

(e) $m < -1/p$. In this case $mp < -1$. Hence $\lim_{x\to\infty} f(x) = \infty$. We infer that

$$\begin{aligned} &\left(\int_x^\infty f^{mp}(t)\sqrt{1+f'^2(t)}\,dt\right)^{1/p}\left(\int_0^x f^{-mp'}(t)\sqrt{1+f'^2(t)}\,dt\right)^{1/p'} \\ &\quad\geq \left(\int_x^\infty f^{mp}(t)|f'(t)|\,dt\right)^{1/p}\left(\int_0^x f^{-mp'}(t)|f'(t)|\,dt\right)^{1/p'} \\ &\quad\geq \left|\int_x^\infty f^{mp}(t)f'(t)\,dt\right|^{1/p}\left|\int_0^x f^{-mp'}(t)f'(t)\,dt\right|^{1/p'} \\ &\quad= \left(\frac{f^{mp+1}(x)}{|mp+1|}\right)^{1/p}\left|\frac{f^{-mp'+1}(x) - f^{-mp'+1}(0)}{-mp'+1}\right|^{1/p'} \\ &\quad= \left|\frac{f^{p'}(x) - f^{-mp'+1}(0)f^{mp'+p'/p}(x)}{|mp+1|^{p'/p}(1-mp')}\right|^{1/p'}. \end{aligned}$$

The last quantity is $O(f(x))$ as $x \to \infty$ for $mp' + p'/p < 0$. Therefore, $\chi_0 = \infty$, which contradicts the hypothesis.

Thus, $m > 0$ and $\lim_{x\to\infty} f(x) = 0$. It may be proved likewise that if $\chi_\infty < \infty$ then $m < 0$ and $\lim_{x\to\infty} f(x) = 0$. This completes the proof of Lemma 3.

Lemma 3 and the preceding considerations imply

THEOREM 2. *Suppose that $p > 1$, f is bounded, $b = \infty$, $j \in \{0, n\}$, and $\Gamma \in \{V_p^j(M), W_p^j(M)\}$. If $d_\Gamma : L_p^j(M) \to L_p^{j+1}(M)$ is normally solvable then $\lim_{x\to\infty} f(x) = 0$.*

REMARK. Assertions 2 and 3 of Lemma 3 imply that in the case of $j = 0$ ($j = n$), d_Γ is normally solvable if and only if $\chi_0 < \infty$ ($\chi_\infty < \infty$). Since $m(0)p = -m(n)p' = n$, it follows that in both cases

$$s_n \int_0^\infty f^n(x)\sqrt{1 + f'^2(x)}\, dx < \infty, \tag{10}$$

where s_n is the volume of the unit n-dimensional sphere. The quantity on the left-hand side of (10) is the volume vol M of the Riemannian manifold M. Therefore, the following assertion holds, whence Theorem 2 is immediate.

THEOREM 2′. *Suppose that $p > 1$, f is bounded, $b = \infty$, $j \in \{0, n\}$, and $\Gamma \in \{V_p^j(M), W_p^j(M)\}$. If $d_\Gamma : L_p^j(M) \to L_p^{j+1}(M)$ is normally solvable then* vol $M < \infty$.

We note that the condition vol $M < \infty$ is insufficient for normal solvability of the operator $d_\Gamma : L_p^j(M) \to L_p^{j+1}(M)$, $j \in \{0, n\}$, as it is examplified by the function $f(x) = (x + 1)^\alpha$ with $\alpha < -1/n$. Clearly, in this case vol $M < \infty$. However, it is easy to see that $\chi_0 = \chi_\infty = \infty$.

2. L_p-cohomology of warped flat fiber bundles

Let $F \to E \xrightarrow{\pi} X$ be a fiber bundle of smooth manifolds with the structure group G. Suppose that the base X and the fiber F are endowed with Riemannian metrics g_X and g_F respectively. We also assume that the manifold X is closed and that the structure group G is a discrete subgroup of the group of smooth isometries of F. A bundle of this kind is called *flat.*

We say that a positive smooth function $f : F \to \mathbb{R}$ is *G-invariant* if

$$(\forall g \in G)(f \circ g = f). \tag{11}$$

For an arbitrary point $e \in E$, consider a chart (U, γ) of the fiber bundle, where U is a neighborhood of the point $x = \pi(e) \in X$ and $\gamma : F \times U \to \pi^{-1}(U)$ is a diffeomorphism. Given a G-invariant function f, we endow $F \times U$ with the metric of the warped product $F \times_f U$ and set

$$g_E(\xi, \eta) = g_{F\times_f U}\big((d_e\gamma^{-1})\xi, (d_e\gamma^{-1})\eta\big), \tag{12}$$

$\xi, \eta \in T_eE$. Equality (12) defines a metric on the tangent space T_eE. This definition does not depend on the choice of (U, γ).

Indeed, let (U', γ') be another chart of our fiber bundle such that $\pi(e) = x \in U'$. Then $\gamma'^{-1} \circ \gamma|_{U\cap U'} : F \times (U \cap U') \to F \times (U \cap U')$ is an isometry, moreover, by the discreteness of G,

$$(\exists g \in G)(\forall x \in U \cap U')((\gamma'^{-1} \circ \gamma)(x) = g).$$

Hence,

$$g_{F\times_f U}\big((d_e\gamma^{-1})\xi, (d_e\gamma^{-1})\eta\big) = g_{F\times_f U'}\big((d_e\gamma'^{-1})\xi, (d_e\gamma'^{-1})\eta\big)$$

and thus, by (11), we have a well-defined Riemannian metric on E we call the *metric of warped fiber bundle with the warping function* f. Using this metric, we can introduce the Sobolev spaces $W_p^j(V)$ of differential forms on open subsets $V \subset E$.

Given $p \in [1, \infty]$, we define a differential sheaf $\mathcal{W}_p^*$ over X as follows. We put $\mathcal{W}_p^j(U) = W_p^j(\pi^{-1}(U))$ for every open set $U \subset X$. The restrictions of $\mathcal{W}_p^*$ are the restrictions of differential forms to open subsets, and the differential $d : \mathcal{W}_p^j \to \mathcal{W}_p^{j+1}$ is induced by the operator of exterior derivation on E. For us, the key property of $\mathcal{W}_p^*$ is the following assertion.

LEMMA 4. *If the function* $1/f$ *is bounded then* $\mathcal{W}_p^*$ *is flabby.*

PROOF. Suppose that A is a closed subset of X and $\theta \in \mathcal{W}_p^j(A)$. We must prove that there exists an element $\omega_0 \in \mathcal{W}_p^j(X)$ such that $\theta = \omega_0|_A$. By Corollary 1 to Theorem 3.3.1 in [8], $\mathcal{W}_p^j(A) = \varinjlim \mathcal{W}_p^j(U)$, the direct limit being taken over all open sets U that contain A, so we have $\theta = \omega|_A$ for some $\omega \in W_p^j(\pi^{-1}(V))$, where V is open and $V \supset A$. Let $h : X \to [0,1]$ be a smooth function which is equal to 1 on A and to 0, on $X \setminus V$, and set $\hat{h} = h \circ \pi$. If $e \in E$ then we put

$$\omega_0(e) = \begin{cases} \hat{h}(e)\omega(e) & \text{if } e \in \pi^{-1}(V), \\ 0 & \text{if } e \notin \pi^{-1}(V). \end{cases}$$

Obviously, $\omega_0 \in L_p^j(E)$. Moreover, $d\omega_0 = \psi_1 + \psi_2$, where

$$\psi_1(e) = \begin{cases} d\hat{h}(e) \wedge \omega(e) & \text{if } e \in \pi^{-1}(V), \\ 0 & \text{if } e \notin \pi^{-1}(V) \end{cases}$$

and

$$\psi_2(e) = \begin{cases} \hat{h}(e) d\omega(e) & \text{if } e \in \pi^{-1}(V), \\ 0 & \text{if } e \notin \pi^{-1}(V). \end{cases}$$

Since $\hat{h}$ is bounded, it follows that $\psi_2 \in L_p^j(E)$. Now, we demonstrate that $\psi_1 \in L_p^j(E)$.

We first note that

$$|\psi_1(e)|_E \leq \text{const}\, |d\hat{h}(e)|_E |\omega(e)|_{\pi^{-1}(V)}.$$

Furthermore, consider a chart (U, γ) of the fiber bundle such that $\pi(e) \in U \subset V$. Denote the projections $F \times U \to F$ and $F \times U \to U$ by pr_F and pr_U respectively. The definitions of the function $\hat{h}$ and the metric on E imply that

$$\begin{aligned} |d\hat{h}(e)|_E &= \left|((\gamma^{-1})^* d(\hat{h} \circ \gamma))(e)\right|_E = \left|d(\hat{h} \circ \gamma)(\gamma^{-1}(e))\right|_{F \times_f U} \\ &= \left|d(h \circ \pi \circ \gamma)(\gamma^{-1}(e))\right|_{F \times_f U} = \left|d(h \circ \operatorname{pr}_U)(\gamma^{-1}(e))\right|_{F \times_f U} \\ &= \frac{1}{f\left(\operatorname{pr}_F(\gamma^{-1}(e))\right)} |dh(\pi(e))|_X. \end{aligned}$$

Since $1/f \leq \text{const}$, from the above it follows that

$$\begin{aligned} |\psi_1(e)|_E &\leq \text{const}\, |d\hat{h}(e)|_E |\omega(e)|_{\pi^{-1}(V)} \leq \text{const}\, |dh(\pi(e))|_X |\omega(e)|_{\pi^{-1}(V)} \\ &\leq \text{const} \sup_{x \in X} |dh(x)|_X |\omega(e)|_{\pi^{-1}(V)}. \end{aligned}$$

We have $\sup_{x\in X} |dh(x)|_X < \infty$ as X is compact. Therefore,

$$|\psi_1(e)|_E \leq \text{const}\, |\omega(e)|_{\pi^{-1}(V)}$$

and, consequently, $\psi_1 \in L_p^{j+1}(E)$. Hence, $\omega_0 \in W_p^j(E)$. The lemma is proved.

Denote by $\mathcal{H}_p^r$ the rth derived sheaf $\mathcal{H}^r(\mathcal{W}_p^*)$ of $\mathcal{W}_p^*$, i.e., the sheaf generated by the presheaf $U \mapsto H^q(\mathcal{W}_p^*(U))$.

The main result of this section can be formulated as follows.

THEOREM 3. *The following assertions hold.*

1. *If the function $1/f$ is bounded then there exists a spectral sequence $\mathbf{E}_*^{**}$ with $\mathbf{E}_2^{qr} = H^q(X; \mathcal{H}_p^r)$ for which $\mathbf{E}_\infty$ is the L_p-cohomology space $H_p^*(E)$ of E.*
2. *If f is bounded then the sheaf $\mathcal{H}_p^r$ is locally constant.*

PROOF. 1. By Lemma 4, $\mathcal{W}_p^*$ is flabby. Therefore, $H^*(X, \mathcal{W}_p^*) = 0$. Moreover, since X is compact, it follows that $\mathcal{W}_p^*(X) = W_p^*(E)$, and thus the cohomology of $(\mathcal{W}_p^*(X), d)$ is $H_p^*(E)$. Now the result is implied by Theorem 4.6.1 of [8].

2. Consider a point $x \in X$. Let U be a neighborhood of x over which the bundle is trivial. Choose an open ball $B_0 \subset U$ that contains x and whose closure $\overline{B}_0$ is included in U. Fix a diffeomorphism $\varphi : \overline{B_0} \to \overline{B}$, where $\overline{B}$ is a closed ball in $\mathbb{R}^n$ $(n = \dim X)$. We prove that the sheaf $\mathcal{H}_p^r|_{B_0}$ is constant.

For an arbitrary $y \in B_0$, consider all kinds of balls $B_{y,d} = B(\varphi(y), d)$ in $\mathbb{R}^n$ of radius d with center $\varphi(y)$ such that $\overline{\varphi^{-1}(B_{y,d})} \subset B_0$ (here and below, the over-line stands for closure). We put $\mathfrak{B}_y = \{\varphi^{-1}(B_{y,d}) \mid \overline{\varphi^{-1}(B_{y,d})} \subset B_0\} \cup \{B_0\}$. The fiber $\mathcal{H}_p^r(y)$ of $\mathcal{H}_p^r$ over y is, by definition, the direct limit $\varinjlim_{V\in\mathfrak{O}(y)} H_p^r(\pi^{-1}(V))$ taken over the direct system $\mathfrak{O}(y)$ of all open neighborhoods of y. Clearly, $\mathfrak{B}_y$ is a direct subsystem of $\mathfrak{O}(y)$; moreover,

$$\varinjlim_{V\in\mathfrak{O}(y)} H_p^r(\pi^{-1}(V)) \cong \varinjlim_{V\in\mathfrak{B}(y)} H_p^r(F \times_f V).$$

Furthermore, if B is a ball in $\mathbb{R}^n$, we have

$$H_p^j(B) = H_p^j(\overline{B}) = \begin{cases} \mathbb{R}, & j = 0, \\ 0, & j \neq 0, \end{cases}$$

where the function $\chi = 1$ generates $H_p^0(B)$. Besides, since $\overline{B_0}$ is compact, φ is a quasi-isometry. Consequently, for every $C_1, C_2 \in \mathfrak{B}_y$ the mapping $\iota^* : H_p^*(C_2) \to H_p^*(C_1)$ induced by the inclusion $\iota : C_1 \to C_2$ is an isomorphism. By the Künneth formula for L_p-cohomology (see [9, 10]), there are natural isomorphisms

$$H_p^r(F \times_f C_i) \cong H_p^r(F, f^{n/p}), \quad i = 1, 2.$$

Therefore, the inclusion $\iota : F \times_f C_1 \to F \times_f C_2$ yields the isomorphism

$$\iota^* : H_p^r(F \times_f C_2) \xrightarrow{\cong} H_p^r(F \times_f C_1) \left(\xrightarrow{\cong} H_p^r(F \times_f B_0) \right).$$

Thus we obtain a canonical identification

$$\mathcal{H}_p^r(y) \cong \mathcal{H}_p^r(x) \cong H_p^r(F, f^{n/p})$$

for each $y \in B_0$, which means that $\mathcal{H}_p^r|_{B_0}$ is constant.

The proof of Theorem 3 is over.

In conclusion, I would like to express my deep gratitude to Professor V. I. Kuz′minov for posing the problems and fruitful discussions. I also would like to thank the members of the Local Committee of "Cruz del Sur" and all the people who made this congress an unforgettable event.

References

1. V. I. Kuz′minov and I. A. Shvedov, *On normal solvability of the operator of exterior derivation on warped products*, Sibirsk. Mat. Zh. **37** (1996), no. 2, 324–337.
2. J. Cheeger, *On the Hodge theory of Riemannian pseudomanifolds*, Proc. Sympos. Pure Math. **36** (1980), 93–146.
3. V. M. Gol′dshtein, V. I. Kuz′minov, and I. A. Shvedov, *On normal and compact solvability of the operator of exterior derivation for homogeneous boundary value conditions*, Sibirsk. Mat. Zh. **28** (1987), no. 4, 82–96.
4. V. M. Gol′dshtein, V. I. Kuz′minov, and I. A. Shvedov, *Normal and compact solvability of linear operators*, Sibirsk. Mat. Zh. **30** (1989), no. 5, 49–59.
5. V. I. Kuz′minov and I. A. Shvedov, *On normal solvability of the exterior differentiation on a warped cylinder*, Sibirsk. Mat. Zh. **34** (1993), no. 1, 85–95.
6. V. I. Kuz′minov and I. A. Shvedov, *On compact solvability of the operator of exterior derivation*, Sibirsk. Mat. Zh. **38** (1997), no. 3, 573–590.
7. A. Baider, *Noncompact Riemannian manifolds with discrete spectra*, J. Diff. Geom. **14** (1979), no. 1, 41–57.
8. R. Godement, *Topologie algébrique et théorie des faisceaux*, Hermann, Paris, 1958.
9. V. M. Gol′dshtein, V. I. Kuz′minov, and I. A. Shvedov, *The Künneth formula for L_p-cohomologies of warped products*, Sibirsk. Mat. Zh. **32** (1990), no. 5, 29–42.
10. K. V. Storozhuk and I. A. Shvedov, *L_p-cohomology of warped products of Lipschitz Riemannian manifolds*, Sibirsk. Mat. Zh. **39** (1998), no. 3, 633–649.

Sobolev Institute of Mathematics, Akademik Koptyug prospect 4, 630090 Novosibirsk, Russia

E-mail address: `yakop@math.nsc.ru`

Contemporary Mathematics
Volume **240**, 1999

Matrices for Fenchel-Nielsen parameters in genus 2

Bernard Maskit

1. Introduction.

In this note we consider the Fenchel-Nielsen space of Riemann surfaces of genus 2, marked by the lengths of three disjoint simple closed geodesics, and twists about these geodesics, where one of the three geodesics is dividing. The other case for genus 2, where all three geodesics are non-dividing, will be considered elsewhere in the general context of an algorithm for determining matrix generators corresponding to Fenchel-Nielsen data on any hyperbolic orbifold with finitely generated fundamental group. We label our basic parameters as $\alpha > 0$, $\beta > 0$, $\gamma > 0$, ρ, σ, τ, where 2γ is the length of the dividing geodesic; 2τ is the twist about the dividing geodesic; 2α is the length and 2ρ the twist about the first non-dividing geodesic; and 2β is the length and 2σ the twist about the second non-dividing geodesic. The twists are all taken to be length of arc of the given geodesic, independent of its length.

We define the Fenchel-Nielsen space to be

$$(\mathbb{R}^+)^3 \times \mathbb{R}^3 = \{\lambda = (\sinh\alpha, \sinh\beta, \sinh\gamma, \sinh\rho, \sinh\sigma, \sinh\tau)\}.$$

For each point $\lambda \in (\mathbb{R}^+)^3 \times \mathbb{R}^3$, we define two auxiliary functions μ and ν (explicit formulae for these appear below); we then write down four matrices, $(\tilde{a}_1, \tilde{b}_1, \tilde{a}_2, \tilde{b}_2) \in (SL(2,\mathbb{R}))^4$, where the corresponding Möbius transformations, a_1, b_1, a_2, b_2, satisfy the usual relation,

$$a_1 b_1 a_1^{-1} b_1^{-1} a_2 b_2 a_2^{-1} b_2^{-1} = 1,$$

and where the entries in these matrices depend algebraically on the point $\lambda \in (\mathbb{R}^+)^3 \times \mathbb{R}^3$. These four transformations generate a discrete group G, so that $\mathbb{H}^2/G$ is the Riemann surface of genus 2 whose Fenchel-Nielsen coordinates are the given ones (this will be made precise below).

We end with a remark concerning a relationship between rational Fenchel-Nielsen coordinates and subgroups of $SL(2,k)$, where k is an algebraic number field. This is related to the work of Buser and Silhol [**1**] in which they relate certain algebraic curves with specific Fenchel-Nielsen data.

1991 *Mathematics Subject Classification.* 30F10.

Research supported in part by NSF Grant DMS 9500557.

In broad outline, the treatment here follows the development given in [**4**], with some modifications that appear in the treatment in [**3**]. However, for the convenience of the reader, the development here is essentially self-contained.

2. Matrices and Möbius transformations.

We will use the following convention throughout. A matrix

$$\tilde{x} = \begin{pmatrix} \alpha & \beta \\ \gamma & \delta \end{pmatrix} \in SL(2, \mathbb{R})$$

determines a unique Möbius transformation

$$x = \frac{\alpha z + \beta}{\gamma z + \delta},$$

and every Möbius transformation has two such representations. We will, without further mention, associate the matrix $\tilde{x}$ with the transformation x.

We denote the trace of the matrix $\tilde{x}$ by $tr(\tilde{x})$. If $|tr(\tilde{x})| > 2$, then the corresponding transformation x is hyperbolic, and has a translation length ξ, defined by $|tr(\tilde{x})| = 2\cosh(\xi/2)$.The hyperbolic transformation x has an axis A_x, which is the hyperbolic line whose (Euclidean) endpoints on the circle at infinity (the extended real axis) are the fixed points of x; one of these fixed points is repelling, the other is attracting. The *positive direction* of the axis A_x goes from the repelling fixed point to the attracting fixed point.

If A_x and A_y are disjoint axes, even at the circle at infinity, then they have a unique common orthogonal, L_{xy}, which we regard as a complete line. We denote the line segment, orthogonal to both A_x and A_y, directed from A_x to A_y as $\hat{L}_{xy}$. Further, we will denote the reflection in L_{xy} by r_{xy}.

In what follows, we will denote the group generated by $a, \ldots$ as $\langle a, \ldots \rangle$.

3. Pants Groups.

A triple of transformations $(a, c, e) \in (PSL(2, \mathbb{R}))^3$ are *geometric generators for a pants group* if the following hold. The transformations a, c and e are all hyperbolic, where the three axes are disjoint, even at the circle at infinity, and they bound a common region, D. When traversing any of these axes in the positive direction, this region D always lies on the right (or always lies on the left). Further, $G = \langle a, c, e \rangle$ has the one defining relation: $eca = 1$.

The axis of a (or c or e, respectively) bounds a half-plane disjoint from D; it is well known that this half-plane is precisely invariant under $\langle a \rangle$ (or $\langle c \rangle$ or $\langle e \rangle$, respectively) in G.

It is also well known (a proof can be found in [**2**]), that if a and c are hyperbolic with disjoint axes, and if these axes are oriented so that the region between them lies on the right (or on the left), then a, c and $e = a^{-1}c^{-1}$ are geometric generators for a pants group if and only if $tr(\tilde{e}) = tr(\tilde{a}^{-1}\tilde{c}^{-1}) < -2$, where $\tilde{a}$ and $\tilde{c}$ have been chosen so that $tr(\tilde{a}) > 2$ and $tr(\tilde{c}) > 2$. In this case, we will also say that the triple of matrices $(\tilde{a}, \tilde{c}, \tilde{e})$ are geometric generators for the corresponding group of transformations $G = \langle a, c, e \rangle$.

For every pants group $G = \langle a, c, e \rangle$, there is a corresponding *extended pants group* $\hat{G} = \langle r_{ac}, r_{ce}, r_{ea} \rangle$. It is well known that $a = r_{ca}r_{ae}$, $c = r_{ec}r_{ca}$, and $e = r_{ae}r_{ec}$.

The fact that we need from the above is that the point of intersection of L_{ce} with A_c lies on A_c half-way (in the hyperbolic sense) between z, the point of intersection of L_{ac} with A_c, and $c(z)$.

4. A fully normalized pair of pants with two equal sizes.

We assume we are given numbers $\alpha > 0$ and $\gamma > 0$; we need to construct matrices $\tilde{a}$, $\tilde{c}$ and $\tilde{e}$, where these geometrically generate a pants group with the following properties: $tr(\tilde{a}) = -tr(\tilde{e}) = 2\cosh\alpha$, and $tr(\tilde{c}) = 2\cosh\gamma$; c has its attracting fixed point at ∞ and its repelling fixed point at 0; and a has both its fixed points positive, with the attracting fixed point smaller than the repelling fixed point, and the product of its fixed points equal to 1. (This normalization can be restated as follows: The fixed points of c are as above; the region between A_a and A_c lies on the right for both axes, and the common orthogonal between A_a and A_c lies on the unit circle.)

We define the matrices $\tilde{a}$ and $\tilde{c}$ by

$$\tilde{a} = \frac{1}{\sinh\mu}\begin{pmatrix} \sinh(\mu-\alpha) & \sinh\alpha \\ -\sinh\alpha & \sinh(\mu+\alpha) \end{pmatrix}, \quad \tilde{c} = \begin{pmatrix} e^{\gamma} & 0 \\ 0 & e^{-\gamma} \end{pmatrix},$$

where $\mu > 0$ is yet to be determined. We note that, as required, c has its fixed points at 0 and ∞, with the attracting fixed point at ∞, and that a has its fixed points at $e^{\pm\mu}$, with the repelling fixed point at $e^{+\mu}$.

We write

$$\tilde{e} = \tilde{a}^{-1}\tilde{c}^{-1} = \frac{1}{\sinh\mu}\begin{pmatrix} e^{-\gamma}\sinh(\mu+\alpha) & -e^{\gamma}\sinh\alpha \\ e^{-\gamma}\sinh\alpha & e^{\gamma}\sinh(\mu-\alpha) \end{pmatrix}.$$

Since we require $tr(\tilde{e}) = -tr(\tilde{a})$, we obtain

$$e^{\gamma}\sinh(\mu-\alpha) = -\sinh(\mu+\alpha). \tag{1}$$

Solving for μ, we obtain

$$\tanh\mu = \tanh\alpha\tanh(\gamma/2). \tag{2}$$

We observe that, with this definition, μ is necessarily positive. We also remark that the quantity μ is related to the distance δ between the axes, A_a and A_c; the relation is given by $\coth\mu = \cosh\delta$, or, equivalently, $\sinh\mu\sinh\delta = 1$.

Let $G_1 = \langle a, c, e\rangle$. Then G_1 is discrete, and $\mathbb{H}^2/G_1$ is a pair of pants; i.e., a sphere with three holes. The axes A_a, A_c, and A_e project to the simple geodesics freely homotopic to the boundary of this surface. Since $tr(\tilde{a}) = -tr(\tilde{e})$, a and e are conjugate in $PSL(2,\mathbb{R})$. We will also need the fact that the one relation in G_1 is that $eca = 1$.

5. Closing the first handle without twist.

Our next goal is to adjoin to G_1 a transformation b which conjugates a into e^{-1} without twist; that is, $bab^{-1} = e^{-1}$, and $A_b = L_{ae}$.

Let M be the perpendicular bisector of $\hat{L}_{ae}$. Then we can write $b \doteq r_M r_a$, where r_a is the reflection in A_a, and r_M is the reflection in M.

Since M is the unique line with the property that reflection in M interchanges A_a and A_e, M is orthogonal to A_c and passes through the point $ie^{\gamma/2}$. Hence we can write the matrix for reflection in M as

$$\tilde{r}_M = \begin{pmatrix} 0 & e^{\gamma/2} \\ e^{-\gamma/2} & 0 \end{pmatrix}.$$

(Observe that matrices for orientation-reversing transformations have negative determinant.)

It was observed in [**3**], and one easily computes, that if

$$\tilde{x} = \begin{pmatrix} \alpha & \beta \\ \gamma & \delta \end{pmatrix} \in SL(2, \mathbb{R}),$$

where x is hyperbolic and $\beta\gamma \neq 0$, then we can write the matrix for the reflection r_x in A_x as

$$\tilde{r}_x = \frac{1}{\sqrt{(\alpha+\delta)^2 - 4}} \begin{pmatrix} \alpha - \delta & 2\beta \\ 2\gamma & \delta - \alpha \end{pmatrix}.$$

Hence we can write

$$\begin{aligned} \tilde{b} &= \frac{1}{\sinh \mu} \begin{pmatrix} 0 & e^{\gamma/2} \\ e^{-\gamma/2} & 0 \end{pmatrix} \begin{pmatrix} -\cosh \mu & 1 \\ -1 & \cosh \mu \end{pmatrix} \\ &= \frac{1}{\sinh \mu} \begin{pmatrix} -e^{\gamma/2} & e^{\gamma/2} \cosh \mu \\ -e^{-\gamma/2} \cosh \mu & e^{-\gamma/2} \end{pmatrix}. \end{aligned}$$

We remark that it is easy to compute, using (1) and/or (2), that $\tilde{b}\tilde{a} = -\tilde{c}\tilde{a}\tilde{b} = \tilde{e}^{-1}\tilde{b}$.

Let $G_2 = \langle a, b, c \rangle$. It is easy to see that G_2 can be constructed from G_1 and b using the second combination theorem; that is, A_a and A_e are disjoint, and they bound disjoint half-planes, H_a and H_e, where H_a (H_e) is precisely invariant under $\langle a \rangle(\langle e \rangle)$ in G_1, and b conjugates a into e^{-1}, while mapping the exterior of H_a onto H_e. We conclude from this that G_2 is discrete and purely hyperbolic; that $\mathbb{H}^2/G_2$ is a torus with one hole, where c is a boundary geodesic, and that the left half-plane is precisely invariant under $\langle c \rangle$ in G_2. It also follows from this combination theorem that G_2 has the presentation:

$$G_2 = \{a, b, c, e | eca = ebab^{-1} = 1\}.$$

6. Twisting the first handle.

If we twist along a, then a and c will be unaffected, but b will be changed. In order to compute the change in b, we first write down the matrix $\tilde{a}(\rho)$, where the element $a(\rho)$ has the same fixed points as a, but $tr(\tilde{a}(\rho)) = 2\cosh \rho$, as compared with $tr(\tilde{a}) = 2\cosh \alpha$. Then

$$\tilde{a}(\rho) = \frac{1}{\sinh \mu} \begin{pmatrix} \sinh(\mu - \rho) & \sinh \rho \\ -\sinh \rho & \sinh(\mu + \rho) \end{pmatrix}.$$

Note that $a(0) = 1$, and $a(\alpha) = a$.

Now, to twist by 2ρ about a, we replace $\tilde{b}$ by $\tilde{b}_\rho = \tilde{b}\tilde{a}(\rho)$. Then

$$\tilde{b}_\rho = \frac{1}{\sinh \mu} \begin{pmatrix} -e^{\gamma/2} \cosh \rho & e^{\gamma/2} \cosh(\mu + \rho) \\ -e^{-\gamma/2} \cosh(\mu - \rho) & e^{-\gamma/2} \cosh \rho \end{pmatrix}.$$

We observe that b_ρ, like b, maps the half-plane H_a onto the exterior of H_e, and conjugates a into e^{-1}. Hence we can replace b by b_ρ in our use of the combination theorem to construct G_2. This yields the new group $G_2(\rho)$ with the presentation:

$$G_2(\rho) = \{a, b_\rho, c, e | eca = eb_\rho ab_\rho^{-1} = 1\}.$$

As above, $G_2(\rho)$ is a purely hyperbolic Fuchsian group representing a torus with one hole, where the left half-plane, with stabilizer $\langle c\rangle$, is a boundary half-plane.

7. Preparing the second pair of pants.

We start by repeating the construction of Sect. 4, but with α replaced by β. Then we also need to replace μ by ν, where

$$\tanh\nu = \tanh\beta\tanh(\gamma/2). \tag{3}$$

This yields three matrices, call them $\tilde{x}$, $\tilde{c}$ and $\tilde{y}$, as follows.

$$\tilde{x} = \frac{1}{\sinh\nu}\begin{pmatrix}\sinh(\nu-\beta) & \sinh\beta\\ -\sinh\beta & \sinh(\nu+\beta)\end{pmatrix}, \quad \tilde{c} = \begin{pmatrix} e^{\gamma} & 0\\ 0 & e^{-\gamma}\end{pmatrix},$$

$$\tilde{y} = \frac{1}{\sinh\nu}\begin{pmatrix}\sinh(\nu-\beta) & e^{\gamma}\sinh\beta\\ -e^{-\gamma}\sinh\beta & \sinh(\nu+\beta)\end{pmatrix}.$$

We want to normalize this group so that the right half-plane is precisely invariant under $\langle c\rangle$. To accomplish this, we represent the reflection r_c in A_c by the matrix

$$\tilde{r}_c = \tilde{r}_c^{-1} = \begin{pmatrix}-1 & 0\\ 0 & 1\end{pmatrix},$$

and write down the following matrices.

$$\tilde{a}' = \tilde{r}_c\tilde{x}\tilde{r}_c^{-1} = \frac{1}{\sinh\nu}\begin{pmatrix}\sinh(\nu-\beta) & -\sinh\beta\\ \sinh\beta & \sinh(\nu+\beta)\end{pmatrix};$$

$$\tilde{c}' = \tilde{c} = \begin{pmatrix} e^{\gamma} & 0\\ 0 & e^{-\gamma}\end{pmatrix};$$

$$\begin{aligned}\tilde{e}' &= \frac{1}{\sinh\nu}\begin{pmatrix}\sinh(\nu-\beta) & -e^{\gamma}\sinh\beta\\ e^{-\gamma}\sinh\beta & \sinh(\nu+\beta)\end{pmatrix}\\ &= \frac{1}{\sinh\nu}\begin{pmatrix}-e^{-\gamma}\sinh(\nu+\beta) & -e^{\gamma}\sinh\beta\\ e^{-\gamma}\sinh\beta & -e^{-\gamma}\sinh(\nu-\beta)\end{pmatrix}.\end{aligned}$$

As above, we have written $\tilde{e}'$ in two separate forms; it follows from (3) that these are in fact equal.

Let $G_3 = \langle a', c, e'\rangle$. Then G_3 is a purely hyperbolic discrete group, where $\mathbb{H}^2/G_3$ is a sphere with three holes; the axes of a', c and e' project to the distinct simple geodesics freely homotopic to the boundary; and G_3 has the presentation:

$$G_3 = \{a', c, e' | e'ca' = 1\}.$$

We note that the left half-plane is precisely invariant under $\langle c\rangle$ in G_2 while the right half-plane is precisely invariant under $\langle c\rangle$ in G_3. However, we need to close the second handle before we adjoin this group to G_2.

8. Closing the second handle without twist.

We repeat the process given in Sect. 5, with the following changes. We first replace μ by ν, and we replace α by β — this yields a matrix $\tilde{z}$, which conjugates $\tilde{x}$ into $\tilde{y}^{-1}$. Then we conjugate $\tilde{z}$ by $\tilde{r}$ to obtain a matrix $\tilde{z}'$, which conjugates $\tilde{a}'$ into $(\tilde{e}')^{-1}$. However we need a transformation conjugating $(e')^{-1}$ to a', so we replace $\tilde{z}'$ by its inverse, which we label as $\tilde{d}$. We obtain

$$\tilde{d} = \frac{1}{\sinh\nu}\begin{pmatrix} e^{-\gamma/2} & e^{\gamma/2}\cosh\nu \\ -e^{-\gamma/2}\cosh\nu & -e^{\gamma/2} \end{pmatrix}.$$

We remark that, by using (3), we can easily compute that $\tilde{d}\tilde{c}\tilde{a}' = -\tilde{a}'\tilde{d}$.

As above, we can use the second combination theorem to conclude that $G_4 = \langle G_3, d\rangle$ is discrete and purely hyperbolic, and represents a torus with one hole. The one hole has the right half-plane, which is precisely invariant under $\langle c\rangle$, as its boundary half-plane.

Also, for future use, we have the presentation:

$$G_4 = \{a', c, d, e' | e'ca' = a'de'd^{-1} = 1\}.$$

9. Twisting the second handle.

We repeat the work of Sect. 6, but with a small difference. We want to twist by 2σ about a', which is conjugate to e'. This twist leaves a', c and e' unchanged, but changes d. We first write down the matrix $\tilde{a}'(\sigma)$, corresponding to the transformation with the same fixed points as a', but with $tr(\tilde{a}'(\sigma)) = 2\cosh\sigma$, as compared with $tr(\tilde{a}') = 2\cosh\beta$. We obtain

$$\tilde{a}'(\sigma) = \frac{1}{\sinh\nu}\begin{pmatrix} \sinh(\nu-\sigma) & -\sinh\sigma \\ \sinh\sigma & \sinh(\nu+\sigma) \end{pmatrix}.$$

As above, we note that $a'(0) = 1$, and $a'(\beta) = a'$. We now write

$$\tilde{d}_\sigma = \tilde{a}'(\sigma)\tilde{d} = \frac{1}{\sinh\nu}\begin{pmatrix} e^{-\gamma/2}\cosh\sigma & e^{\gamma/2}\cosh(\nu-\sigma) \\ -e^{-\gamma/2}\cosh(\nu+\sigma) & -e^{\gamma/2}\cosh\sigma \end{pmatrix}.$$

After we replace d by d_σ, we obtain the group $G_4(\sigma) = \langle a', c, d_\sigma, e'\rangle$. We can obtain $G_4(\sigma)$ from G_3 and d_σ by using the combination theorem exactly as we obtained G_4. We conclude that $G_4(\sigma)$ is discrete and purely loxodromic; $\mathbb{H}^2/G_4(\sigma)$ is a torus with one hole; the right half-plane, which is precisely invariant under $\langle c\rangle$, is a boundary half-plane corresponding to the hole; and $G_4(\sigma)$ has the presentation:

$$G_4(\sigma) = \{a', c, d_\sigma, e' | e'ca' = a'd_\sigma e'd_\sigma^{-1} = 1\}.$$

10. Amalgamating the two groups without twist.

We have already observed that the left, respectively, right, half-plane is precisely invariant under $\langle c\rangle$ in $G_2(\rho)$, respectively, $G_4(\sigma)$; hence we can apply the first combination theorem to amalgamate these two groups. We obtain that $G' = \langle G_2(\rho), G_4(\sigma)\rangle = \langle a, a', b_\rho, c, d_\sigma, e, e'\rangle$ is discrete and purely hyperbolic; it represents a closed Riemann surface of genus 2, and has the presentation:

$$G' = \{a, a', b_\rho, c, d_\sigma, e, e' | eca = eb_\rho ab_\rho^{-1} = e'ca' = a'd_\sigma e'd_\sigma^{-1} = 1\}. \tag{4}$$

11. Doing it with a twist.

In order to describe the twist about L_c, we consider the common orthogonals L_{ac} and $L_{a'c}$. We say that the twist about L_c is 2τ, if the distance, measured in the positive direction along A_c, between L_{ac} and $L_{a'c}$ is 2τ.

We can accomplish the twist by 2τ about L_c by keeping $\tilde{a}$, $\tilde{e}$, and $\tilde{b}_\rho$ unchanged, and conjugating $\tilde{a}'$, $\tilde{e}'$ and $\tilde{d}_\sigma$ by

$$\tilde{c}(\tau) = \begin{pmatrix} e^{\tau} & 0 \\ 0 & e^{-\tau} \end{pmatrix}.$$

This changes $\tilde{a}'$ to $\tilde{a}'_\tau = \tilde{c}(\tau)\tilde{a}'\tilde{c}(\tau)^{-1}$, $\tilde{e}'$ to $\tilde{e}'_\tau = \tilde{c}(\tau)\tilde{e}'\tilde{c}(\tau)^{-1}$, and $\tilde{d}_\sigma$ to $\tilde{d}_{\sigma\tau} = \tilde{c}(\tau)\tilde{d}_\sigma\tilde{c}(\tau)^{-1}$. We explicitly write down these matrices.

$$\tilde{a}'_\tau = \frac{1}{\sinh\nu}\begin{pmatrix} \sinh(\nu-\beta) & -e^{2\tau}\sinh\beta \\ e^{-2\tau}\sinh\beta & \sinh(\nu+\beta) \end{pmatrix};$$

$$\tilde{e}'_\tau = \frac{1}{\sinh\nu}\begin{pmatrix} \sinh(\nu-\beta) & -e^{2\tau+\gamma}\sinh\beta \\ e^{-2\tau-\gamma}\sinh\beta & \sinh(\nu+\beta) \end{pmatrix};$$

$$\tilde{d}_{\sigma\tau} = \frac{1}{\sinh\nu}\begin{pmatrix} e^{-\gamma/2}\cosh\sigma & e^{2\tau+\gamma/2}\cosh(\nu-\sigma) \\ -e^{-2\tau-\gamma/2}\cosh(\nu+\sigma) & -e^{\gamma/2}\cosh\sigma \end{pmatrix}.$$

As above, the group $G_4(\sigma,\tau) = \langle a'_\tau, c, d'_{\sigma\tau}\rangle$ is discrete and purely hyperbolic; it represents a torus with one hole, where the right half-plane, which is precisely invariant under $\langle c\rangle$, is a boundary half-plane.

We can again use the first combination theorem to amalgamate $G_2(\rho)$ to $G_4(\sigma,\tau)$. We thus obtain our final group $G = \langle a, a'_\tau, b_\rho, c, d_{\sigma\tau}, e, e'_\tau\rangle$. We have shown that G is discrete and purely hyperbolic, and that $\mathbb{H}^2/G$ is a closed surface of genus 2.

Finally, as in equation (4), G has the presentation:

$$G = \{a, a'_\tau, b_\rho, c, d_{\sigma\tau}, e, e'_\tau \mid eca = eb_\rho ab_\rho^{-1} = e'_\tau ca'_\tau = a'_\tau d_{\sigma\tau}e'_\tau d_{\sigma\tau}^{-1} = 1\}. \tag{5}$$

12. A standard basis.

Finally, we express our result in terms of a standard basis. We observe that we can define a standard basis, satisfying the usual relation, by setting $a_1 = b_\rho$, $b_1 = a$, $a_2 = d_{\sigma\tau}$ and $b_2 = (e'_\tau)^{-1}$. It follows from (5) that, with these definitions, $c = a_1b_1a_1^{-1}b_1^{-1} = (a_2b_2a_2^{-1}b_2^{-1})^{-1}$.

Then, finally, our matrices for these generators are:

$$\tilde{a}_1 = \tilde{b}_\rho = \frac{1}{\sinh\mu}\begin{pmatrix} -e^{\gamma/2}\cosh\rho & e^{\gamma/2}\cosh(\mu+\rho) \\ -e^{-\gamma/2}\cosh(\mu-\rho) & e^{-\gamma/2}\cosh\rho \end{pmatrix};$$

$$\tilde{b}_1 = \tilde{a} = \frac{1}{\sinh\mu}\begin{pmatrix} \sinh(\mu-\alpha) & \sinh\alpha \\ -\sinh\alpha & \sinh(\mu+\alpha) \end{pmatrix};$$

$$\tilde{a}_2 = \tilde{d}_{\sigma\tau} = \frac{1}{\sinh\nu}\begin{pmatrix} e^{-\gamma/2}\cosh\sigma & e^{2\tau+\gamma/2}\cosh(\nu-\sigma) \\ -e^{-2\tau-\gamma/2}\cosh(\nu+\sigma) & -e^{\gamma/2}\cosh\sigma \end{pmatrix};$$

$$\tilde{b}_2 = (\tilde{e}'_\tau)^{-1} = \frac{1}{\sinh\nu}\begin{pmatrix} \sinh(\nu+\beta) & e^{2\tau+\gamma}\sinh\beta \\ -e^{-2\tau-\gamma}\sinh\beta & \sinh(\nu-\beta) \end{pmatrix}.$$

13. Statement of Results.

For convenience in exposition, we choose a base surface. Let $\alpha_0 = \beta_0 = \gamma_0$ be defined by $\cosh(\alpha_0) = \cosh(\beta_0) = \cosh(\gamma_0) = 2$, and let $\rho_0 = \sigma_0 = \tau_0 = 0$. Let G_0 be the group generated by the corresponding matrices $\tilde{a}_1^0, \tilde{b}_1^0, \tilde{a}_2^0, \tilde{b}_2^0$. We remark that it was shown in [**3**] that $\mathbb{H}^2/G_0$ is a normal subgroup of the $(2,4,6)$-triangle group; i.e., it is the algebraic curve: $w^2 = z^6 - 1$.

THEOREM 13.1. *Let $\alpha > 0$, $\beta > 0$, $\gamma > 0$, ρ, σ and τ be given real numbers. Define the quantities μ, respectively, ν, by (2) and (3), respectively, and define the matrices $\tilde{a}_1$, $\tilde{b}_1$, $\tilde{a}_2$ and $\tilde{b}_2$ by the formulae in Sect. 12. Then*
(i) *the group, $G = G(\alpha, \beta, \gamma, \rho, \sigma, \tau) = \langle a_1, b_1, a_2, b_2 \rangle$ is discrete and purely hyperbolic, and the surface $S = \mathbb{H}^2/G$ is closed and of genus 2;*
(ii) *the mapping Φ defined by*

$$\lambda = (\sinh\alpha, \sinh\beta, \sinh\gamma, \sinh\rho, \sinh\sigma, \sinh\tau) \mapsto (a_1, b_1, a_2, b_2)$$

defines an algebraic diffeomorphism between $(\mathbb{R}^+)^3 \times \mathbb{R}^3$ and the Teichmüller space of closed Riemann surfaces of genus 2, described as the identity component of the space of discrete faithful representations of G_0 into $PSL(2, \mathbb{R})$, modulo conjugation.

PROOF. We have already proven statement (i) by using combination theorems. For every $\lambda \in (\mathbb{R}^+)^3 \times \mathbb{R}^3$, there is the obvious isomorphism from G_0 onto G, given by $a_1^0 \mapsto a_1$, etc. One easily sees, using the combination theorems, that this isomorphism is induced by an orientation-preserving homeomorphism of the closed upper half-plane. Hence we have a well defined map $\Phi\colon (\mathbb{R}^+)^3 \times \mathbb{R}^3 \to \mathcal{DF}_0(G_0, SL(2, \mathbb{R}))$, the identity component of the space of discrete faithful representations of G_0 into $SL(2, \mathbb{R})$, modulo conjugation.

It is essentially immediate from the definitions of the matrices that Φ is algebraic. In order to establish the unicity of Φ, it suffices to show that the six parameters are intrinsically defined by the generators. To this end, we first note that $a, b_\rho, c, d_{\sigma\tau}, e, a'_\tau, e'_\tau$ are an equivalent set of generators for G, so it suffices to observe that the parameters $(\alpha, \beta, \gamma, \rho, \sigma, \tau)$ can be intrinsically defined in terms of these generators.

We first note that $tr(a) = 2\cosh\alpha$, $tr(a'_\tau) = 2\cosh\beta$, and $tr(c) = 2\cosh\gamma$.

Further, the distance, measured along A_e, between the image under b_ρ of the beginning point of $\hat{L}_{ae}$, and the endpoint of $\hat{L}_{ae}$, is exactly 2ρ. Similarly, the distance on $A_{a'_\tau}$ between the beginning point of $\hat{L}_{a'_\tau e'_\tau}$, and the $d_{\sigma\tau}$ image of its endpoint, is exactly 2σ. Finally, we have already mentioned that the distance, measured along A_c, between the points of intersection with L_{ac} and with $L_{a'_\tau c}$ is exactly 2τ. These distances are of course all measured in the positive sense on the respective axes; thus allowing ρ, σ and τ to take on all values.

The above remarks show equally well that Φ^{-1} is well defined and differentiable on all of $\mathcal{DF}(G_0, SL(2, \mathbb{R}))^+$. □

14. Final remark.

We have shown that the generators, and hence the whole Fuchsian group, lies in some $PSL(2, k)$, where k is a number field, if and only if the parameters $\sinh\alpha, \sinh\beta, \sinh\gamma, \sinh\rho, \sinh\sigma, \sinh\tau$ are all algebraic. It follows that these parameters are all algebraic for any one such homotopy basis if and only if they are all algebraic for every such homotopy basis. In particular, it follows that the

Teichmüller modular group acts on this space of Fenchel-Nielsen coordinates as a group of algebraic diffeomorphisms.

We remark that the above phenomena occur in the more general context of an arbitrary hyperbolic orbifold with finitely generated fundamental group; this will be pursued elsewhere.

References

[1] P. Buser and R. Silhol. Geodesics, periods and equations of real hyperelliptic curves. to appear.

[2] J. Gilman and B. Maskit. An algorithm for 2-generator fuchsian groups. *Mich. Math. J.*, 38:13–32, 1991.

[3] B. Maskit. New parameters for fuchsian groups of genus 2. *Proc. Amer. Math. Soc.* to appear.

[4] B. Maskit. Explicit matrices for fuchsian groups. *Cont. Math.*, 169:451–466, 1994.

Mathematics Department, The University at Stony Brook, Stony Brook NY 11794-3651

E-mail address: bernie@math.sunysb.edu

Contemporary Mathematics
Volume **240**, 1999

Arithmetic Fuchsian groups of signature $(0; e_1, e_2, e_3, e_4)$ with $2 \leq e_1 \leq e_2 \leq e_3,\ e_4 = \infty$

T. Nakanishi, M. Näätänen, and G. Rosenberger

Let $G = \langle A, B, C, D;\ A^{e_1} = B^{e_2} = C^{e_3} = D^{e_4} = ABCD = 1\rangle$, $2 \leq e_1 \leq e_2 \leq e_3 \leq e_4 \leq \infty$, be a Fuchsian group of signature $(0; e_1, e_2, e_3, e_4)$. For $e_4 = \infty$, we say D is parabolic and write $D^\infty = 1$, analogously for e_1, e_2, e_3. Let tr $A = \alpha \geq 0$, tr $B = \beta \geq 0$, tr $C = \gamma \geq 0$ and tr $D = \delta \geq 0$ (we use, with slight ambiguity, tr E for the trace of E, $E \in$ PSL $(2, \mathbb{R})$, and normalize $\alpha, \beta, \gamma, \delta$ to be non–negative.) For discreteness reasons, we then have $x = -\text{tr } BC > 2$, $y = -\text{tr } AC > 2$, and $z = -\text{tr } AB > 2$. We define numbers a, b, c and d, depending only on the traces A, B, C and D by $a = \alpha\delta + \beta\gamma$, $b = \beta\delta + \alpha\gamma$, $c = \gamma\delta + \alpha\beta$ and $d = \alpha\beta\gamma\delta + \alpha^2 + \beta^2 + \gamma^2 + \delta^2 - 4$. The Teichmüller space $\mathcal{T} = \mathcal{T}(e_2, e_2, e_3, e_4)$ for the signature $(0; e_1, e_2, e_3, e_4)$, as the space of conjugacy classes in PGL$(2, \mathbb{R})$ for this signature, is represented by

$$\{(x, y, z) \in \mathbb{R}^3_+ \mid x > 2,\ y > 2,\ z > 2\}$$

fulfilling

$$x^2 + y^2 + z^2 - xyz + ax + by + cz + d = 0. \tag{1}$$

Here x, y and z represent the traces of the products as above. The equation (1) is obtained from tr $(ABC)^2 = \delta^2 - 2$ by using several times the equation tr $AB +$ tr $(BA)^{-1} =$ tr A tr B for $A, B \in$ SL$(2, \mathbb{R})$, see [5]. Recall that the Teichmüller space has dimension 2 as a real algebraic set.

The transformations

$$T_1 : (x, y, z) \mapsto (yz - x - a, y, z)$$
$$T_2 : (x, y, z) \mapsto (x, xz - y - b, z)$$
$$T_3 : (x, y, z) \mapsto (x, y, xy - z - c)$$

generate a group which, as a subgroup of the Teichmüller modular group, acts discontinuously on $\mathcal{T}$. Using the transformations T_i above, we find in finitely many steps a solution (x, y, z) of (1), satisfying the inequalities $2 < x \leq \frac{1}{2}(yz - a)$,

1991 *Mathematics Subject Classification.* Primary 30F35; Secondary 20H10.
The first author was supported by Grant-in-Aid for Scientific Research (No. 09640176), Ministry of Education, Science and Culture of Japan.
The second author was supported by the Academy of Finland.
The third author was supported by the University of Helsinki.

$2 < y \leq \frac{1}{2}(xz - b)$, and $2 < z \leq \frac{1}{2}(xy - c)$. Such a solution is called here a *local fundamental solution.* Depending on the orders e_i, $i = 1, \ldots, 4$, automorphisms of G, which induce non-trivial permutations of x, y, z, might exist. Hence, if the equation (1) is symmetric in two or all three of x, y, z, then, in addition, in a local fundamental solution two or all three of x, y, z may be ordered.

Our aim is to classify – up to conjugacy in $PGL(2, \mathbb{R})$ – the arithmetic Fuchsian groups of signature $(0; e_1, e_2, e_3, e_4)$ with $2 \leq e_1 \leq e_2 \leq e_3 \leq e_4 = \infty$. This is done by using the equation (1). In other words, we want to find those (ordered if possible) local fundamental solutions (x, y, z), of the related equation (1), which correspond to an arithmetic Fuchsian group G of signature $(0; e_1, e_2, e_3, e_4)$, $2 \leq e_1 \leq e_2 \leq e_3 \leq e_4 = \infty$. Such a triple is called a *describing triple* (x, y, z). Recall that G is determined up to conjugacy by $\alpha, \beta, \gamma, \delta, x, y$ and z, see [7].

The motivation for this classification comes firstly from our interest to find the integer solutions of (1) with $a, b, c, d \in \mathbb{N}$. and secondly from Takeuchi's classification of the subgroups of the modular group with signature $(0; e_1, e_2, e_3, e_4)$ [11].

K. Takeuchi's characterization of arithmetic Fuchsian groups [10] has been reformulated in [4] as

Theorem 1. Let Γ be a Fuchsian group of the first kind. Then Γ is an arithmetic Fuchsian group if and only if both of the following conditions hold

(i) $k_1 = \mathbb{Q}(\mathrm{tr}\ g \mid g \in \Gamma)$ is a totally real number field and $\mathrm{tr}\ \Gamma \subset \mathcal{O}_{k_1}$, where $\mathrm{tr}\ \Gamma = \{\mathrm{tr}\ g \mid g \in \Gamma\}$ and $\mathcal{O}_{k_1}$ the ring of integers in k_1

(ii) for every $\mathbb{Q}$–isomorphism $\sigma : k_1 \to \mathbb{R}$ such that $\sigma \mid k_2 \neq \mathrm{id}$, where $k_2 = \mathbb{Q}((\mathrm{tr}\ g)^2 \mid g \in \Gamma)$, then

$$|\sigma(\mathrm{tr}\ g)| < 2 \ \text{ for all } \ g \neq 1. \quad \square$$

If Γ is an arithmetic Fuchsian group having parabolic elements, then $k_2 = \mathbb{Q}$ and Γ is commensurable with the modular group $PSL(2, \mathbb{Z})$. From this it follows that there are only finitely many conjugacy classes of arithmetic Fuchsian groups of signature $(0; e_1, e_2, e_3, e_4)$ having parabolic elements. Γ being commensurable with $PSL(2, \mathbb{Z})$ means that it is conjugate to a subgroup of finite index in some $\Gamma_1(n)$, where

$$\Gamma_1(n) = \text{normalizer of } \ \Gamma_0(n) \ \text{ in } \ PSL(2, \mathbb{R}) \ \text{ and}$$

$$\Gamma_0(n) = \{\begin{pmatrix} a & b \\ c & d \end{pmatrix} \in PSL(2, \mathbb{Z}) \mid c \equiv 0 \bmod n\} \ \text{ for } \ n \ \text{ square-free [1].}$$

If Γ is a subgroup of some $\Gamma_1(n)$, then

$$\mu(\Gamma_1(n)) = \frac{\pi}{3} \prod_{p|n} \frac{p+1}{2} \leq \mu(\Gamma),$$

where μ denotes the non–Euclidean volume. For a fixed Γ this gives restrictions for n; if for instance $\mu(\Gamma) \leq 2\pi$, then $1 \leq n \leq 14$. The structure of $\Gamma_1(n)$ is well understood, see [3]. Let n be a positive square–free integer. Then $\Gamma_1(n)$ can only have maximal finite subgroups of orders 2,3,4 or 6. Furthermore, $\Gamma_1(n)$ has

(a) exactly one conjugacy class of maximal parabolic subgroups

(b) exactly one conjugacy class of maximal finite subgroups of order 4 if $n = 2p_2 \cdots p_t$ with each prime $p_i \equiv 1 \pmod 4$, and no element of order 4 otherwise
(c) exactly one conjugacy class of maximal finite subgroups of order 6 if $n = 3p_2 \cdots p_t$ with each prime $p_i \equiv 1 \pmod 3$, and no element of order 6 otherwise
(d) exactly one conjugacy class of maximal finite subgroups of order 3 if $n = p_1 p_2 \cdots p_t$ with each prime $p_i \equiv 1 \pmod 3$, and no conjugacy class of maximal finite subgroups of order 3 otherwise.

Moreover, if $n = 2p_2 \cdots p_t$ with each prime $p_i \equiv 1 \pmod 4$, then $\Gamma_1(n)$ is isomorphic to a subgroup of finite index in $\Gamma_1(2)$ which has signature $(0; 2, 4, \infty)$, and if $n = 3p_2 \cdots p_t$ with prime $p_i \equiv 1 \pmod 3$, then $\Gamma_1(n)$ is isomorphic to a subgroup of finite index in $\Gamma_1(3)$ which has signature $(0; 2, 6, \infty)$.

For $n = p_1, \ldots, p_t$, $t \geq 0$, the p_i being pairwise different primes, the genus g of $\Gamma_1(n)$ and the number of conjugacy classes of maximal finite groups of order 2 in $\Gamma_1(n)$ can now be easily computed using the following two equations

$$\mu(\Gamma_0(n)) = 2^t \mu(\Gamma_1(n))$$
$$\mu(\Gamma_1(n)) = \tfrac{\pi}{3} \prod_{i=1}^{t} \tfrac{p_i+1}{2},$$

using the structural knowledge on $\Gamma_0(n)$, see [8], and the above facts for $\Gamma_1(n)$. A table of the signatures of $\Gamma_1(n)$ for $1 \leq n < 50$ is given in [3].

Let G be a Fuchsian group of signature $(0; e_1, e_2, e_3, \infty)$, with $2 \leq e_1 \leq e_2 \leq e_3$. Our aim is to decide if G is arithmetic and to give a complete list of conjugacy classes (in $PGL(2, \mathbb{R})$ of arithmetic Fuchsian groups of this signature in terms of x, y, z introduced above. There are only finitely many conjugacy classes of arithmetic Fuchsian groups of such a signature, and the conjugacy class of G is given by $\alpha, \beta, \gamma, \delta, x, y$ and z, see [6]. For such a signature, $k_2 = \mathbb{Q}((\text{tr } g)^2 \mid g \in G) = \mathbb{Q}$, hence we have to consider the cases $\delta = 2$, $\alpha, \beta, \gamma \in \{0, 1, \sqrt{2}, \sqrt{3}, 2\}$ and $x = \sqrt{p}x_0$, $y = \sqrt{q}y_0$, $z = \sqrt{r}z_0$, where $x_0, y_0, z_0 \in \mathbb{N}$ and $p, q, r \in \mathbb{N}$ are square free. The restriction $1 \leq p, q, r < 50$ follows from the fact that $\mu(G) \leq 4\pi$ and G is conjugate to a subgroup of some $\Gamma_1(n)$. Moreover, the structural tables for the signatures of $\Gamma_1(n)$, $1 \leq n < 50$ are available, see [3]. We employ parallelly two different methods. Firstly, a result of D. Singerman [9] can be used to check if a group G of signature $(0; e_1, e_2, e_3, \infty)$ is conjugate to a subgroup of some $\Gamma_1(n)$:

Theorem 2. Let Γ be a Fuchsian group of the first kind with the canonical presentation

$$\Gamma = \langle x_1, \ldots, x_r, p_1, \ldots, p_s, a_1, b_1, \ldots, a_g, b_g;$$
$$x_1^{m_1} = \cdots = x_r^{m_r} = \prod_{i=1}^{g} [a_1, b_i] \prod_{j=1}^{r} x_j \prod_{k=1}^{s} p_k = 1\rangle,$$

$2 \leq m_i < \infty$ for all i, that is a Fuchsian group of the first kind with the canonical signature $(g; m_1, \ldots, m_r; s)$. Then Γ contains a subgroup Γ' of the first kind and index N with the respective canonical signature $(g'; n_{11}, n_{12}, \ldots, n_{1\rho_1}, \ldots, n_{r_1}, n_{r_2}, \ldots, n_{r\rho_r}; s')$, $2 \leq n_{ij} < \infty$ for all pairs (i, j), if and only if

(a) There exists a finite permutation group H transitive on N points, and an epimorphism $\theta : \Gamma \to H$ satisfying the following conditions:

(i) The permutation $\theta(x_j)$ has precisely ρ_j cycles of lengths less than m_j, the lengths of these cycles being $\frac{m_j}{n_{j1}}, \dots, \frac{m_j}{n_{j\rho_j}}$,

(ii) if we denote the number of cycles in the permutation $\theta(p_k)$ by $\delta(p_k)$, then $s' = \sum_{k=1}^{s} \delta(p_k)$.

(b) $\frac{\mu(\Gamma')}{\mu(\Gamma)} = N$. □

Secondly, we consider for each of the 35 combinations for α, β and γ the equation (1) for the possibility of having solutions of the type $x = \sqrt{p}x_0,\ y = \sqrt{q}y_0,\ z = \sqrt{r}z_0$ with the restrictions and inequalities described above. We consider in details four examples:

Example 1. Let G be a Fuchsian group of signature $(0; \infty, \infty, \infty, \infty)$. Then $a = b = c = 8$ and $d = 28$, hence the equation (1) is now

$$x^2 + y^2 + z^2 - xyz + 8x + 8y + 8z + 28 = 0.$$

On one side, up to isomorphism, G occurs as a subgroup of $\Gamma_1(n)$ only for $n \leq 6$. On the other side, the equation cannot have a solution of the type $x = \sqrt{p}x_0,\ y = \sqrt{q}y_0$ and $z = \sqrt{r}z_0$ with $p \geq 2$, $q \geq 2$ or $r \geq 2$. In this case the equation is symmetric in x, y and z, hence we can look for solutions $x, y, z \in \mathbb{N}$ with $2 < x \leq y \leq z \leq \frac{1}{2}(xy - 8)$. From the equation we get

$$y \leq z = \frac{1}{2}\,[xy - 8 - \sqrt{(xy-8)^2 - 4(x^2 + y^2 + 8x + 8y + 28)}],$$

and, hence,

$$xy^2 \leq x^2 + 2y^2 + 8x + 16y + 28,$$

which yields $xy^2 < 16y^2$, that is, $3 \leq x \leq 15$ because $x \leq y$.

The Reduce–Computerprogram by Michael Niermann (Dortmund), which is enclosed, gives the triples

$$(x, y, z) = (3, 18, 23), (4, 10, 16), (6, 6, 14), (7, 7, 7).$$

For each triple there is a realization of G as an arithmetic Fuchsian group, see also [11].

Example 2. Let G be a Fuchsian group of signature $(0; 4, \infty, \infty, \infty)$. Then $a = b = c = 4 + 2\sqrt{2},\ d = 10 + 8\sqrt{2}$ and (1) cannot have a solution of the type $x = \sqrt{p}x_0,\ y = \sqrt{q}y_0,\ z = \sqrt{r}z_0$ with restrictions as above, hence no arithmetic Fuchsian group appears in this signature.

Example 3. Let G be a Fuchsian group of signature $(0; 2, 4, 6, \infty)$. Then $a = \sqrt{6},\ b = 2\sqrt{2},\ c = 2\sqrt{3}$ and $d = 5$. By Singerman's subgroup theorem and Maclachlan's list G cannot be realized as a subgroup of any $\Gamma_1(n)$, and, hence G is not an arithmetic Fuchsian group. Recall that a subgroup of $\Gamma_1(n)$ cannot have an element of order 4 and an element of order 6 at the same time for any n.

Example 4. Let G be a Fuchsian group of signature $(0; 2, 3, 6, \infty)$. Then $a = \sqrt{3},\ b = 2,\ c = 2\sqrt{3}$ and $d = 4$. The equation (1) gets now the form

$$x^2 + y^2 + z^2 - xyz + \sqrt{3}x + 2y + 2\sqrt{3}z + 4 = 0.$$

Here we have necessarily $x = \sqrt{3}x_0$, $y \in \mathbb{N}$ and $z = \sqrt{3}z_0$, $x_0, z_0 \in \mathbb{N}$. There is no symmetry for x, y and z, hence we have to distinguish as in Example 3. This leads to $2 < x < 6$, $2 < y < 6$ or $2 < z < 6$, and the Reduce–program now gives only one possible triple, $(x, y, z) = (3\sqrt{3}, 8, 2\sqrt{3})$. For this triple there is a realization of G as a subgroup of $\Gamma_1(3)$, hence G is an arithmetic group. □

The other possible signatures are handled in an analogous manner. We used in addition the Reidermeister-Shreier-method, see [2], for the respective groups $\Gamma_1(n)$ to have a second calculation for the number of conjugacy classes.

To summarise, we get

Theorem: The following list completely describes the conjugacy classes of arithmetic Fuchsian group of signature $(0; e_1, e_2, e_3, \infty)$ with $2 \le e_1 \le e_2 \le e_3 \le \infty$ in terms of the signature and the describing triple (x, y, z):

Signature	x	y	z
$(0; 3, \infty, \infty, \infty)$	3	14	18
	4	8	13
	5	6	12
	6	6	7
$(0; 3, 3, \infty, \infty)$	3	11	14
	5	5	10
	6	6	5
$(0; 3, 3, 3, \infty)$	5	5	5
$(0; 2, 3, \infty, \infty)$	3	8	10
	4	5	8
	5	6	4
$(0; 2, 2, 3, \infty)$	4	4	3
	$2\sqrt{3}$	$2\sqrt{3}$	5
$(0; 2, 3, 3, \infty)$	5	4	4
$(0; 4, 4, \infty, \infty)$	$4\sqrt{2}$	$4\sqrt{2}$	8
	$2\sqrt{2}$	$14\sqrt{2}$	16
$(0; \infty, \infty, \infty, \infty)$	3	18	23
	4	10	16
	6	6	14
	7	7	7
$(0; 2, \infty, \infty, \infty)$	3	10	13
	4	6	10
	5	5	7
$(0; 2, 2, \infty, \infty)$	3	6	7
	4	4	6
	5	5	3
	$3\sqrt{2}$	$3\sqrt{2}$	4
	$2\sqrt{5}$	$3\sqrt{5}$	3
$(0; 2, 2, 2, \infty)$	3	3	3
	$2\sqrt{2}$	$2\sqrt{2}$	4
	$\sqrt{6}$	$2\sqrt{3}$	$3\sqrt{2}$
	$\sqrt{5}$	$2\sqrt{5}$	5

Signature	x	y	z
$(0;2,4,\infty,\infty)$	$3\sqrt{2}$	$4\sqrt{2}$	6
	$2\sqrt{2}$	$10\sqrt{2}$	9
$(0;2,4,4,\infty)$	8	$3\sqrt{2}$	$3\sqrt{2}$
$(0;2,6,\infty,\infty)$	$2\sqrt{3}$	$7\sqrt{3}$	6
$(0;3,6,6,\infty)$	11	$3\sqrt{3}$	$3\sqrt{3}$
$(0;2,3,6,\infty)$	$3\sqrt{3}$	8	$2\sqrt{3}$

Remark: In a case where x, y and z are all integers we find always a realization of a Fuchsian group with the corresponding signature $(0; e_1, e_2, e_3, e_4)$ as a subgroup of the modular group $\mathrm{PSL}(2,\mathbb{Z})$. In the following tables we give for each case which comes in question a related subgroup of $\mathrm{PSL}(2,\mathbb{Z})$ in terms of generating matrices A, B and C. To get these tables we started with a suitable normalised parabolic element $\begin{pmatrix} 1 & u \\ 0 & 1 \end{pmatrix}$: therefore the orders of indices in a signature are different from the standard ones $(0; e_1, e_2, e_3, e_4)$ with $e_1 \le e_2 \le e_3 \le e_4$. Also we use the notation $(a, b; c, d) = \begin{pmatrix} a & b \\ c & d \end{pmatrix}$.

TABLE 1. $(\infty,\infty,\infty,\infty)$

$-trBC$	$-trCA$	$-trAB$	A	B	C
3	18	23	(-9,-4;25,11)	(1,-1;0,1)	(1,0;5,1)
4	10	16	(-5,-6;6,7)	(1,-3;0,1)	(1,0;2,1)
6	6	14	(-3,-8;2,5)	(1,-8;0,1)	(1,0;1,1)
7	7	7	(1,-1;0,1)	(1,0;9,1)	(-2,-1;9,4)

TABLE 2. $(\infty,\infty,\infty,2)$

$-trBC$	$-trCA$	$-trAB$	A	B	C
3	10	13	(-5,-12;3,7)	(1,-5;0,1)	(1,0;1,1)
4	6	10	(-3,-4;4,5)	(1,-3;0,1)	(1,0;2,1)
5	5	7	(-2,-1;9,4)	(1,-1;0,1)	(1,0;7,1)

TABLE 3. $(\infty,\infty,2,2)$

$-trBC$	$-trCA$	$-trAB$	A	B	C
3	6	7	(1,-3;0,1)	(1,0;3,1)	(-1,-1;2,1)
4	4	6	(1,-4;0,1)	(1,0;2,1)	(-1,-2;1,1)
5	5	3	(1,-5;0,1)	(1,0;1,1)	(-2,-5;1,2)

TABLE 4. $(2,2,2,\infty)$

$-trBC$	$-trCA$	$-trAB$	A	B	C
3	3	3	(0,-1;1,0)	(1,-1;2,-1)	(1,-2;1,-1)

TABLE 5. $(\infty, \infty, \infty, 3)$

$-trBC$	$-trCA$	$-trAB$	A	B	C
3	14	18	(-7,-16;4,9)	(1,-5;0,1)	(1,0;1,1)
4	8	13	(-4,-5;5,6)	(1,-3;0,1)	(1,0;2,1)
5	6	12	(-3,-8;2,5)	(1,-7;0,1)	(1,0;1,1)
6	6	7	(-2,-1;9,4)	(1,-1;0,1)	(1,0;8,1)

TABLE 6. $(\infty, \infty, 3, 3)$

$-trBC$	$-trCA$	$-trAB$	A	B	C
3	11	14	(1,-4;0,1)	(1,0;4,1)	(-1,-1;3,2)
5	5	10	(1,-6;0,1)	(1,0;2,1)	(-1,-3;1,2)
6	6	5	(1,-1;0,1)	(1,0;7,1)	(-2,-1;7,3)

TABLE 7. $(3, 3, 3, \infty)$

$-trBC$	$-trCA$	$-trAB$	A	B	C
5	5	5	(0,1;-1,1)	(-4,7;-3,5)	(-2,7;-1,3)

TABLE 8. $(\infty, 2, 3, \infty)$

$-trBC$	$-trCA$	$-trAB$	A	B	C
10	8	3	(1,-3;0,1)	(0,-1;1,0)	(-4,-7;3,5)
8	4	5	(1,-5;0,1)	(0,-1;1,0)	(-2,-7;1,3)
4	5	6	(1,-6;0,1)	(0,-1;1,0)	(-1,-3;1,2)

TABLE 9. $(\infty, 2, 2, 3)$

$-trBC$	$-trCA$	$-trAB$	A	B	C
3	4	4	(1,-4;0,1)	(0,-1;1,0)	(-1,-2;1,1)

TABLE 10. $(\infty, 2, 3, 3)$

$-trBC$	$-trCA$	$-trAB$	A	B	C
4	4	5	(1,-5;0,1)	(0,-1;1,0)	(-1,-3;1,2)

We wish to thank Michael Niermann for placing the Reduce-program at our disposal.

% REDUCE-program

```
procedure nst(xmax,n,m,k,a,b,c,d);

begin
scalar x,y,z,erg,diskr;
  on rounded;
  erg:={};
    for x:=floor(2/sqrt n)+1:xmax do
    << y:=floor(sqrt(n/m))*x;
      while -(k*m*n)**0.5*x*y**2+a*n*x+b*m*y+c*k*y+d+k*y**2+m*y**2+
              n*x**2
geq 0
        do
        << z:=
    if (diskr:=-2*sqrt(k*m*n)*c*k*x*y-4*a*k*n*x-4*b*k*m*y+c**2*k**2-
              4*d*k+k*m*n*x**2*y**2-4*k*m*y**2-4*k*n*x**2) geq 0
    then (-sqrt(diskr)+sqrt(k*m*n)*x*y - c*k)/(2*k)
            else 1/2;
          if z - floor z < 0.0000001
            then erg:=append(erg,{{x,y,z,"**",n,m,k}});
     y:=y+1
        >>
    >>;
    return erg
end;
```

References

[1] H. Helling: Bestimmung der Kommensurabilitätsklasse der Hilbertschen Modulgruppe; Math. Z. 92 (1966), 269–280.

[2] R.C. Lyndon and P.E. Schupp: Combinatorial group theory; Springer–Verlag 1977.

[3] C. Maclachlan: Groups of units of zero ternary quadratic forms; Proc. of the Royal Society of Edinburgh 88A (1981), 141–157.

[4] C. Maclachlan and G. Rosenberger: Two generator arithmetic Fuchsian groups, Math. Proc. Camb. Phil. Soc. 93 (1983), 383–391.

[5] T. Nakanishi and M. Näätänen: Areas of Two-dimensional Moduli Spaces: Reports of the Department of Mathematics, University of Helsinki, Preprint 158, October 1997.

[6] W. Plesken and G. Rosenberger: Simultaneous conjugation in quaternion algebras; Results in Mathematics 25 (1995), 120–124.

[7] G. Rosenberger; Some remarks on a paper of A.F. Bearden and P.C. Waterman about strongly discrete subgroups of $SL_2(\mathbb{C})$; J. London Math. Soc. 27 (1983), 39–42.

[8] B. Schoeneberg: Elliptic Modular Functions; Springer–Verlag 1974.

[9] D. Singerman: Subgroups of Fuchsian groups and finite permutation groups; Bull. London Math. Soc. 2 (1970), 319–323.

[10] K. Takeuchi: A Characterization of arithmetic Fuchsian groups; J. Math. Soc. Japan 27 (1975), 600-612.

[11] K. Takeuchi: Subgroups of the modular group with signature $(0; e_1, e_2, e_3, e_4)$; Saitama Math. J. 14 (1996) 55–78.

Graduate School of Mathematics, Nagoya University, Furo-cho, Chikusa-ku, Nagoya 464-8602, Japan
E-mail address: tosihiro@math.nagoya-u.ac.jp

Department of Mathematics, University of Helsinki, P.O. Box 4, FIN–00014 University of Helsinki, Finland
E-mail address: marjatta.naatanen@helsinki.fi

Universität Dortmund, 44221 Dortmund, Germany
E-mail address: rosenberger@mathematik.uni-dortmund.de

Contemporary Mathematics
Volume **240**, 1999

Geometric and holomorphic moduli for extremal Riemann surfaces

J. R. Quine

ABSTRACT. The theory of Hermite-type extremals for Riemann surfaces is discussed in relation to examples contained in a genus 3 curve family.

1. Introduction

This paper gives a short sketch of the theory of Hermite-type extremals for Riemann surfaces and techniques for finding them. As an example, extremals in the curve family KFT, so named in a paper of Rodriguez and Gonzales-Aguilera [**22**], are discussed. The family is named for the classical curves in the family, the Klein and the Fermat curves, and for singular tetrahedra in the family. These extremals give new examples of surfaces in the family which, unlike the Klein and Fermat curves, are not distinguished by extra automorphisms in addition to the group S_4. There are two extremals, the Schmutz Schaller and the Bavard extremals, the first defined in terms of geometric parameters and the second in terms of holomorphic parameters. The relationship between these two parametrizations is unknown except in the cases of the classical curves, and so the "distance" between these curves is not known, or even if they are different. The problem that emerges from this discussion is to find for KFT analytic relationships between the length and twist parameters and the Jacobian parameters defined below. Another interesting problem is to describe the action of the modular group on the length and twist parameters analogous to the action of $\Gamma_0(4)$ on the plane zA_3 in Siegel's space $\mathcal{H}_3$, as described in [**22, 5**].

2. Holomorphic and geometric moduli

2.1. Geometric moduli. The fundamental group of an orientable genus g surface can be represented in terms of generators and relations as

$$F_g = \{a_1, \cdots, a_g, b_1, \cdots, b_g \mid [a_1, b_1][a_2, b_2] \cdots [a_g, b_g] = 1\}.$$

Such a presentation will be called standard, and the generators a_j and b_j a standard set of generators. This is just one of many Fuchsian group types [**15, 16, 9**]; others are discussed below in connection with curve families.

1991 *Mathematics Subject Classification.* Primary 30F10 Secondary 14H15.
Key words and phrases. Klein curve, extremal Riemann Surface, Jacobian, length spectrum, Hermite problem, moduli space, symplectic lattice, curve family.

From the point of view of hyperbolic geometry, a marked compact Riemann Surface of genus g is given by a discrete representation of F_g into the group of sense preserving isometries of the hyperbolic plane, $PSL(2,\mathbb{R})$, and the space of such representations modulo the action of $PSL(2,\mathbb{R})$ is the Teichmüller space,

$$\mathcal{T}_g \subset PSL(2,\mathbb{R})\backslash \mathrm{Hom}\,(F_g, PSL(2,\mathbb{R})).$$

If $\rho : F_g \to PSL(2,\mathbb{R})$ is such a representation and $\gamma \in \mathrm{Aut}\,(F_g)$ then $\rho \circ \gamma$ is also in $\mathcal{T}_g$ and this gives the action of $\mathrm{Aut}\,(F_g)$ on $\mathcal{T}_g$. The action is trivial for an inner automorphism and so is defined for $\Gamma_g = \mathrm{Out}(F_g)$. The moduli space for compact genus g surfaces is the space

$$\mathcal{M}_g = \mathcal{T}_g/\Gamma_g.$$

The modular group Γ_g permutes the conjugacy classes of F_g.

The images $\rho\, a_j$ and $\rho\, b_j$ of the generators, suitably normalized, are a set of geometric parameters for $\mathcal{T}_g$. These are the Fricke coordinates ([**17**] section 2.5) and they can be related, with some difficulty, to the Frenchel-Nielsen length and twist coordinates for a pants decomposition [**2**].

The surface S_ρ associated to the representation ρ is the upper half-plane modulo the image of F_g,

$$S_\rho = \rho(F_g)\backslash \mathbb{H}.$$

2.2. Holomorphic moduli. The Jacobian gives a different parametrization of the moduli space, not easily related to the geometric one. The Jacobian can be thought of as a map into the Siegel upper half-plane [**12**],

$$\mathrm{Jac} : \mathcal{T}_g \to \mathcal{H}_g,$$

giving a marked symplectic lattice for each marked Riemann Surface. The lattice can be thought of as the integral homology $\mathrm{H}_1(S,\mathbb{Z})$ given as a sublattice of the complex dual of the space of holomorphic one-forms. It becomes a symplectic lattice when endowed with the Hermitian norm induced by integration over S of the wedge product of a form with its conjugate. Suitably normalized, the imaginary part of the inner product is the intersection form on cycles.

The action of the modular group on $\mathrm{H}_1(S,\mathbb{Z})$ can be described more explicitly. By abelianization, given a set of standard generators for F_g, there is a map ab : $F_g \to \mathbb{Z}^g \oplus i\mathbb{Z}^g$ given by $a_j \to e_j$, $b_j \to ie_j$, where e_j, $j = 1, \dots, g$ is the standard basis for $\mathbb{R}^g$. The elements $ab\, f$, $f \in F_g$, can be thought of as cycles in the integral homology group. When no confusion arises, the cycles $ab\, a_j$ and $ab\, b_j$ will be referred to simply as a_j and b_j respectively. This gives a map

$$Sp\, : \mathrm{Out}(F_g) \to Sp\,(2g,\mathbb{Z})$$

defined by the commutative diagram

$$\begin{array}{ccc} F_g & \xrightarrow{\;\gamma\;} & F_g \\ {\scriptstyle ab}\big\downarrow & & \big\downarrow{\scriptstyle ab} \\ \mathbb{Z}^g \oplus i\mathbb{Z}^g & \xrightarrow[Sp\,[\gamma]]{} & \mathbb{Z}^g \oplus i\mathbb{Z}^g \end{array}$$

and the map Sp also gives the commutative diagram

$$\begin{array}{ccc} \mathcal{T}_g & \xrightarrow{[\gamma]} & \mathcal{T}_g \\ \text{Jac}\big\downarrow & & \big\downarrow\text{Jac} \\ \mathcal{H}_g & \xrightarrow[Sp\,[\gamma]]{} & \mathcal{H}_g \end{array}$$

For $Z = X + iY$ in $\mathcal{H}_g$, the gram matrix, or matrix of inner products with respect to the standard basis of cycles, is given by

$$\begin{pmatrix} Y^{-1} & Y^{-1}X \\ XY^{-1} & Y + XY^{-1}X \end{pmatrix} + iJ$$

where

$$J = \begin{pmatrix} 0 & I_g \\ -I_g & 0 \end{pmatrix} \quad .$$

It follows that the gram matrix for the a cycles is Y^{-1}

3. Extremal surfaces

3.1. Length and Jacobi spectra. The length spectrum of a Riemann Surface is a function on the set of conjugacy classes $\mathcal{C}(F_g)$ of the fundamental group. If the surface is given by $\rho \in \mathcal{T}_g$ then the trace of an element $f \in F_g$, $\operatorname{tr} f$, is constant on the conjugacy class $[f]$. This trace will be denoted $\operatorname{tr}[f]$. The lengths are defined in terms of the traces by

$$\ell\,[f] = \frac{1}{2}\operatorname{arccosh}\frac{1}{2}|\operatorname{tr}\rho\,[f]\,|.$$

The map from $\mathbb{Z}^{2g}$ to $\mathbb{R}$ defined by norms of integral combinations of a standard basis of cycles will be referred to as the Jacobi spectrum.

3.2. Hermite-type extremals. A Hermite-type extremal is based on the classical Hermite problem for lattices, to find $\max\min N(v)$ where N denotes the norm, or length squared, and where the minimum is taken over all non-zero vectors in the lattice and the maximum is taken over the space of all lattices of dimension n and determinant 1, $SO(n)\backslash SL(n,\mathbb{R})/SL(n,\mathbb{Z})$. A good survey of classical extremals can be found in [**7**]. The problem can be generalized to other situations where there is an infinitely indexed set of norms N or lengths ℓ which are permuted under the action of a modular group [**4, 5, 20, 23**]. For example, in the case of symplectic lattices the space is $U(g)\backslash Sp\,(2g,\mathbb{R})/Sp\,(2g,\mathbb{Z})$, and the modular group is $Sp\,(2g,\mathbb{Z})$ which permutes the norms indexed as a function of $\mathbb{Z}^{2g}$. In the case of a Riemann Surface the lengths indexed as a function on $\mathcal{C}(F_g)$ are permuted under the modular group Γ_g.

The indices corresponding to minimum length or norm are called the systoles. Since generally $\ell\,([a]) = \ell\,([a^{-1}])$, this pair of systoles is counted once.

A few extremals for the classical Hermite problem are Jacobians of Riemann surfaces. The dimension 6 extremal $A_6^{(2)}$ discovered by Barnes [**1**] is, properly normalized, the Jacobian of the Klein curve. Another extremal for a subset of the classical lattices was later discovered to be a Jacobian: the dimension 6 extremal $M(E_6)$ for isodual lattices discovered by Conway and Sloane is the Jacobian of the Picard-Wiman curve [**7, 19, 4**].

Generalization of classical results of Voronoi show that for the existence at p of a local maximum for the Hermite problem for length functions ℓ defined on a space $\mathcal{T}$, the following two conditions must hold for the set $\mathcal{S}$ of systoles [**4**]:

1. the set of differentials $d\ell_s(p)$, $s \in \mathcal{S}$, affinely spans the tangent space $T(\mathcal{T})(p)$
2. there exist real constants $c_s > 0$, $s \in \mathcal{S}$, such that $\sum_{\mathcal{S}} c_s d\ell_s(p) = 0$.

As a consequence of the first condition, the number of systoles must be greater than the real dimension of the moduli space.

4. Symmetries

In order to have a Hermite-type extremal on $\mathcal{M}_g$ it follows from the Voronoi conditions that there must be $6g - 5$ systoles or more. To find examples of such extremal it is useful to have families of curves where the spectrum has multiplicity. This is the case for families with fixed automorphism group. Such families are well known in genus 2 and 3 [**13**].

4.1. Genus 3 families, KFT. The genus 3 families of curves with automorphisms can be described in terms of Fuchsian group types. Most common are genus 0 type groups. The general genus 0 type will be denoted $(n_1, n_2, \ldots, n_k)$ given in terms of generators and relations by

$$(n_1, n_2, \ldots, n_k) = \{x_1, \ldots, x_k \mid x_1^{n_1} = x_2^{n_2} = \cdots = x_k^{n_k} = x_1 x_2 \ldots x_k = 1\}.$$

The Picard family [**10**] is associated with the group (3,3,3,3,3) while the family KFT [**22, 21**], or S_4 family, is associated with the group $(2, 2, 2, 3)$. If F_g is a normal subgroup of G, the group G describes a group of automorphisms of a family of surfaces of genus g. In this case the automorphism group is G/F_g and the group G acts on F_g by conjugation, giving G/F_g as a subgroup of $\Gamma_g = \mathrm{Out}(F_g)$.

A representation in $PSL(2,\mathbb{R})\backslash\mathrm{Hom}\,(F_g, PSL(2,\mathbb{R}))$ is obtained by restricting an element of $PSL(2,\mathbb{R})\backslash\mathrm{Hom}\,(G, PSL(2,\mathbb{R}))$. A count of free variables and constraints shows that this space is parameterized by $2k - 6$ real parameters. An element ρ in $\mathrm{Hom}\,(G, PSL(2,\mathbb{R}))$ can be specified by choosing an appropriate fundamental polygon. A type of fundamental polygon is described by what Macbeath calls a surface symbol [**15**] but there are many other types. A different type for the $(2, 2, 2, 3)$ is described in [**22**] and software to construct these in general is provided in [**14**].

For the example $(2, 2, 2, 3)$, a topological construction shows that there is an S_4 action on a genus 3 surface whose quotient map is a branched covering to S^2 which is branched over 4 points. The order of the branching over the points is 2,2,2 and 3. In other words, the quotient map is S^2 with 4 orbifold points of type 2,2,2, and 3. One way to see the S_4 action is to think of the genus 3 surface as the "filled in" Cayley graph of S_4 given in terms of generators and relations by

$$\{x_1, x_2, x_3, x_4 \mid x_1^2 = x_2^2 = x_3^2 = x_4^3 = (x_2x_1)^2 = (x_1x_3)^3 = x_1x_2x_3x_4 = 1\}.$$

This presentation can be realized by the map

$$x_1 \to (12) \quad x_2 \to (12)(34) \quad x_3 \to (23) \quad x_4 \to (234) \tag{1}$$

from $(2, 2, 2, 3)$ to S_4 whose kernel is $N\langle (x_2x_1)^2, (x_1x_3)^3\rangle$, the smallest normal subgroup containing the group generated by $(x_2x_1)^2$ and $(x_1x_3)^3$. Since this is a universal covering map, the kernel is F_3. Thus F_3 is a normal subgroup of $(2, 2, 2, 3)$.

4.2. Fuchsian uniformization of KFT. Finding a discrete representation in

$$\mathrm{Hom}\,((2,2,2,3), PSL(2,\mathbb{R}))$$

requires finding elements $\rho\, x_j$ in $PSL(2,\mathbb{R})$ satisfying the relations

$$(\rho\, x_1)^2 = (\rho\, x_2)^2 = (\rho\, x_3)^2 = (\rho\, x_4)^3 = (\rho\, x_1)(\rho\, x_2)(\rho\, x_3)(\rho\, x_4) = 1.$$

One way to find such a representation is to choose a fundamental domain for $\rho\,(2,2,2,3)\backslash\mathbb{H}$ in the form of a five sided polygon with vertices $CB'A'AB$ when followed clockwise around it. Choose the polygon under the constraints that the interior angles at the vertices are respectively $2\pi/3$, β, γ, $\pi-\gamma$, $\pi-\beta$ and the following sides are equal in length:

$$|CB| = |CB'|, \quad |AB| = |A'B'|\,.$$

Let D be the midpoint of AA'. The generators are then

$$\begin{aligned} \rho\, x_1 &= \text{rotation an angle } \pi \text{ about D} \\ \rho\, x_2 &= \text{rotation an angle } \pi \text{ about A} \\ \rho\, x_3 &= \text{rotation an angle } \pi \text{ about B} \\ \rho\, x_4 &= \text{rotation an angle } 2\pi/3 \text{ about C}\,. \end{aligned} \tag{2}$$

One useful feature of such a choice of fundamental polygon that 6 images of one polygon under the action of six elements of $\rho\, G$ (including the identity) form a hexagon with interior angles alternating γ and $\pi-\gamma$. Under the action of ρF_3 this becomes a pair of pants (or Y piece) with 3 boundary geodesics all of length $4|AB|$. Four of these are glued together with a single twist parameter to give the genus 3 surface $\rho\, F_3\backslash\mathbb{H}$. (For more information on length and twist parameters for pants decompositions, see [**2**]). The action of S_4 on the surface permutes the pants and in fact is isomorphic to the permutation group on the pants. In this way the surface can be thought of as a tetrahedron whose vertices are pairs of pants, since the symmetry group of a tetrahedron is the permutation group on its vertices. If the action of the group on the pants is given by (2) then the permutation representation is given by (1). Also the four Y pieces can be thought of as cosets of a dihedral group D_3 in S_4.

Thus given the combinatorial joining of the Y pieces as described above, the family KFT is defined by two geometric parameters, a length ℓ and a twist α. P. Schmutz [**23**] defines the family in a different way: by joining two surfaces of type (0;4) with tetrahedral symmetry and whose boundary geodesics have the same length by a single twist parameter along the four joined boundaries. These lengths and twists will not in general be the same as the ℓ and α described above.

The family KFT can also be described in terms of Jacobians, and this gives a holomorphic parametrization of the family. This point of view is described in [**22, 21**]. Each curve in the family has Jacobian given by a point zA_3 in $\mathcal{H}_3$ where A_3 is the matrix

$$A_3 = \begin{pmatrix} 2 & 1 & 1 \\ 1 & 2 & 1 \\ 1 & 1 & 2 \end{pmatrix},$$

and $z \in \mathbb{H}$. The matrix A_3 is a gram matrix for the face-centered cubic lattice [**6**] generated by vectors $(1,0,1)$, $(1,1,0)$, and $(0,1,1)$. Its dual has gram matrix

$$A_3{}^{-1} = \frac{1}{4}\begin{pmatrix} 3 & -1 & -1 \\ -1 & 3 & -1 \\ -1 & -1 & 3 \end{pmatrix}.$$

The modular group Γ_3 induces an action of $\Gamma_0(4)$ on $z \in \mathbb{H}$ and the quotient gives a moduli space for the family in terms of Jacobians. Thus z is referred to as a holomorphic parameter for the family.

5. Two genus 3 extremals

In genus 3, curve families with automorphism have been studied for a long time, but genus three extremals have only recently been identified. The search for extremals for the Jacobian is inspired by [**3**]. Extremals have also been studied by Jenny [**11**]. Two extremals will be described here for the $(2,2,2,3)$ family, KFT. These extremals are especially interesting because they are new surfaces different from classically known ones like the Klein and the Picard surfaces. The extremal for the length spectrum was discovered by Schmutz [**23**] and the extremal for the Jacobian by Bavard [**5**].

5.1. The Schmutz Schaller extremal for the length spectrum. Consider the action of S_4 on $\mathcal{C}(F_3)$ given by $(2,2,2,3)$ acting on the kernel of (1) by conjugation. Let

$$a' = x_1x_3, \quad b' = x_2x_1, \quad c' = x_2x_3 = b'a'.$$

Under the map (1) the images of these are (124), (34), and (1243) respectively. It follows that

$$a = a'^3, \quad b = b'^2, \quad c = c'^4$$

are in F_3. Under the mapping ρ these group elements are mapped to hyperbolic transformations with axes given by hyperbolic lines determined by the segments BD, AD, and AB respectively on the boundary of the fundamental polygon. In addition let

$$d = (x_1x_4)x_1(x_1x_4)^{-1}x_3 = a'b'c'a'.$$

The element $\rho\, d$ is hyperbolic with axis determined by the points D and $\rho(x_1x_4)\,D$. It maps to the identity under (1) and therefore is in F_3.

The elements in the orbits under $S_4 \subset \Gamma_3$ of the conjugacy classes of a, b, c and d in union with their inverses can be counted:

$$|S_4([a] \cup [a^{-1}])| = 4$$
$$|S_4([b] \cup [b^{-1}])| = 6$$
$$|S_4([c] \cup [c^{-1}])| = 3$$
$$|S_4([d] \cup [d^{-1}])| = 12.$$

It helps to visualize these on a tetrahedron, for example, the conjugacy class $[b] \cup [b^{-1}]$ corresponds to the edges of the tetrahedron and $[a] \cup [a^{-1}]$ to the triangles. The computation of the number of orbits can be done by computing the order of the stabilizer in $(2,2,2,3)/F_g = S_4$ of the various systoles. For example, the stabilizer of $[a] \cup [a^{-1}]$ is the 6 element group $\langle x_1, x_3 \rangle$ and the stabilizer of $[b] \cup [b^{-1}]$ is the 4 element group $\langle x_1, x_2 \rangle$

The extremal surface can be expressed in terms of these. In either case, from the Voronoi conditions, at least 13 systoles are needed for an extremal surface. If the systoles are among these conjugacy classes, $[b]$ or $[d]$ must be systoles. The Schmutz Schaller extremal $T(x|y)$ is defined by the equations

$$\ell\,[a] = \ell\,[b] = \ell\,[d]$$

with 22 systoles. The length of the systole is about $2\operatorname{arccosh}(3.632)$. By comparison, the Klein surface is given by

$$\ell\,[c] = \ell\,[b] = \ell\,[d]$$

with 21 systoles of length about $2\operatorname{arccosh}(3.648)$.

5.2. The Bavard extremal for the Jacobian. The discussion for this example is similar to the one for the Schmutz Schaller example except that the Jacobi spectrum is considered and the function considered is the norm N considered as a functions of cycles in $\mathrm{H}_1(S, \mathbb{Z})$.

The symplectic lattice describing the Jacobian of this surface is easily described by giving a corresponding point in $\mathcal{H}_g$, in this case $Z = ((3 + \sqrt{15}i)/8)\, A_3$. The real part of the corresponding gram matrix with respect to cycles a_1, a_2, a_3, b_1, b_2, b_3 is

$$\frac{1}{\sqrt{15}} \begin{pmatrix} 8{A_3}^{-1} & 3I \\ 3I & 3A_3 \end{pmatrix} .$$

The a cycles generate a copy of the face-centered cubic lattice. The vectors of minimum norm are given by the conjugacy classes of b_1 with 6 elements, of $a_1 - b_1$ with 12 elements, and of a_1 with 4 elements, for a total of 22 systoles. The minimal norm is $6/\sqrt{15}$, and

$$N(a_1) = N(b_1) = N(a_1 - b_1) = 6/\sqrt{15}.$$

By comparison, the Klein curve is given by $Z = ((3 + \sqrt{7}i)/8)\, A_3$ and the corresponding gram matrix is

$$\frac{1}{\sqrt{7}} \begin{pmatrix} 8{A_3}^{-1} & 3I \\ 3I & 2A_3 \end{pmatrix}$$

and the vectors of minimum norm are given by the conjugacy classes of b_1 with 6 elements, of $a_1 - b_1$ with 12 elements, and of $a_1 + a_2 - b_1 - b_2 + b_3$ with 3 elements for a total of 21 systoles. The minimal norm is $4/\sqrt{7}$ and

$$N(a_1 + a_2 - b_1 - b_2 + b_3) = N(b_1) = N(a_1 - b_1) = 4/\sqrt{7}.$$

The systoles for the Jacobi and for the length spectra correspond in a one-to-one fashion for the Klein curve [**18**]. The systoles for the Schmutz Schaller and Bavard curves seem to correspond also, and the method of construction is similar, but probably they are not the same curve. For a comparison of these two curves it will be necessary to determine the correspondence between the geometric and the holomorphic parameters for the KFT family.

References

[1] E. S. Barnes, The perfect and extreme senary forms *Canadian Journal of Mathematics*, **9**(1957), 235-242.

[2] P. Buser, *Geometry and Spectra of Compact Riemann Surfaces*, Birkhäuser, Boston, 1992.

[3] P. Buser and P. Sarnak, On the period matrix of a Riemann surface of large genus (with an appendix by J. H. Conway and N. J. A. Sloane), *Invent. Math.* **117**(1994), 27-56.

[4] C. Bavard, Systole et invariant d'Hermite, *J. reine angew. Math.*, **482**(1997), 93-120.

[5] C. Bavard, Familles hyperboliques de réseaux symplectique, *Prépublication du laboratoire de Mathématique Pure de Bordeaux*, **59**(1997).

[6] J. H. Conway and N. J. A. Sloane *Sphere Packings, Lattices and Groups*, Springer-Verlag, 1988.

[7] J. H. Conway and N. J. A. Sloane, Low dimensional lattices III, perfect forms *Proc. R. Soc. Lond. A* **418**(1988), 43-80.

[8] J. H. Conway and N. J. A. Sloane, On lattices equivalent to their duals, *Journal of Number Theory*, **48**(1994), 373-382.

[9] H.S.M. Coxeter and W. O. J. Moser, *Generators and Relations for Discrete Groups*, Springer-Verlag, New York, 1980.

[10] R. P. Holzapfel, *Geometry and Arithmetic Around Euler Partial Differential Equations*, Reidel, 1986.

[11] F. W Jenny, Ueber den ersten Eigenwert des Laplace-Operators auf ausgewälten Beispielen kompacter Riemannscher Flächen, *Comment. Math. Helvetici* **59**(1984), 193-203.

[12] H. Klingen, *Introductory Lectures on Siegel Modular Forms*, Cambridge University Press, 1990.

[13] I. Kuribayashi and A. Kuribayashi, Automorphism groups of compact Riemann surfaces of genus three and four, *Journal of Pure and Applied Algebra*, **65**(1990), 277-292.

[14] C. O'Ryan Lira, Unifpack: A package for studying Riemann surface with symmetry, http://www.geom.umn.edu.

[15] A. M. Macbeath, The classification of non-euclidean plane crystallographic groups, *Can. J. Math.*, **19**(1967), 1192-1205.

[16] A. M. Macbeath, On a theorem of Hurwitz, *Glasgow Math. J.*, **5**(1961), 90-96.

[17] S. Nag, *The Complex Analytic Theory of Teichmüller spaces*, John Wiley & Sons, 1988.

[18] J. R. Quine, Systoles of two extremal Riemann surfaces, *Journal of Geometric Analysis*, **6**(1996).

[19] J. R. Quine, Jacobian of the Picard curve, *Contemporary Mathematics Volume 201*, 1997, 33-41.

[20] J. R. Quine and P. L. Zhang, Extremal symplectic lattices, *Israel Journal of Math.*, to appear.

[21] J. F. X. Ries, The splitting of some Jacobi varieties using their automorphism groups, in *Contemporary Mathematics Volume 201*, 1997, 81-124.

[22] R. Rodriguez and V. Gonzalez-Aguilera, Fermat's quartic curve, Klein's curve and the tetrahedron, in *Contemporary Mathematics Volume 201*, 1997, 43-62.

[23] P. Schmutz, Riemann surfaces with shortest geodesic of maximal length, Geometric and Functional Analysis, **3**(1993) 564-631.

DEPARTMENT OF MATHEMATICS, FLORIDA STATE UNIVERSITY, 208 LOVE BUILDING, TALLAHASSEE FL, 32306-4510

E-mail address: quine@math.fsu.edu